QUADRUPOLE ION TRAP
MASS SPECTROMETRY

SECOND EDITION

CHEMICAL ANALYSIS

A SERIES OF MONOGRAPHS ON ANALYTICAL CHEMISTRY AND ITS APPLICATIONS

Series Editor
J. D. WINEFORDNER

VOLUME 165

QUADRUPOLE ION TRAP MASS SPECTROMETRY

Second Edition

RAYMOND E. MARCH
Trent University, Peterborough, Ontario, Canada

JOHN F. J. TODD
University of Kent, Canterbury, UK

WILEY-INTERSCIENCE

A JOHN WILEY & SONS, INC., PUBLICATION

Library of Congress Cataloging-in-Publication Data:

March, Raymond E.
    Quadrupole ion trap mass spectrometry / Raymond E. March, John F. J. Todd.—2nd ed.
        p. cm.—(chemical analysis ; v. 165)
    Rev. ed of: Quadrupole storage mass spectrometry. c1989.
    Includes bibliographical references and index.
    ISBN-13 978-0-471-48888-0
    ISBN-10 0-471-48888-7 (cloth : alk. paper)
1. Mass spectrometry.   I. Todd, John F. J.   II. Title.   III. Series.

    QD96.M3M335 2005
    543'.65—dc22   2004020701

Printed in the United States of America

10   9   8   7   6   5   4   3   2   1

CONTENTS

PREFACE

The advent of commercial quadrupole ion trap instruments has been followed by a wide range of applications of such instruments for mass spectrometric studies and investigations facilitated, at least in part, by the publication in 1989 of *Quadrupole Storage Mass Spectrometry*. This book was intended as a primer, an elementary text-book, to explain briefly and concisely the basic operation of the quadrupole ion trap. On the basis of many laudatory comments (the inclusion of which is precluded by surfeit of modesty and paucity of space) from graduate students who had been assisted in their researches by this primer, we concluded that the primer had been a success. In the same year, the award of the Nobel Prize in Physics, in part, to Wolfgang Paul and Hans Dehmelt recognized the invention of the quadrupole (or Paul) ion trap by Wolfgang Paul and Hans Steinwedel.

It was a great privilege for us to be invited, in 1991, to undertake the preparation of a research monograph on quadrupole ion trap mass spectrometry for CRC Press. The principal objective of this monograph was to present an account of the development and theory of the quadrupole ion trap and its utilization as an ion storage device, a reactor for ion/molecule reactions, and a mass spectrometer. A secondary objective was to expand the reader's appreciation of ion traps from that of a unique arrangement of electrodes of hyperbolic form (and having a pure quadrupole field) to a series of ion traps having fields with hexapole and octopole components; furthermore, the reader was introduced to the practical ion trapping device in which electrode spacing had been increased. It was gratifying to discover that so many of our colleagues were willing to contribute to such a monograph. All of the research groups and individuals invited initially to contribute to the monograph accepted their invitations with enthusiasm that matched, if not exceeded, that which they displayed for ion trap research. Their enthusiasm and hard work had not gone unrewarded, in that significant advances were made during the early 1990s to our understanding of ion trap behavior, new traps, trapped ion trajectory control, and trapped ion behavior. The net result was a burgeoning of new material beyond that anticipated for the original monograph and in excess of that accommodated normally in a single book.

In 1995, not one, but three, volumes of *Practical Aspects of Ion Trap Mass Spectrometry* were published. These volumes were comprised of a total of 32 chapters from eighteen of the leading ion trap research laboratories in the world.

In 2002, it appeared as though *Quadrupole Storage Mass Spectrometry* had gone out of print: despite a number of searches on both sides of the Atlantic Ocean, not a single copy could be found! Following discussions with Robert Esposito of John Wiley & Sons, it was agreed in February 2003 that some additional copies would be printed and that consideration should be given to the preparation of a second edition. We agreed to submit an outline for a second edition because the quadrupole ion trapping field had expanded so rapidly during the 1990s and into the twenty-first century. Parenthetically, two copies of the book were located serendipitously in a warehouse near Toronto's Pearson International Airport while one of us (R. E. M.) was waiting for an aircraft. The proposal for a second edition included an updated history of the development of ion trapping devices, a revised treatment of the theory of ion trapping in quadrupole fields, and an introduction of new instruments that employed quadrupole fields for the confinement of gaseous ions. John Wiley & Sons accepted the advice of Dominic Desidario to the effect that a second edition of *Quadrupole Storage Mass Spectrometry* would be an asset to the Wiley analytical list.

Quadrupole Ion Trap Mass Spectrometry is a tightly focused primer directed particularly to relative newcomers to the field of ion trap mass spectrometry. The aim of this book is to present a primer on the theory of ion containment in a quadrupole field and to discuss the relevant dynamics of ion motion as a background to discussions of the development and performance of new instruments that have become available *in this* century. Our understanding of the behavior of both two- and three-dimensional quadrupole fields is now much more advanced with the result that new ion trapping devices of greatly improved sensitivity and speed have been developed. The discussions of new instruments and new forms of existing instruments are up to date as of this time. We have been able to present the latest aspects of these instruments through the ready cooperation of scientists in the field and manufacturers of the instruments.

The approach to this book may be described as that of a "pseudo-single author" in that we have shared the writing but, at each stage, every chapter has been subjected to intense discussion back and forth until complete final agreement has been achieved. In this manner, the reader is presented with a text that has a high degree of clarity of expression so as to promote ready comprehension. While this book is *not* an attempt to review recent research in quadrupole ion trap mass spectrometry, it does give numerous references to the literature.

Quadrupole Ion Trap Mass Spectrometry is composed of nine chapters. Chapter 1, A Historical Review of the Early Development of the Quadrupole Ion Trap, is presented as a prelude to the theory, dynamics, computer modeling, operational aspects, and applications of the quadrupole ion trap in its various forms (including the recently introduced "linear ion trap") that are discussed in succeeding chapters. We first examine the early historical development of the ion trap leading up to its emergence as a commercial instrument. The aim is to provide the reader with a fairly general account containing a level of explanation sufficient for the understanding of the story as it unfolds. Chapter 2 deals with the theory of quadrupole instruments. A new unified step-by-step theoretical treatment of the mathematics of ion containment in quadrupole fields is presented gently at a level that should be readily comprehensible to graduates in Chemistry that have an elementary knowledge of calculus and physics. This presentation of the theoretical treatment or derivation flows through three stages. First, the general expression for the potential is modified by the Laplace condition. Second, it is recognized that the structure of each quadrupolar device is constrained by the requirement that the hyperbolic electrodes share common asymptotes. Third, the derivation is based on a demonstration of the equivalence of the force acting on an ion in a quadrupole field and the force derived from the Mathieu equation; this equivalence permits the application of the solutions of Mathieu's equation to the confinement of gaseous ions. For the quadrupole ion trap, no assumption is made "a priori" concerning the ratio r_0/z_0.

The entire treatment given in Chapter 2 is based on the assumption that we are examining the behavior of a single ion in an infinite, ideal quadrupole field in the total absence of any background gas. Clearly in most applications a single ion is rarely of great value and background gases cannot be excluded totally; in reality, therefore, the theory of ion containment must be considered along with the dynamical aspects of ion behavior. Chapter 3 deals with the dynamics of ion trapping and includes discussion of the effects arising from the presence of other trapped ions (i.e., space charge), the existence of higher order (nonquadrupole) field components, and collisional effects, both with the background and with deliberately added neutral gases. Furthermore, in practice it is obviously necessary to create ions within the quadrupole field or to inject them from an external source and, in most applications, to eject them for detection. Other possible experiments may make use of the ability to excite resonantly the motion of ions mass selectively through the application of additional potentials oscillating with frequencies that match one or more of the secular frequencies. All these topics are considered in an integrated approach, in which first we consider a simplifying approximation to the theory of containment that assumes that the motion of the ions can be described in terms of their being trapped within a "pseudopotential well." This approximation allows us, in turn, to develop models that predict the maximum number of ions that may be stored within an ion trap and to make estimates of the kinetic energies of ion motion. From here we examine the influence of higher order field components and the effects of "nonlinear resonances" on the motion of the ions, together with such phenomena as "black holes" and mass shifts. The reader is introduced to the action of nonlinear resonances in a quadrupole field and how such resonances have been harnessed to improve instrument performance.

Chapter 4 is concerned with the simulation of ion trajectories in the quadrupole ion trap. The ready availability of software programs for the calculation of ion trajectories in quadrupole devices provides an enormously useful tool that complements experimental investigations. Here, we present an introduction to the theory underlying the various types of ion trajectory simulations that may be carried out, from ion trajectories in a pure quadrupole field in the absence of collisions with neutral particles to relatively complex simulations of resonantly excited ion trajectories in nonideal ion traps in the presence of buffer gas. Not only does this discussion cover the basic elements of ion trajectory simulation but the pros and cons of different approaches serve to highlight the opportunities that are available for ion trap research and the necessary compromises that must be made in the creation of a simulation.

Chapters 5–8 are concerned with new types and new forms of quadrupole instruments. In Chapter 5, the linear quadrupole ion trap mass spectrometer is introduced. Since 1983, the two-dimensional quadrupole mass filter has been eclipsed to some extent by the high versatility and performances of successive generations of three-dimensional quadrupole ion traps. Yet the development of two-dimensional quadrupole instruments has continued apace. Two areas of development of such quadrupole instruments are discussed here: the first area is the development of linear ion traps wherein ions are confined within a two-dimensional quadrupole field and subsequently ejected mass selectively; the second area concerns the development of ion tunnels for the efficient transmission of ions in a pressure region. These two areas are interrelated because of the successful combination of electrospray (ESI) ionization and quadrupole devices despite the wide disparity in ambient pressures. The award of the Nobel Prize in Chemistry for 2002 to Koichi Tanaka and John Fenn "for their development of soft desorption ionization methods for mass spectrometric analyses of biological macromolecules" is a testament to the eminence of mass spectrometry as an important analytical tool for the investigation of biological molecules. Ion tunnels are required for the efficient transmission of ions from the ESI at atmospheric pressure to the linear ion trap wherein the pressure is some seven to eight orders of magnitude lower. The absence of an axial trapping field in the linear ion trap facilitates ready admission of ions.

Chapter 6 is devoted to a discussion of the cylindrical ion trap mass spectrometer, where the cylindrical ion trap has been derived as a simplified version of a quadrupole (or Paul) ion trap to affect confinement of charged species. The general direction of research into cylindrical ion traps has been the study of small ion traps, described as miniature ion traps, for the confinement of a single ion or for applications as sophisticated as tandem mass spectrometry. In a further development, multiple cylindrical ion traps have been mounted in an array.

In 1983, the ion trap detector was marketed as a mass detector for the examination of the effluent from a gas chromatograph. The instrument offered such simplicity of operation as to obviate the requirement for personnel with long-term mass spectrometric experience. Chapter 7 is devoted to a discussion of gas chromatography/mass spectrometry. The commercial success of the quadrupole ion trap as a mass detector for a gas chromatograph has permitted the continued development of this instrument. These developments consist of the operation of the ion trap as a tandem mass spectrometer in conjunction with positive-ion chemical ionization, including selective reagent chemical ionization, negative-ion chemical ionization, selected ion monitoring, injection of externally generated ions into the ion trap, multiresonance ion ejection techniques, and variable mass resolution.

In Chapter 8 we examine the combination of ion trap mass spectrometry/liquid chromatography and discuss recent developments in the field of tandem mass spectrometry applied to ESI of polar compounds such as proteins and peptides. The evolution of these instruments has been rapid and their performances have been demonstrated particularly in the areas of metabolite structural elucidation and peptide sequence determination. Two principal manufacturers of instruments have combined electrospray with a quadrupole ion trap; ThermoFinnigan introduced the LCQ™ instrument and Bruker/Franzen introduced the Esquire™ ion trap mass spectrometer. Recently, Finnigan has introduced the LCQ Deca XP MAX mass spectrometer while Bruker Daltonics has introduced a high-capacity ion trap, the Esquire (HCT) mass spectrometer, which claims to open a new bioanalytical dimension of ESI/ion trap performance. Both instruments perform rapid metabolite identification and employ multiple stages of mass selectivity (MS^n) in combination with liquid chromatography (LC) on the time scale of elution of peaks from LC. The HCT instrument offers substantially greater ion storage capacity and features ultrahigh-performance scan modes for protein sequencing and metabolite research.

In addition, Shimazu has introduced a digital ion trap (DIT), also in combination with ESI, in which ions are trapped by a digital waveform. In the DIT, the trapping rectangular waveform is applied to the ring electrode of a nonstretched quadrupole ion trap.

Chapter 9 is entitled An Ion Trap Too Far? The Rosetta Mission to Characterize a Comet. At precisely 07:17 Greenwich Mean Time (GMT) on Tuesday March 2, 2004, an Ariane-5 rocket carrying the Rosetta "comet chaser" was launched at Kourou in French Guyana: the mission, to characterize the comet Churyumov-Gerasimenko, otherwise known as "67P," having a cross-sectional area roughly equal to that of a major airport. The purpose of this chapter is to give an account of this highly unusual application of the ion trap and, in particular, to explore some of the technical and design considerations of a system that is fully automated yet is not due to reach its sample until 2014! The Orbiter's payload includes 11 experiments, all controlled by different research groups. The Lander carries a further nine experiments, including a drilling system to take samples of subsurface material. Described here is the MODULUS experiment where the name "MODULUS" stands for methods of determining and understanding light elements from unequivocal stable isotope compositions. It was concocted by Professor Colin Pillinger, FRS, and his coinvestigators of the Planetary and Space Sciences Research Institute at the Open University in the United Kingdom in honor of Thomas Young, the English physician turned physicist who was the initial translator of the Rosetta Stone and whose name is best known by the measure of elasticity, "Young's modulus."

Our thanks are due to many people who have assisted us in the successful completion of a manuscript. We thank the editorial staff, Robert Esposito, Heather Bergman, and Rosalyn Farkas of John Wiley & Sons for their ready cooperation, advice, and encouragement. One of the greatest pleasures that we enjoyed while preparing this manuscript was the extraordinary degree of cooperation that we received from colleagues in the quadrupole ion trapping community and the alacrity with which they responded to our requests. A good example of such cooperation is afforded by the difficulty we faced in attempting to reproduce Dr. Randall D. Knight's "simple geometric calculation shows that the asymptote of the $r_0 = \sqrt{2}z_0$ trap bisects, at high values of r and z, the ring electrode–end-cap electrode gap"; see Chapter 2, Appendix. Eventually, we decided to approach Dr. Knight and to request his geometric calculation. Alas, the calculation could not be found but, to his credit, Dr. Knight responded readily to our request; so readily that we suspect he spent much of his President's Day holiday weekend at the task. The calculation shown in the Appendix to Chapter 2 may or may not be the same as that carried out some 20 years previously, but it is similar. The calculation is certainly not simple and, while it is not difficult, it is somewhat devious!

In preparing this manuscript and the three volumes of *Practical Aspects of Ion Trap Mass Spectrometry*, we have been the fortunate recipients of encouragement and support from Professor R. Graham Cooks, of Purdue University. We acknowledge gratefully the contributions from Professor Cooks' laboratory concerning the miniaturization of cylindrical ion traps, both singly and in arrays. We have great affection for the cylindrical ion trap and we applaud the progress in this field since the observation of our first mass spectrum obtained with a cylindrical ion trap in 1978. We thank also Professor Cooks and Zheng Ouyang for a prepublication copy of the manuscript concerning the conceptual evolution of the rectilinear ion trap from the linear ion trap and the cylindrical ion trap. In addition, we thank Professor Cooks for permission to reproduce Figure 5.6; this figure, which was created by Professor Cooks, is a visual summary of this book. We appreciate the cooperation of Dr. James W. Hager for helpful discussions and for supplying copies of figures showing the performance of the linear ion trap. Throughout the quadrupole ion trap conferences held at La Benerie, near Paris, in the 1990s and in the preparation of Volume 1 of *Practical Aspects of Ion Trap Mass Spectrometry*, we have enjoyed the fellowship and support of Dr. Jochen Franzen, of Bruker Daltonics. We acknowledge gratefully his continuing support by way of figures and material that we have used in the discussion of the high-charge ion trap in Chapter 8. An exciting addition to the complement of quadrupole instruments is the digital ion trap, and we thank Dr. Li Ding, of Shimazu, for prepublication material and figures that we have used in our discussion of this new instrument. The authors are greatly indebted to Dr. Simeon J. Barber, of the Planetary and Space Sciences Research Institute, Open University, Milton Keynes, United Kingdom, for his considerable help in preparing the material for Chapter 9.

We acknowledge gratefully the support of Scott Thompson in maintaining operation of REM's computer. We thank our wives, Kathleen and Mavis, respectively, for their unfailing support during preparation of the manuscript, for the many cups of tea, and for their tolerance of our work habits and our mercifully few idiosyncrasies. Finally, we thank each other. Each of us could, perhaps, have prepared a manuscript alone, but this manuscript is undoubtedly superior because of the cooperation between us with the result that you, the readers, are the beneficiaries; furthermore, we actually enjoyed the cooperative experience!

RAYMOND E. MARCH
JOHN F. J. TODD
Peterborough, Ontario and Canterbury, Kent
February 2005

PREFACE TO FIRST EDITION

The announcement by Finnigan MAT, in 1983, of a novel scanning technique for the ion trap detector, by which a mass spectrum of stored ions could be acquired, heralded a new era in mass spectrometry. The quadrupole ion trap or QUISTOR, described by its inventors in 1953 as "still another electrode arrangement," had survived as an object of curiosity in only a dozen laboratories around the world. Characterization of the quadrupole ion trap developed slowly in the 1960s and 1970s; its behavioral and operational modes were explored, while simulation studies using numerical phase-space methods for trajectory mapping of ions within a trapped ion ensemble were carried out to determine velocity and spatial distributions. The ion trap detector was presented as a concatenation of the Mathieu equation from which the stability diagram was derived, a simple electrode structure for the ion trapping device, toleration (or requirement) of an appreciable background pressure, a facile voltage scan for mass-selective trajectory instability, compact and stable electronics, software appropriate for device operation in a pulsed mode, and a relatively inexpensive microprocessor by which complete instrumental control is achieved.

The achievement of precise control of both ion trajectory and collision number within the quadrupole ion trap lay in the method of mass-selective ion ejection developed by Finnigan MAT. To the QUISTOR devotee the announcement of a great improvement in ion trap performance was enormously exciting. Yet the excitement was not only for the rather narrow reason of expediting research in gaseous ion chemistry and physics, important though this may be, but also for the much wider possibility of making mass spectral information readily available at greatly reduced cost. There is now abundant evidence of the application to the health services of mass spectrometric techniques with concomitant high sensitivity and resolution for toxicological studies; studies of metabolism and incipient disease; environmental problems; the quality of food, well water, and materials; forensic sciences; and so forth. Thus the advent of the ion trap detector permits a much greater use of mass spectrometric techniques not only in the technically advanced countries but also in those countries which are technically less advanced. The effects of quadrupole storage mass spectrometry through utilization of the ion trap detector will be considerable.

With the ready acceptance of the ion trap detector, the need for a monograph to supplement the material supplied by the manufacturer was soon realized. The excellent text *Quadrupole Mass Spectrometry* by Dr. Peter H. Dawson of the National Research Council of Canada was out of print and only second-hand copies were available. As Dr. Dawson did not intend to prepare a revised edition, we decided to prepare a primer, an elementary textbook, to explain briefly and concisely the basic operation of the manila-colored "black-box" which was the ion trap detector. After further consideration and discussion with the Editor, James L. Smith of John Wiley & Sons, Inc., it was agreed that a somewhat lengthier monograph would be appropriate to the Chemical Analysis Series.

We present here the story of the Paul radio frequency trap, which became known as the quadrupole ion trap (or store), the QUISTOR, and the ion trap detector. Although it is but one of a family of quadrupole devices, this account is restricted to the quadrupole ion trap except for the theoretical treatment. Here, due to the strong familial relationship, we present an introduction to the theory of operation of the quadrupole mass filter en route to developing quadrupole ion trapping theory.

This monograph is composed of seven chapters, references, a bibliography, and a listing of all patents and theses pertaining to the quadrupole ion trap. Chapter 1, a historical review written by J. F. J. Todd, is a review of the development and characterization of the quadrupole ion trap from the original patent filed by Paul and Steinwedel in 1953 to the introduction of, initially, the ion trap detector and then the ion trap mass spectrometer by Finnigan MAT. The historical review assumes a knowledge of basic quadrupole ion trapping theory, rather than attempting to include a brief outline of the theory of ion containment sufficient perhaps for making connective linkages in the story but somewhat daunting to the novice. In Chapter 2 we present a detailed treatment of the theory of ion containment with the quadrupole ion trap, tracing its genesis from the quadrupole mass filter and emphasizing the importance of the stability diagram. We have strived to present an unambiguous, step-by-step mathematical treatment of quadrupole theory which will encourage the reader to persevere through to an understanding of the significance of the stability diagram. The explanatory notes and worked examples will, we hope, reassure the reader of progress being made.

The third and fourth chapters deal with the physics and chemistry, respectively, of the ion trap, though division of the subject matter is somewhat arbitrary. A major part of Chapter 3 is devoted to the applications of numerical and phase-space methods to ion trajectory calculations and velocity and spatial distributions of an ion ensemble. In Chapter 4 emphasis is accorded to infrared multiphoton dissociation of gaseous ions and to the confinement of externally generated ions. In the treatment of the former, we demonstrate the versatility of the quadrupole ion trap for ion synthesis, ion isolation, collisional focusing of the ion cloud, laser irradiation of the focused ion cloud, control of collision number and collision partner, and trapping of photoproduct ions with subsequent mass analysis. These examples illustrate applications of the quadrupole ion trap for isomer differentiation and the determination of the frequency dependence of photodissociation. We are of the opinion that important applications of ion trapping in the future will involve the capture of externally generated ions, and hence the state of the art of this technique is described in some detail.

For Chapter 5, a thorough review is given of the theory and utilization of cylindrical ion traps. These devices, which are of simple fabrication and structure, retain many of the ion storage and behavioral aspects of the quadrupole ion trap and have been applied to studies of RF spectroscopy, metastable ions, and cluster ions. Other types of ion traps are reviewed briefly.

Chapters 6 and 7 are devoted to the Finnigan MAT ion trap detector and ion trap mass spectrometer, respectively. The derivation of ion trapping parameters leading to development of the stability diagram is repeated in condensed form for two reasons: First, much of the subsequent discussion is based upon treatment of the working point within the stability diagram or at a boundary; and second, this concise derivation should be comprehended at this stage; if not, then a revisitation of Chapter 2 is highly recommended. In our opinion, the ion trap detector (and mass spectrometer) will find extensive application in the fields of medical biochemistry; thus we have discussed in some detail the contribution of S.-N. Lin and R. M. Caprioli on measurement of urinary organic acids and their new small sample volume procedure. This unpublished work was a solicited contribution in this important area. In Chapter 7, following discussion of both tandem mass spectrometry in its various facets and automatic reaction control, four distinct aspects of ion trap operation are examined; these are performance comparisons with other instruments, the application of DC and RF voltages, Fourier transform mass spectrometry, and negative ion studies. The work of J. E. P. Syka and W. J. Fies, Jr., on quadrupole Fourier transform mass spectrometry is quoted almost verbatim as communicated to us. The final part of Chapter 7 is devoted to an extensive discussion of trapped negative ions. As the literature in this area is sparse, we have relied heavily on the pioneering work of McLuckey, Glish, and Kelley.

Behind the writing of every story such as this one there is the story of the writers, which in this case is a happy one. Upon returning from Yale University, J. F. J. Todd, who had been intrigued by two papers on ion storage in three-dimensional quadrupole fields (by P. H. Dawson and N. R. Whetten, published in the *Journal of Vacuum Science*), set out to build an apparatus (which was known later as a QUISTOR-quadrupole mass filter combination) for the study of metastable ions. However, construction of this "electronic test tube" revealed such a myriad of possibilities for the study of gaseous ion chemistry that Todd and his co-workers were side-tracked for several years of very fruitful research, and it was C. Lifshitz and her co-workers who demonstrated so elegantly, much later, the application of a cylindrical ion trap to metastable ion studies. In the study of vibrationally excited N_2 formed in the fast and highly exothermic reaction of NO with N· atoms by monitoring electron impact induced fluorescence from $N_2{}^+$, R. E. March observed the effects of competing ion/molecule reactions. At this stage the decision was made to devote a sabbatical leave to an apprenticeship in mass spectrometry in the laboratory of J. Durup at Orsay, France. The last three months of this leave were spent at Canterbury, England, in the laboratory of J. F. J. Todd. March and Todd had been undergraduates together at Leeds University, and a fruitful and enjoyable continuing collaboration got under way. It has been a great personal pleasure for us to have John Todd associated with this monograph; the Trent-Kent connection is alive and well. R. F Bonner, who had been a research student with John Todd at that time, accepted a postdoctoral position with March at Trent University; Ron Bonner's contributions to the development and pursuit of quadrupole ion trapping at Trent University were considerable and are acknowledged gratefully here. Just prior to Ron Bonner's departure for Rockefeller University, R. J. Hughes joined the laboratory, and his association with this laboratory has continued with only minor interruptions. It has been a privilege and an enormous personal pleasure to have been associated with Richard Hughes over this period and to collaborate with him in the preparation of this monograph.

This monograph represents the confluence of the contributions of many people, which we gladly acknowledge here. Dr. Peter H. Dawson not only supplied us with papers and a list of publications but also gave us the initial encouragement to take on this task and maintained an active interest throughout. The scientists and engineers of Finnigan MAT responsible for the development of the ion trap detector and ion trap mass spectrometer, particularly Dr. Paul F. Kelley, Dr. George C. Stafford, Jr., and Dr. John E. P. Syka, supplied us with a large collection of Finnigan Corporation artwork and mass spectra pertaining to the above instruments. We are deeply appreciative of their support. Dr. Richard M. Caprioli and Shen-Nan Lin supplied us with unpublished work on the measurement of urinary organic acids and the details of a new small sample volume procedure which enhanced the recovery of water-soluble urinary organic acids. We also wish to thank the following people: Dr. Richard A. Yost for graciously supplying figures relating to the dynamic range of the ion trap detector. Dr. Anthony O'Keefe for papers of the late Professor Bruce A. Mahan. Our friends in Marseille, Dr. Fernande Vedel, Professor Jacques André, Dr. Michel Vedel, Dr. Georges Brincourt, Dr. Alili Abdelmalek (Malek), S. Mahmood Sadat Kiai, Dayyoub Nazir, Yves Zerega, André Teboul, and Robert Catella, for their hospitality to R.E.M., les diners les Vedel, le tennis, and their comments on the lecture series on the ion trap detector. Professor Hans G. Dehmelt, Dr. N. Rey Whetten, Professor Chava Lifshitz, Professor Ronald G. Brown, Professor R. Graham Cooks, Professor Dr. Karl-Peter Wanczek, Professor Michel Desaintfuscien, Dr. Earl C. Beaty, Professor Hans A. Schuessler, and Dr. Connie O. Sakashita for supplying us with reprints, lists of publications, and encouragement. James L. Smith, Senior Editor with John Wiley & Sons, Inc., and Dr. James D. Winefordner appreciation for their patience, encouragement, and assistance.

We are indebted to our friends at Trent. The care and extra effort of those who assisted in the preparation and typing of the manuscript are acknowledged with thanks: Bonnie MacKinnon, who stayed with us throughout; Dorothy Sharpe; Connie Bartley; Margaret Nolan; and Neil Snider. In the preparation of the figures we are particularly grateful to Gregory K. Koyanagi, who not only redrew many of the figures to achieve a measure of uniformity, but also recalculated all of the figures which contained stability diagrams and iso-β lines. We thank Jim Tomlinson for assistance with computer software, and Louis Taylor for his photographic assistance.

To Dr. Adam W. McMahon and Dr. Alex B. Young, who undertook to read through the manuscript with a critical eye, we offer our grateful thanks.

We thank our families, whom we rejoin with pleasure, for their patience, tolerance, and love.

RAYMOND E. MARCH
RICHARD J. HUGHES
JOHN F .J. TODD
Peterborough, Ontario
August 1988

NOMENCLATURE

3-D	Three-dimensional
A	Ampere
A	A potential applied between electrodes of opposing polarity
A	Time-dependent collisional damping term
A, B, C, D	Regions on the stability diagram
A, B, C, D	Stages in a mass-selective instability mode scan
A′, B′, C′, D′	Stages in a mass-selective instability mode scan
Å	Ångstrom unit
$(A_{CID})_{cong}$	Fragment ion signal intensity per pg of congener injected
AC	Alternating current
ACQUIRE	Advanced control of the quadrupole ion trap for isotope ratio experiments
$(A_{EI})_{cong}$	Ion signal intensity per pg of congener injected
AGC	Automatic gain control
ALICE	Ultraviolet imaging spectrometer
a_m	Characteristic curves of a cosine-type function of order m
amol	Attomole
amu	Atomic mass unit
amu/s	Mass scan rate
A_n, A_n^0	Weighting factors, arbitrary coefficients
APXS	Alpha proton X-ray spectrometer
ARC	Automatic reaction control
a, b	Constants in the general equations for electrode surfaces
a_0	Trapping parameter for the digital ion trap
a_0, b_1	Characteristic curves
a_x	Acceleration of an ion in the x direction
a_x, a_y, a_z, a_r, a_u	Trapping parameter in the x, y, z, r, and u directions, respectively

B	Magnetic field intensity
BAD	Boundary-activated dissociation
b_m	Characteristic curves of a sine-type function of order m
c	Speed of light
C	A fixed potential applied to all the electrodes so as to float the device
$C_{2n,u}$	Coefficient of ion motion amplitude
CAD	Collision-activated dissociation
CCD	Charge-coupled device
CCLRC	Council for the Central Laboratory of the Research Councils
CDMS	Control and data management system
CE	Capillary electrophoresis
CF	Collision factor
CI	Chemical ionization
CID	Collision-induced dissociation
CIT	Cylindrical ion trap
ÇIVA/ROLIS	Rosetta lander imaging system
C_0	Constant proportional to AC voltage
CONSERT	Comet nucleus sounding experiment by radiowave transmission
COSAC	Cometary sampling and composition experiment
COSIMA	Cometary secondary ion mass analyser
CRM	Charge residue model
CW	Chemical warfare
d	Positive voltage fraction of an asymmetric rectangular waveform
Da	Mass in daltons
DAC	Digital-to-analog converter
DC	Direct current
DIT	Digital ion trap
$\bar{D}_{DIT}$	Pseudopotential well depth for the digital ion trap
DMMP	Dimethyl methyl phosphonate
DMNB	Diaminonitrobenzene
DNT	Dinitrotoluene
DOS	Disk operating system
$\bar{D}_r$	Pseudopotential well depth acting along the r direction
d_{ring}	Distance between a point on the ring electrode and the asymptote
d_{endcap}	Distance between a point on an end-cap electrode and the asymptote
d_x	Half-distance between the x electrodes
d_y	Half-distance between the y electrodes
$\bar{D}_z$	Pseudopotential well depth acting along the z direction
e, e_0	Electronic charge
e	Base of natural logarithms

E	Ion kinetic energy
E	Electric potential
EBE	Triple sector instrument with electrostatic (E) and magnetic (B) sectors
E_C	Relative kinetic energy of colliding particles in center-of-mass frame
ECD	Electron capture dissociation
EI	Electron ionization
E_k	Kinetic energy
$E_{k,max}$	Maximum kinetic energy
$E_{k,min}$	Minimum kinetic energy
E_L	Total kinetic energy of colliding species in laboratory frame
E_{max}	Maximum kinetic energy
$E(\text{total})$	Effective total ion kinetic energy
$\langle E(z) \rangle$	Average ion kinetic energy in the z direction
ESI	Electrospray ionization
ESQUIRE	Bruker ion trap mass spectrometer
eV	Electronvolt, unit of energy
f	Frequency expressed in Hz
$\mathbf{F}$	Force acting upon an ion
FAST	Forced asymmetric trajectory
FC-43	Mass calibration compound (see PFTBA)
fg	Femtogram
FIM	Field interpolation method
$(F_{iso})_{cong}$	Sum of the fractional abundances of two mass-selected ions
FM	Flight module
fmol	Femtomole
FNF	Filtered noise field
FWHM	Full width at half maximum
g_i	Degeneracy of the population of level i
GIADA	Grain impact analyser and dust accumulator
GC	Gas chromatography
GC/MS	Gas chromatography combined with mass spectrometry
GC/MS/MS	Gas chromatography combined with tandem mass spectrometry
GMT	Greenwich mean time
h	Signal height
h	Hamiltonian function
$h^{(2)}$	Hamiltonian function representing kinetic and potential energies
h'	Hamiltonian function representing a perturbing potential
h_∞	Signal height at infinite time
H_6CDD	Hexachlorodibenzo-p-dioxin
H_7CDD	Heptachlorodibenzo-p-dioxin

HCT	High-ion-capacity (or high-charge) ion trap
$\dot{h}_0$	Initial rate of increase of signal height
HPLC	High-performance liquid chromatography
HRGC	High resolution gas chromatography
HRMS	High resolution mass spectrometry
Hx	Hexapole rod assembly
Hz	hertz
i	Square root of minus one
i	Electron beam current
ICC	Ion charge control
ICR	Ion cyclotron resonance
ICDR	Ion cyclotron double resonance
i.d.	Inner diameter
I_e	Electron beam intensity
IEM	Ion evaporation model
IRMPD	Infrared multiphoton dissociation
ISIS	Integrated system for ion simulation
IT	Ion tunnel
ITD	Ion trap detector
ITD 700	Finnigan MAT ion trap detector
ITD 800	Finnigan MAT ion trap detector
ITMS	Finnigan MAT ion trap mass spectrometer
ITS-40	Finnigan quadrupole ion trap instrument
ITSIM	Ion Trajectory SIMulation program
IQ_1	First interquadrupole aperture
IQ_2	Second interquadrupole aperture
IQ_3	Third interquadrupole aperture
IUPAC	International Union of Pure and Applied Chemistry
J	joule
j	An integer
k	Boltzmann constant
k	An integer
k_1, k_2, etc.	Rate constants
kHz	Kilohertz
kJ	Kilojoule
kV	Kilovolt
L	Ionization path length
l	An integer
l_A	Length of the SCIEX linear ion trap
l_B	Length of the Thermo Finnigan linear ion trap
LC	Liquid chromatography

LC/MS	Liquid chromatography combined with mass spectrometry
LCQ Deca 3-D	Liquid Chromatograph/Thermo Finnigan quadrupole ion trap
LHS	Left hand side
LMCO	Low-mass cutoff
LIT	Linear ion trap
m	Ion mass
m	An integer
m_a to m_b	Range of variation of parameter m
m, m'	Slopes of asymptotes
m_1, m_2, etc.	Masses of ions
M_1, M_2, etc.	Masses of ions
MA	Program for direct integration of the Mathieu equation
MALDI	Matrix-assisted laser desorption ionization
mbar	Millibar
Mbyte, MB	Megabyte
MDS	Medical Diagnostic Services
meV	Millielectronvolt
MFI	Multifrequency irradiation
MFP	Mean-free-path
MHz	Megahertz
m_i	Relative ion mass
micro-CIT	Cylindrical ion trap with radius <1 mm
MIDAS	Micro-imaging dust analysis system
MIMS	Membrane introduction mass spectrometry
mini-CIT	Miniature cylindrical ion trap
MIRO	Microwave instrument for the Rosetta orbiter
m_n	Mass of background neutral atom or molecule
m_0	Mass/charge ratio of the "genuine" ion
$m/\Delta m$	Mass resolution
MODULUS	Methods Of Determining and Understanding Light elements from Unequivocal Stable isotope compositions
mol	Mole
MOSFET	Metal-oxide-semiconductor field-effect transistor
MOWSE	MOlecular Weight SEarch
mPa	Millipascal
MPAe	Max-Planck-Institut für Aeronomie
MRFA	A tetrapeptide, methionine-argenine-phenylalanine-alanine
MRM	Multiple-reaction monitoring
ms	Millisecond
MS	Mass spectrometry
MS	Mass scan
MS-30	Kratos double-focusing mass spectrometer
MS/MS	Tandem mass spectrometry involving one stage of mass selection followed by ion activation and mass analysis

MS/MS/MS	Tandem mass spectrometry involving two stages of mass selection, each followed by ion activation, and mass analysis following the second stage of ion activation
MS^3	MS/MS/MS
MS^n	Tandem mass spectrometry involving (n-1) stages of mass selection, each followed by ion activation, and mass analysis following the final stage of ion activation
mTorr	Millitorr
MUPUS	MUlti-PUrpose Sensor for surface and subsurface science
MW, M	Molecular weight
m/z	Mass/charge ratio
n	Order of the multipole
n	Number of parameters
n	An integer
n_r	An integer
n_z	An integer
N	Order of the resonance
N	Total number of ions in an ensemble
N	Ions per cm^3
$N_{2D,A}$	Ion capacity of the SCIEX linear ion trap
$N_{2D,B}$	Ion capacity of the Thermo Finnigan linear ion trap
N_{3D}	Ion capacity of the 3-D quadrupole ion trap
N_∞, N_{max}	Maximum ion density
NB	Nitrobenzene
N_{cong}	Number of molecules per pg of congener
ng	Nanogram
NG	Nitroglycerine
n_i	Number of ions having energy ε_i
NICI	Negative ion chemical ionization
NIST	National Institute of Standards and Technology
nmol	Nanomole
ns	Nanosecond
NT	Nitrotoluene
o	An integer
o-, m-, and p-	Ortho, meta, and para
O_8CDD	Octachlorodibenzo-p-dioxin
o.d.	Outer diameter
OSIRIS	Optical, spectroscopic and infrared remote imaging system
p	An integer
p	Pressure of neutral molecules
p	Number of protons
P	A potential in the ion trap

P	Absolute probability
P	Any ion property
P	Probability of an ion suffering a collision
Pa	Pressure in pascals
PCB	Polychlorinated biphenyl compound
PCDD	Polychlorodibenzo-p-dioxin
P_5CDD	Pentachlorodibenzo-p-dioxin
PCDF	Polychlorodibenzofuran
P_5CDF	Pentachlorodibenzofuran
PDMS	Poly(dimethylsiloxane)
PETN	Pentaerythritol-tetranitrate
PFTBA	Perfluorotri-n-butylamine
pg	Picogram
P_{inf}	Informing power
pmol	Picomole
P_n	Legendre polynomial of order n
P_n	Pixel number n
PICI	Positive ion chemical ionization
PTFE	Polytetrafluoroethylene
q_0	Trapping parameter for the digital ion trap
Q_0	RF-only quadrupole ion guide
Q_1	First quadrupole mass analyzer
Q_2	Second quadrupole mass analyzer
Q_C	Quadrupole collision cell
QIT	Quadrupole ion trap
QITMS	Hypothetical amino acid pentamer Q-I-T-M-S
QM	Qualification model
QMF	Quadrupole mass filter
QUISTOR	QUadrupole Ion STORe
q_x, q_y, q_z, q_r, q_u	Trapping parameter in the x, y, z, r, and u directions, respectively
$q_{x,max}$	Maximum q_x value for stable ion trajectories
r	Field radius of a rod array
r_0	Radius of the inscribed circle of the rod array
r_0, r_1, r_2	Radial dimensions of Beaty's readily machinable ion trap
r_1	Radius of ring electrode of a cylindrical ion trap
R^2	Measure of the linearity of a series of points
$r_{2D,A}$	Radius of the inscribed circle of the SCIEX linear ion trap
$r_{2D,B}$	Radius of the inscribed circle of the Thermo Finnigan linear ion trap
r_{3D}	Radius of the ring electrode of a quadrupole ion trap
rad	Radian
r_c	Radius of ring electrode of cylindrical ion trap
RDX	Cyclotrimethylenetrinitramine, an explosive compound

RF	Radio-frequency
$R(g)$	Constant resolution of a capillary GC column
RIT	Rectilinear ion trap
$R(m)$	Mass resolution
r_{max}	Maximum radial excursion
r_{min}	Minimum radial excursion
ROMAP	Rosetta lander magnetometer and plasma monitor
ROSINA	Rosetta orbiter spectrometer for ion and neutral analysis
RPC	Rosetta Plasma Consortium
R_{sample}	Ratio of given isotopes in a sample
RSI	Radio science investigation
$R_{standard}$	Ratio of given isotopes in a standard
$[S]$	Sample concentration
S_i	Number of measurable steps for a given quantity
SCIEX	SCIentific EXport
SD2	Sample drill and distribution system
SESAME	Surface electrical, seismic and acoustic monitoring experiments
SFI	Single-frequency irradiation
SFM	Secular-frequency modulation
SI	International system of units
SID	Surface-induced dissociation
SIMION	ION and electron optics SIMulation program
SIS	Selected-ion storage
SMA	Shape memory alloy
$S(m)$	Ion intensity range
S/N	Signal/noise ratio
SPQR	Simulation Program for Quadrupolar Resonance
STP	Standard temperature and pressure
SWIFT	Stored waveform inverse Fourier transform
t_n	Time after n time increments, each of duration Δt
T	Temperature
T	Time in primary and secondary mission phases
T	Period of asymmetric rectangular waveform
T_4CDD	Tetrachlorodibenzo-p-dioxin
T_4CDF	Tetrachlorodibenzofuran
t_d	Delay of midpoint of DC pulse
Th	thomson, a measure of the mass/charge ratio
Th/s	Mass/charge ratio scanning rate in thomsons per second
TIC	Total ion current
TNT	Trinitrotoluene
Torr	Pressure in torr
t_r	Retention time

TSQ	Triple-stage quadrupole mass spectrometer
T_{SWF}	Period of the square-wave potential
t_w	Peak width
u	One of the coordinates x, y, and z
u	Unit atomic mass
u_{max}	Maximum displacement in the u direction
$\dot{u}$	Velocity in the u direction
$\langle \dot{u}_{max} \rangle$	Average of the maximum ion velocity in the u direction
U	DC voltage
U'	Amplitude of the DC component of the tickle potential
$u_1(\xi)$, $u_2(\xi)$	Two linear independent solutions
UHV	Ultra-high vacuum
v	An integer equal to or greater than zero
V	Volt
V_1	Positive voltage component of a rectangular waveform
V_2	Negative voltage component of a rectangular waveform
V'	Amplitude of the RF component of the tickle voltage
V_{AC}	Amplitude of AC voltage
v_C	Velocity of the center of mass
V_d	Amplitude of DC pulse
V_f	Final amplitude of RF voltage
V_{FAE}	Potential applied to field-adjusting electrode
VG	Vacuum Generators
v_i	Ion velocity
V_i	Initial amplitude of RF voltage
VIRTIS	Visible and infrared mapping spectrometer
v_n	Neutral particle velocity
$V_{0\text{-}p}$	Zero-to-peak amplitude of an oscillating potential
$V_{p\text{-}p}$	Peak-to-peak amplitude of an oscillating potential
v_r, v_R	Relative velocity of an ion and a neutral gas atom
V_{ring}	Amplitude of the RF potential applied to the ring electrode of a CIT
V_{RF}	Amplitude of the RF voltage
V_{rms}	Root-mean-square voltage
$v_{x,t}$	Velocity of an ion in the x direction at time t
W	watt
w_d	Width of DC pulse
$\overline{W}_z$	Kinetic energy of an ion oscillating in a pseudopotential well of depth $\overline{D}_z$
z_{endcap}	Height of a point on an end-cap electrode above the radial axis
z_{ring}	Height of a point on the ring electrode above the radial axis

z	Secular displacement along the z axis
z_0	Half the separation distance of the endcap electrodes in a QIT
z_0, z_1, z_2	Axial dimensions of Beaty's readily machinable ion trap
z_1	Half the separation distance of the end-cap electrodes in a CIT
ZE	Zero enrichment
z_{max}	Maximum axial excursion
z_{min}	Minimum axial excursion
α	Angle of the asymptote to the radial plane
α	Polarizability
α	Ratio of number of ions detected to number of ions formed
α_0	Electronic polarizability
$\beta_x, \beta_y, \beta_z, \beta_r, \beta_u$	Secondary trapping parameter in the x, y, z, r, and u directions, respectively
γ	RF phase angle
Γ, Γ'	Constants of integration
δ	Differential isotope ratio
δ	Micromotion associated with trajectory ripple
∇	First differential
∇^2	Second differential
$\Delta a_r, \Delta a_z$	Shifts in the a values
δ_m	Mass displacement
δ_m	Smallest distinguishable increment in the parameter m
Δm	Width of a mass peak measured at half-height
$\Delta(m/e)$	Mass shift
$\Delta_r H$	Standard enthalpy change of reaction
Δt	Integration step size
ΔU	Effective DC voltage increment
ε	Emittance of the beam ellipse
$\varepsilon_i, \varepsilon_j$, etc.	Energy of an allowed level
ε_0	Permittivity of a vacuum
$\varepsilon(\%)$	Percentage trapping efficiency
$(\eta_{CID})_{cong}$	CID efficiency for each chlorocongener
θ	Azimuth
λ, σ, γ	Weighting constants for the x, y, and z directions, respectively
μ	Characteristic exponent
μ	Reduced mass of ion and target
μA	Microamp
μL	Microliter
μm	Micrometer
μ_{max}	Maximum excursion
μs	Microsecond
ξ	A dimensionless quantity equal to $\Omega t/2$
ξ_0	Initial phase of RF potential
ξ_0'	Initial phase of the tickle potential

$\rho,\ \theta,\ \phi$	Spherical polar coordinates
ρ_{max}	Theoretical space charge-limited ion density
σ	Cross-sectional area
σ_{cong}	Congener electron impact ionization cross section
τ	Half period of the secular oscillation
$\bar{\tau}_i$	Mean ion lifetime
ϕ	Potential
ϕ	Elevation
ϕ_{aux}	Auxiliary potential
ϕ_{endcap}	Potential applied to the endcap electrode(s)
ϕ_i	Repulsive electrostatic potential within an ion ensemble
ϕ_0	Electric potential difference between rod pairs
$\phi_0,\ \phi_{ring}$	Potential applied to the ring electrode
$\phi_{x\ pair},\ \phi_{y\ pair}$	Electric potential applied to a rod pair
$\phi_{x,y,z}$	Potential at a coordinate position $(x,\ y,\ z)$
ϕ_0	Electric potential difference between rod pairs
$\Phi_0^{\ E}$	Potential applied in phase to the endcap electrodes
$\Phi_0^{\ R}$	Potential applied to the ring electrode
ψ	Trapping potential
$\omega_{u,n}$	Secular frequency in direction u and of order n
ω_T	Tickle frequency (See Ω')
Ω	Radial frequency of the RF potential
Ω'	Frequency of the RF component of the tickle potential

1

THE HISTORICAL REVIEW OF THE EARLY DEVELOPMENT OF THE QUADRUPOLE ION TRAP

1.1. INTRODUCTION

The three-dimensional radio-frequency (RF) quadrupole ion trap (QUISTOR), which forms the subject of this book, is only one of a family of devices that utilize path stability as a means of separating ions according to the ratio mass/charge-number (m/z). The later chapters are concerned with the theory, dynamics, computer modeling, operational aspects, and applications of the quadrupole ion trap in its various forms (including the recently introduced "linear ion trap"), but as a prelude to these more detailed considerations, we examine first the early historical development of the ion trap leading up to its emergence as a commercial instrument. The aim is to provide the reader with a fairly general account containing a level of explanation sufficient for the understanding of the story as it unfolds.

The original public disclosure, filed in 1953, of the quadrupole ion trap, which it described as "still another electrode arrangement," is to be found in the same patent [1] as that in which Paul and Steinwedel, working at the University of Bonn, first described the operating principle of the quadrupole mass spectrometer. However, the same ideas were also put forward in that year by Post and Heinrich [2] for a "mass spectrograph using strong focusing principles" and by Good for "a particle containment device" [3]. Evidently these proposals were stimulated through the publication by Courant et al. in the previous year [4] of the theory of strong focusing of charged particle beams using alternating gradient quadrupole magnetic fields. Yet the strong-focusing technique had been discovered two years earlier by N. C. Christofilos [see 5], an electrical engineer working in Athens, Greece. Despite his applications for patents and submission of a report on his work to the University of California Radiation Laboratory, his work was overlooked at that time. The principle of using strong-focusing fields for mass analysis was recognized by Wolfgang Paul [6] and his colleagues at the University of Bonn, and the first detailed account of the operation of a quadrupole ion trap appeared in the thesis of Berkling [7] in 1956.

1.2. PRINCIPLES OF OPERATION

The geometry of the quadrupole mass spectrometer is shown in Figure 1.1a, which is reproduced from the original patent [1]. The analyzer consists of a parallel array of four rod electrodes mounted in a square configuration. The ideal geometry (see below) dictates that each electrode should be hyperbolic in cross section, but, in practice, for ease of manufacture, round cylindrical rods often are employed, with the spacing optimized to approximate the ideal electric field [8]. The field within the analyzer is created by coupling opposite pairs of rods together and applying RF and direct-current (DC) potentials between the pairs (see Figure 1.1b [9]). Ions created within the source are injected through the parallel array, and under the influence of the fields they describe complex trajectories. Some of these trajectories are *unstable* in that they tend toward infinite displacement from the center so that the ions are lost, for example, through collision with an electrode. Ions that are successfully transmitted through the analyzer are said to possess *stable* trajectories, and these are recorded on the detection system. For a given interelectrode spacing $2r_0$, the path stability of an ion with a particular value of m/z depends on the amplitude of the RF drive potential (V), the magnitude of its frequency Ω, and the ratio of the amplitudes of the RF and DC (U) potentials. When $U = 0$, a wide band of m/z values is transmitted, and as the value of the ratio U/V is increased, the resolution increases so that at the stability limit only a single value of m/z corresponds to a stable trajectory,

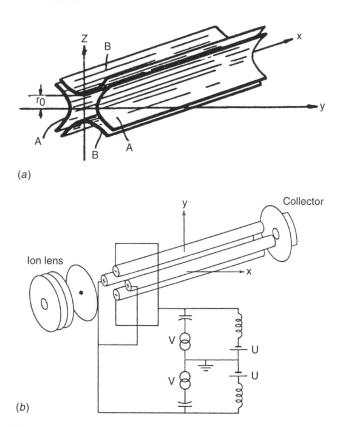

Figure 1.1. The quadrupole mass spectrometer, now known as the quadrupole mass filter. (*a*) The arrangement of four electrodes A, A, B, B of hyperboloidal shape which serve to create a cylindrically symmetric field, the electrodes being arranged at a distance r_0 from the *x* axis. (*b*) Schematic arrangement of the quadrupole mass filter. (Reproduced with permission from P. F. Knewstubb, *Mass Spectrometry and Ion-Molecule Reactions,* Cambridge University Press, Cambridge, 1969. Figure 1.1 from *Quadrupole Storage Mass Spectrometry*, by R. E. March, R. J. Hughes, J. F. J. Todd, Wiley-Interscience, 1989.)

resulting in the transmission and collection of ions of a single mass/charge ratio. In this way the quadrupole mass spectrometer acts as a mass filter, rather than as an energy or momentum spectrometer, and is referred to hereafter as a *quadrupole mass filter*. A mass spectrum may be generated by scanning the values of U and V with a fixed U/V ratio and constant drive frequency or by scanning the frequency and holding U and V constant. An introductory description of the operation of the quadrupole mass filter is to be found in Chapter 2 and in the standard text by Dawson [10].

The RF quadrupole ion trap is related directly to the quadrupole mass filter in that it can be visualized as being a solid of revolution generated by rotating the hyperbolic rod electrodes about an axis perpendicular to the *z* axis and passing through the centers of two opposing rods. This rotation of the electrodes results in one pair of rods joining up to form a doughnut-shaped ring electrode and the other two forming end-cap electrodes which are moved closer together, as illustrated in an early model of the ion trap shown in Figure 1.2a [11]; a cross-sectional view of the electrode structure is shown in Figure 1.2b. The system is axially symmetric, and for ideal field geometry within the trap the surfaces should again be hyperbolic. The field is generated by applying the RF and DC voltages between the ring electrode and the pair of end-cap electrodes, but, as will be seen later, it is generally more convenient to maintain the end-cap electrodes at ground potential and simply supply power to the ring electrode. The resulting field geometry and the potential ϕ at any point (x, y, z) within the device are given in Chapter 2. The internal radius of the ring electrode is now r_0.

From a consideration of the force $\mathbf{F}$ acting upon an ion, the equations of motion in the x, y, and z directions may be derived [see Eqs. (2.43)–(2.77)]. The absence of cross terms of the type xy, yz in the derived equations of motion means that the components of motion in each of the mutually perpendicular directions may be treated independently. At this stage, however, we pause to note that whereas the equations of motion are identical for x and y, there is a reversal of sign in the equation governing motion in the z-direction and a factor of 2 has been introduced. The opposite sign reflects the fact that the phase of the RF potential applied to the ring (x, y) is 180° out of phase with that experienced in the axial (z) direction, and the factor 2 results from the application of the Laplace condition that the rate of change of field gradient should be uniform throughout the trap volume ($\nabla^2 \phi = 0$). The form of the potential relates to the most commonly used form of trap geometry, in which we set $r_0^2 = 2z_0^2$, where $2z_0$ is the closest distance between opposing end-cap electrodes; a general form of the field potential has been described by Knight [12], and in this present work the *Knight formulation* has been used extensively throughout.

Each of the equations of ion motion is an example of the Mathieu equation [see Eq. (2.33)] [13, 14], the general form of which is characterized by the parameters a_u and q_u. The subscript u refers to motion in either the radial (xy) plane or the axial (z) direction due to the cylindrical symmetry of the quadrupole ion trap. The parameters a_r, a_z, q_r, and q_z are related to the experimental variables U, V, m/z, r_0, and Ω. The parameters a_u and q_u are functions of U and V, respectively, which refer to the amplitudes of the potentials developed between the ring and end-cap electrodes.

A fundamental property of the Mathieu equation is that the values of the parameters a_u and q_u determine whether the solutions are stable, that is, whether the displacement passes periodically through zero, or whether the displacement increases without limit to infinity (i.e., unstable). These two conditions are described in terms of *stability diagrams* plotted in (a_u, q_u) space, and for the ion trap we must consider two sets of overlapping diagrams such as to represent simultaneous stability of motion in the radial and axial directions. Various overlapping regions may be obtained, but the one of most direct relevance to the operation of the trap is that closest to the origin in (a_u, q_u) space. Only those ions having a_u, q_u coordinates lying within the four boundaries of the stability diagram will remain in stable trajectories within the device. The stability region is subdivided by the so-called iso-β_r and iso-β_z lines, and these define the characteristics of the

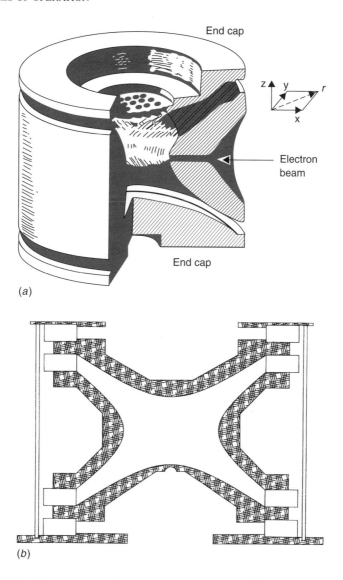

Figure 1.2. The three-dimensional quadrupole ion trap with rotational symmetry about the z axis. (*a*) The electrode structure required to produce three-dimensional rotationally symmetric quadrupole fields used in quadrupole ion trap. Note that with this early ion trap the electrons were injected through the ring electrode; on modern instruments the electrons or externally created ions are normally injected through one or more holes in an end-cap electrode. (Reproduced from P. H. Dawson and N. R. Whetten, The three-dimensional quadrupole ion trap, *Naturwissenschaften* **56** (1969), 109–112. Copyright Springer-Verlag (GmbH).) (*b*) Cross-sectional view of quadrupole ion trap electron structure. (Figure 1.2 from *Quadrupole Storage Mass Spectrometry,* by R. E. March, R. J. Hughes, J. F. J. Todd, Wiley-Interscience, 1989.)

ion trajectories in the radial and axial directions, respectively (see Figure 2.15). The derivation of the parameter β_u, which is related to a_u and q_u, is to be found in Chapter 2, and it is sufficient here to note how the nature of the ion motion may change quite dramatically as the values of β_r and β_z are altered. The importance of these effects in terms of the operation of the ion trap will become evident at a later stage.

1.3. UTILIZATION OF THE QUADRUPOLE ION TRAP

So far we have considered only the underlying principles of the means by which ions may be stored within the quadrupole ion trap. We now proceed to trace the various early applications of the device, concentrating mainly on its use as a mass spectrometer or as an ion storage source employed in conjunction with an external mass analyzer. However, there are other avenues of development, in particular its use for studying the spectroscopy of trapped ions, and while this topic is essentially peripheral to the theme of this book, several useful concepts concerning the physics of the system have emerged from this work. For its use as a mass spectrometer one clearly has to provide means both for the creation of ions and for their detection, and it is the development of the latter which has characterized the history of the ion trap over the past 50 years. Indeed, the milestones in the development of the ion trap for mass spectrometric applications may be divided into three distinct periods, as suggested in Table 1.1.

1.3.1. Early Mass-Selective Modes of Operation

The first methods employed for the detection of ions were based on the principle of *mass-selective detection*, in which the presence of ions in the trap was recorded through sensing the motion of the ions by means of circuitry connected between the end-cap electrodes. This was followed by *mass-selective storage*, in which the positive ions were ejected through holes in the end-cap electrodes into an electron multiplier; this method was extended further by interposing a mass analyzer between the trap and the detector to give a tandem arrangement that permitted the external mass analysis of stored ions. It is perhaps significant that up to this stage ion trap mass spectrometers were not actually available as commercial instruments, probably reflecting the complexity of the systems and the fact that they did not appear to offer any appreciable advantage over other types of analytical mass analyzers. More recently, we have seen the use of *mass-selective axial ejection* as the means of generating mass spectra, and it is this technique that has led to the dramatically increased interest in ion trap instruments; this topic is considered in greater detail at the end of this chapter and, indeed, forms the main subject of this book.

1.3.1.1. Mass-Selective Detection   This means of mass-selective ion detection was presented briefly by Paul and Steinwedel in the original ion trap patent [1], wherein they indicated that, while ions which possess unstable trajectories and thus impinge upon the electrodes represent an ohmic charge in the high-frequency circuit, ions with stable orbits are inductive charges since they do not contribute to the flow of current. The stable ions add to the inductive load of the system, and their presence may thus be detected by means of power-measuring devices.

TABLE 1.1. Milestones in the Mass Spectrometric Development of the Quadrupole Ion Trap

Year	Milestone
	Mass-Selective Detection
1953	First disclosure (Paul and Steinwedel)
1959	Storage of microparticles (Wuerker, Shelton, and Langmuir)
1959	Use as a mass spectrometer (Fischer)
1962	Storage of ions for RF spectroscopy (Dehmelt and Major)
	Mass-Selective Storage
1968	Ejection of ions into an external detector (Dawson and Whetten)
	Use as a mass spectrometer (Dawson and Whetten)
1972	Combination of QUISTOR with quadrupole mass filter for analysis of ejected ions (Todd, Lawson, and Bonner)
	Characterization of the trap, chemical ionization (CI), ion/molecule kinetics, etc. (Todd et al.)
1976	Collisional focusing of ions (Bonner, March, and Durup)
1978	Selective ion reactor (Fulford and March)
1979	Resonant ejection of ions (Armitage, Fulford, Hoa, Hughes, March, Bonner, and Wong)
1980	Use as GC detector (Armitage and March)
1982	Multiphoton (IR) dissociation of ions (Hughes, March, and Young)
	Mass-Selective Ejection
1984	Disclosure of ion trap detector (ITD) (Stafford, Kelley, Syka, Reynolds, and Todd)
1985	Ion trap mass spectrometer (ITMS) (Kelley, Stafford, Syka, Reynolds, Louris, and Todd)
1987	MS/MS, CI, photo dissociation, injection of ions, mass-range extension, etc. Fourier transform quadrupole ion trap (Syka and Fies)
1984, 1988	Deliberate addition of contributions from nonlinear field using stretched geometry and nonhyperbolic electrode surfaces (Kelley, Stafford, Syka, Taylor, Franzen)
1989	Extension of mass/charge range via resonant ejection (Kaiser, Louris, Amy, Cooks)
1990	High-resolution mode of operation (Schwartz, Syka, Louris)
1994, 1998	Linear ion traps (Schwartz, Senko, Syka, Hager)
1997	Use of ion/molecule reactions in isotope ratio measurements (Barber, Wright, Morse, Pillinger. Kent, Todd)
2002	Digital ion traps (Ding)

The precise means of achieving this nondestructive ion detection was described in the pioneering publications by Paul et al. [15] and by Fischer [16] and made use of the circuit shown in Figure 1.3. The main source of RF power was a 500-kHz generator coupled to the ring electrode, upon which was superimposed a DC voltage which could

References pp. 25–33.

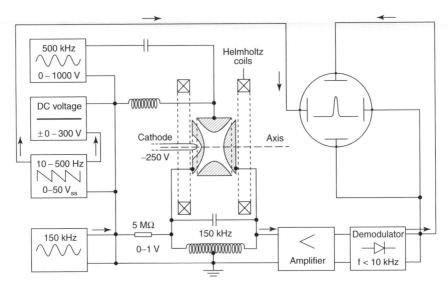

Figure 1.3. Circuit used by Fischer [16] for detection of ions using damping of an auxiliary RF circuit tuned to the fundamental frequency of ion motion. (Reproduced by permission of General Electric Corporate Research and Development. Figure 1.3 from *Quadrupole Storage Mass Spectrometry*, by R. E. March, R. J. Hughes, J. F. J. Todd, Wiley-Interscience, 1989.)

be swept with a sawtooth waveform. Ions were created continuously by an electron beam injected through one of the end-cap electrodes (and collimated by means of Helmholtz coils), and the secular motion of the ions along the axis of the trap was then detected by the resonant absorption of power from an auxiliary generator oscillating at 150 kHz, the output of which was applied across a 5-$M\Omega$ resistor and across half the pure resistance of a resonator. For resonance to occur, the a_u, q_u coordinates of the ions were slowly swept through the $\beta_z = 0.6$ line on the stability diagram. At resonance, the voltage developed was proportional to the resistance of the resonator and inversely proportional to the attenuation, such that the presence of ions led to the appearance of a y deflection on an oscilloscope display. Fischer [16] succeeded in recording a mass spectrum of krypton; however, the mass range and resolution of the instrument were severely limited. A further drawback was that the heavier ions were present in the trap during the detection of the lighter species, but not vice versa, so that the conditions under which the ions were detected changed during the scan. Yet another problem was that through the use of continuous ionization the ion density was almost certainly at the space charge limit of saturation, although Fischer suggested that at lower pressures the resolving power should increase through the increased mean collision time with neutral species. The lowest detectable partial pressure was reported to be $\sim 3 \times 10^{-6}$ Pa, equivalent to $\sim 2 \times 10^4$ ions cm^{-3} in the trap.

Some eight years after Fischer's work, Rettinghaus [17] described an alternative means for operating the ion trap utilizing mass-selective detection. The trap was fabricated with spherical electrode surfaces ($r_0 = 12$ mm, $z_0 = 8.5$ mm), and ions were created at low pressure (3×10^{-7} Pa) by injecting a long-duration (5s) pulse of electrons through an aperture in the ring electrode. The main drive frequency was 1.6 MHz, and the "detection" and "comparison" circuits were first balanced in the absence of ions in the trap. The spectrum was then generated by scanning the RF amplitude V along the $a_z = 0$ line (i.e., zero applied DC potential, $U = 0$ or the q_z axis) such that when the operating point for each m/z value crossed the $\beta_z = 0.5$ line the secular motion of the ions oscillating at 0.41 MHz came into resonance with the detection circuit. It was noted that with a longer ionization time the relative intensities of the higher masses decreased, and this was ascribed by Rettinghaus to the effects of space charge causing discrimination against these species. Space charge is in fact a very important consideration in ion trap operation, since it limits the ion concentration which may be achieved. As first noted by Fischer [16], the presence of this charge acts so as to defocus the ions in all directions and is manifested as a shift in the boundaries of the stability diagram. The effects of space charge are considered in more detail in Chapter 3.

Another feature observed by Rettinghaus [17] in the mass spectra of background gases was the appearance of a peak at m/z 29. This peak was ascribed to the species COH^+ formed through ion/molecule reactions occurring between CO^+ and hydrocarbon molecules present in the background. Evidence for this mechanism was presented in the form of a plot of the signal intensity due to m/z 29 divided by the ratio of the sum of the signal intensities due to m/z 29 plus m/z 28; this ratio showed an approximately sixfold increase with storage time over a range of 0–7 min and is reproduced in Figure 1.4. This is the first recorded observation of ion/molecule reactions within an ion trap; such processes have been the subject of much subsequent study. Other secondary effects occurring within quadrupole ion traps were noted by

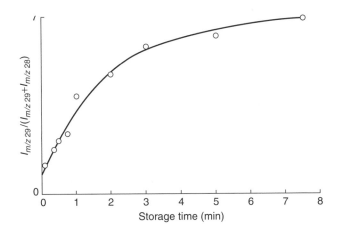

Figure 1.4. Ratio of signal intensity due to m/z 29 to sum of signal intensities due to m/z 28 and m/z 29 as function of storage time [17]. (Reproduced by permission of Birkhauser Verlag from *Zeitschrift für Angewandte Physik*. Figure 1.7 from *Quadrupole Storage Mass Spectrometry*, by R. E. March, R. J. Hughes, J. F. J. Todd, Wiley-Interscience, 1989.)

References pp. 25–33.

Burnham and Kleppner [18] at about this time (1968), but no detailed account of their work seems to have appeared.

The main aim of Rettinghaus' investigation was to examine the potential of the ion trap for use as a partial-pressure gauge in ultrahigh vacuum systems, and he reported that it was possible to detect partial pressures as low as 10^{-11} Pa, corresponding to approximately four ions in the trap. The longest trapping time (half-life of the ion concentration) observed was about 20 min. However, the mass discrimination observed proved difficult to control, and there seemed to be little commercial advantage to be gained over the quadrupole residual gas analyzers that were being manufactured at that time.

Until relatively recently this appeared to be the end of the history of quadrupole ion trap operation using in situ mass-selective detection, but lately there has been renewed interest in this approach. Syka and Fies [19, 20] demonstrated the feasibility of implementing Fourier transform techniques to transient ion image currents by producing mass spectra obtained in this manner from an ion trap. Essentially, the scheme is to excite the resonances of ions having a broad range of m/z values at a specific value of β_z and then to perform a Fourier transformation on the output from the receiving circuit coupled between the end-cap electrodes. Studies to date indicate that this mode of detection does not offer any advantages over other mass spectrometric modes of operating the ion trap; the relatively prolonged image current decay transients require low pressure, thereby negating a strength of the ion trap which is to yield mass spectra of high quality at elevated pressures. Goeringer et al. [21] employed image current detection to demonstrate multiple remeasurement of the same population of ions held in an ion trap at a pressure of $\sim 10^{-4}$ Torr; the efficiency of ion remeasurement was >99%. The potential of broadband Fourier transform ion trap mass spectrometry has been explored also by Cooks and co-workers [22, 23].

1.3.1.2. Mass-Selective Storage

The middle to late 1960s saw a major renewal of interest in the ion trap and, at this stage, the history of its development really begins to follow two parallel branches. On the one hand, spectroscopists, led by H. G. Dehmelt at the University of Washington, Seattle, saw the RF quadrupole ion trap as a means of enabling a wide variety of gas-phase spectroscopic experiments to be performed with simple atomic and molecular ions. This early work was summarized by Dehmelt in two important reviews [24, 25], and these have acted as significant stimuli for further work on the physical characterization of the ion trap as an ion source and as a mass spectrometer. More recent work [26–30] has been reviewed by Schuessler [31] and Wineland et al. [32, 33]. The other branch of development, as a mass analyzer, originates from the realization by Dawson and Whetten [34–36] that ions could be ejected efficiently from the trap through holes in one of the end-cap electrodes onto the first dynode of an electron multiplier or into a "channeltron" and thus be detected externally, thereby avoiding many of the difficulties associated with mass-selective detection.

In order to use the ion trap as a mass spectrometer, the method of mass-selective storage was developed. The idea follows closely the operating principle of the quadrupole mass filter and involves selecting a working region of the stability diagram where

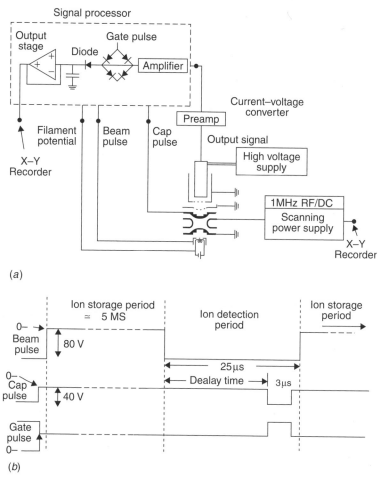

Figure 1.5. (*a*) Circuit employed by Dawson and Whetten [37] for detection of ions by mass-selective ion storage. (*b*) Timing sequence for the circuit. (Reproduced by permission of General Electric Corporate Research and Development. Figure 1.8 from *Quadrupole Storage Mass Spectrometry,* by R. E. March, R. J. Hughes, J. F. J. Todd, Wiley-Interscience, 1989.)

only ions with a single value of *m/z* possess stable trajectories and hence are stored. The circuit and associated timing sequence employed by Dawson and Whetten [37] are shown in Figures 1.5*a* and 1.5*b*, respectively. Ions were created by admitting a beam of electrons from a filament via a gating electrode through holes in one of the end-cap electrodes. This ionization pulse lasted ~5 ms, after which the ions were stored for a delay time of 25 μs before being extracted into the multiplier by means of a pulse applied to the other end-cap electrode. Mass selection occurred during the delay time, and to generate a mass spectrum the amplitudes of the DC and RF fields

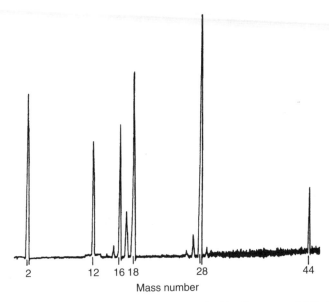

Mass number

Figure 1.6. Mass spectrum of background gases acquired by Dawson and Whetten [37] at a total pressure of ca. 10^{-5} Pa. (Reproduced by permission of General Electric Corporate Research and Development. Figure 1.9 from *Quadrupole Storage Mass Spectrometry,* by R. E. March, R. J. Hughes, J. F. J. Todd, Wiley-Interscience, 1989.)

were scanned slowly at constant U/V. With a total duty period of ~5 ms the maximum repetition frequency was approximately 200 Hz, with the signal duration being in 3-μs bursts. Thus, to avoid the loss of intensity that would otherwise result from the averaging of the output and to eliminate the detection of the excess of ions present during the storage and detection periods (see Figure 1.5b), a "boxcar detector" system was incorporated into the signal-processing circuit. Compared to the earlier mass spectra produced employing mass-selective detection, the quality of the data obtained with this new method of operation was excellent. An example of the recording of peaks in the mass spectrum of background gases is shown in Figure 1.6.

The idea of detecting ions by ejecting them from the trap really represents a watershed in the development of the device for mass spectrometric applications, and within a short space of time various other groups of workers, including Harden and Wagner [38, 39], Dawson and Lambert [40], Mastoris [41], and Sheretov and co-workers [42–48], reported further developments and refinements of this method.

1.3.2. Ion Loss Processes

The sensitivity of any mass spectrometer is determined directly by the fraction of the ions formed from a given quantity of sample and which, after analysis, reaches the detector. For mass analyzers based upon principles involving beam transport, for example, magnetic sector, time of flight, or quadrupole, this is clearly related to the transmission characteristics of the analyzer. With the ion trap the corresponding parameter is the rate of ion loss during the period between creation and detection, and it is hardly surprising, therefore, that considerable attention has been paid to the study of this phenomenon. Ion loss may occur through a number of different processes.

1.3.2.1. Unstable Trajectories   There are essentially two types of unstable trajectories:

1. Intrinsically (mathematically) unstable trajectories, where the a_u, q_u coordinates for a given ionic species equate with a working point that lies outside the stability boundary. The trajectory is unbounded, so that the ion is rapidly removed from the trap; this mechanism is, of course, the means by which mass-selective storage is achieved wherein unwanted ions are ejected from the ion trap.
2. Quasi-unstable trajectories, where the mathematical conditions for stability exist but nevertheless the ion is lost because the limit of excursion of the ion exceeds the internal dimensions of the device. Such an ion might, for example, have been formed very near one of the electrode surfaces and/or have a significant initial velocity.

1.3.2.2. Interactions   The occurrence of ion/neutral molecule collisions and ion/ion interaction processes may lead to the charged species developing unstable orbits. A number of different kinds of effects are important here: for example, depending upon the nature of the species, ion/neutral collisions may variously lead to damping, elastic scattering, inelastic scattering, charge transfer, and ion/molecule reactions.

In scattering collisions one can visualize an ion with a stable trajectory being suddenly deflected such that its new situation is effectively one of having an unfavorable initial starting position and/or initial velocity; in addition there may be a transfer of kinetic energy (i.e., velocity) from one coordinate direction to another so that the original simple assumption relating to the independence of the three components of motion no longer applies. Such collisions have been the subject of a number of studies [49–52] in which it has been shown that the relative masses of the ion and the neutral species are critical in determining whether the ion trajectory is destabilized (when the neutral mass is heavier than that of the ion) or stabilized (when the neutral mass is less than that of the ion). Indeed, the latter effect plays a crucial role in improving the performance of the modern ion trap mass spectrometers [53] through the influence of momentum-moderating collisions with helium buffer gas, which, by reducing the kinetic energies of the ions and causing the trajectories to "collapse" toward the center of the trap, improves dramatically both the sensitivity and resolution of the device. The theory, dynamics, and modeling of the ion trap mass spectrometer are discussed in detail in Chapters 2, 3 and 4, respectively. Collisional cooling has also been important in the uses of RF quadrupole ion traps for the study of the spectroscopy of trapped ions.

In the case of charge transfer, an ion in a stable trajectory is removed and a new one, probably with effectively zero initial velocity, is created. Whether or not this new species will remain stable again depends upon the precise operating conditions at the time of the event as well as upon the m/z value of the product ion. Provided the conditions are chosen correctly, charge transfer does not necessarily lead to ion loss; for example, low-pressure kinetic studies on argon–methane mixtures have shown that within experimental error the rate of loss of $Ar^{+\cdot}$ is exactly balanced by the rate of formation of CH_3^+ and CH_2^+ [54, 55]. March and co-workers [56, 57] reported on detailed experiments and the theoretical modeling of systems such as $Ar^{+\cdot} + Ar$, where it has been shown that such interactions can again lead to the migration of ions toward the center of the trap, thus improving the storage efficiency of the device. The influence of collisions on the motion of trapped ions is considered in greater detail in Section 3.4.

The problems associated with ion/ion scattering are probably less easy to discuss and may be visualized as giving rise to two different kinds of effects. On the one hand, at the microscopic level of individual collisions, a pair of like-charged ions will repel one another, possibly leading to either or both of them developing unstable trajectories. This effect will clearly become more important as the concentration of ions within the trap is increased, for example, by lengthening the ion creation period. Attempts have been made to quantify the process of ion/ion scattering in terms of the kinetic approach described below. On the other hand, the theory of ion containment within the trap, described above and in Chapter 2, is based on the assumption that there is only a single ion in the trap. As the ion concentration increases, the trapping potentials are modified by a defocusing effect due to the space charge, which has the effect of modifying the locations of the boundaries of the stability diagram. The importance of space charge perturbation was first recognized by Fischer [16] and again has been examined in more detail by other workers. A more detailed account of the effects of space charge is included in Chapter 3.

Reference was made to the idea that ion scattering could be considered in terms of a kinetic approach. The first such approach was employed by Fischer [16], who determined a *mean ion lifetime* $\bar{\tau}_i$ by observing how the signal height h varied as the ionization time was increased. Thus over a period of 5–20 μs (depending upon the electron beam current) the signal was found to first increase linearly and then reach saturation, corresponding to level h_∞. The equation

$$\bar{\tau}_i = \frac{h_\infty}{\dot{h}_0} \tag{1.1}$$

then gave a value for $\bar{\tau}_i$, where $\dot{h}_0$ is the rate of increase of the signal with time over the initial linear period.

The problem was considered subsequently by Dawson and Whetten in terms of a rate law analogous to those employed in chemical kinetics, and this approach was also adopted by other workers [39, 58, 59]. Consider the situation where the trap contains N ions cm^{-3} then during the creation period the rate of change of N with time must represent the difference between the rate of creation of ions and the rate of their loss:

$$\frac{dN}{dt} = k_1 p - (k_2 N^2 + k_3 Np) \tag{1.2}$$

where p is the pressure of neutral molecules, k_1 is a rate constant for ion creation, k_2 the rate constant for loss by ion/ion scattering, and k_3 the rate constant for loss by ion/neutral scattering. In practice, a typical plot of N versus t is obtained by increasing progressively the length of the ionization period, allowing a short period for the quasi-unstable ions to be rejected, and then ejecting the ions into the detector by pulsing an end-cap electrode.

To evaluate the rate constants, we can simplify Eq. (1.2) by imposing certain sets of conditions. Thus over the initial linear period we can put $N \cong 0$ so as to obtain

$$\frac{dN}{dt} = k_1 p \tag{1.3}$$

from which k_1 may readily be found for a given value of p. Similarly, if we work under saturation conditions ($N = N_\infty$), then we have that $dN/dt = 0$, so Eq. (1.2) becomes

$$k_1 p = k_2 N_\infty^2 + k_3 N_\infty p. \tag{1.4}$$

Operating at relatively high pressure allows us to assume that $k_3 p \gg k_2 N$, so that k_3 may be found from

$$k_3 = \frac{k_1}{N_\infty}. \tag{1.5}$$

The value of k_2 may now be obtained using a curve-fitting routine on the first derivative of the buildup plot of N versus t. An alternative approach, which is more precise, is to monitor the decay in the value of N after the electron beam has been switched off. Thus

$$\frac{dN}{dt} = -(k_2 N^2 + k_3 p). \tag{1.6}$$

Further details of these two methods and comparisons of different sets of results have been published elsewhere [55, 58]. A typical set of data for $Ar^{+\cdot}$ ions published by Todd and co-workers [60] is given in Table 1.2, from which one can deduce the following rates of ion loss:

$$k_2 N^2 = 2.9 \times 10^{10}\,\mathrm{cm^{-3}\,s^{-1}} \quad \text{(ion/ion scattering)}$$

$$k_3 N p = 4.2 \times 10^9\,\mathrm{cm^{-3}\,s^{-1}} \quad \text{(ion/neutral scattering)}$$

Thus under these conditions it would appear that ion/ion scattering is slightly the more dominant effect, although ion/neutral scattering will become progressively more important as the value of N falls. Dawson et al. [61] deduced that in their trap ion/neutral scattering was the major ion loss mechanism above 10^{-6} Pa in the mass-selective storage mode, whereas Harden and Wagner concluded that the corresponding pressure was $\sim 10^{-4}$ Pa [39].

1.3.2.3. Nonlinear Resonances   So-called nonlinear resonances constitute another mechanism for ion loss. Because of deliberate or unintentional imperfections in the quadrupole field due, for example, to nonideal electrode spacing or nonperfect surfaces, there may exist higher order terms for the expression for the potential in addition to the ideal second-order form of the potential. These effects were first considered by von Busch and Paul [62] in relation to the behavior of the mass filter, and a detailed application to the ion trap has been presented by Whetten and Dawson [63]. An in-depth treatment of this topic is beyond the scope of this brief history, but essentially what happens is that at certain values of β_r and/or β_z the ion motion comes into resonance with the applied field, resulting in unstable trajectories. In Figure 1.7

TABLE 1.2. Typical Ion Loss Parameters for Ar$^{+\cdot}$ from an Ion Trap

r_0	1.0 cm
Pressure	32 mPa
Drive frequency	0.762 MHz
V	140 V (zero to peak)
q_z	0.59
N_∞	1.04×10^7 cm^3
k_1	1.05×10^{13} cm^3 Pa^{-1} s^{-1}
k_2	2.7×10^{-4} cm^3 s^{-1}
k_3	1.3×10^5 Pa^{-1} s^{-1}

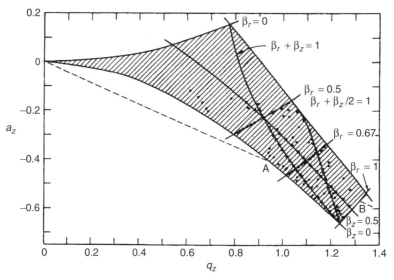

Figure 1.7. Stability diagram for quadrupole ion trap showing resonance lines caused by third- and fourth-order distortion. The points represent experimental measurements. The principal resonance dips fall on the theoretically predicted curves. (Reproduced, with permission, from P. H. Dawson, *Quadrupole Mass Spectrometry and Its Applications*, Elsevier Science Publishers B.V.)

are shown the predicted resonance lines for third- and fourth-order distortions of the ion trap. Evidently a number of these lines converge at the lower apex, and this has been found to correlate well with the extensive peak splitting which may sometimes he observed when the device is operated in the mass-selective storage mode. Further detailed consideration of nonlinear resonances is given in Chapters 3, 7, and 8.

1.3.2.4. Self-Emptying   A final means of ion loss from the low-pressure trap, which is even less well characterized than that of nonlinear resonances, is "self-emptying." Thus both Fischer [16] and Whetten and Dawson [63] observed that at high ion concentrations (continuous ionization) the ion signal was oscillatory. Fischer attributed this to the effects of space charge causing a shift in the position of the resonance lines within the stability diagram so that ions were removed, leading to a reduction in the space charge and relaxation of the resonance line back to its former position. Because of the continuous creation of ions, this cycle was then repeated, with a frequency of 50–1000 Hz depending upon the operating conditions. In the experiments described by Whetten and Dawson it was found that when the (a_z, q_z) working points were close to the $\beta_z = \frac{1}{2}$ or $\frac{2}{3}$ lines, the output of the detector exhibited an oscillatory character, the period of which ($\sim 6 \times 10^{-4}$ s) was inversely proportional to the ionization rate (which was changed by varying the gas pressure or the electron current) and was proportional to the square of the frequency of the applied RF power. The phenomenon was explained on the basis of the idea that the ions are able to gain additional energy from plasma oscillations of the charge cloud.

1.4. THE LOW-PRESSURE QUISTOR–QUADRUPOLE COMBINATION

The ability of the ion trap to store ions with a wide range of m/z values in the RF-only mode prompted Todd and co-workers to utilize the low-pressure trap as the ion source for a quadrupole mass filter in a series of experiments designed originally to study the fragmentation behavior of metastable ions. Thus, in this early form of tandem instrument, ions could be created by electron ionization, trapped for a defined period of time (along with any product ions from metastable decay reactions), and then pulse ejected into the quadrupole mass filter for mass analysis. This development led to the creation of the name QUISTOR (QUadrupole Ion STORe) [64], and this tandem instrument has proved to be an excellent vehicle for characterizing the physical aspects of ion trap operation as well as for studying the kinetics of ion/molecule reactions. Ionization was effected by pulsing in a beam of electrons through a hole in the ring electrode such that after a defined storage time the application of appropriate pulses to either or both of the end-cap electrodes caused the ion packet to be ejected from the ion trap into the quadrupole mass filter. The "detection pulse" corresponded to the opening of the amplifier gate of a "boxcar" detector circuit whose function was to eliminate spurious background signals arising from unstable ions ejected during the creation and storage periods and to retrieve the signal pulses, which would otherwise be averaged out over the complete duty cycle. A typical repetition frequency was 100 Hz. Depending upon the particular experiment being conducted, the quadrupole mass filter could be set to transmit a specific value of m/z or scanned slowly to generate a mass spectrum of the ejected ions.

Early results showed that despite the presence of RF voltages applied to the QUISTOR, the trap behaved like a conventional ion source when used for the determination of the ionization energy of helium [58]. Ion/molecule reactions could readily be investigated by recording the mass spectra of the ejected ions as a function of the storage time, and it was found that with a pressure of $\sim 10^{-2}$ Pa of methane and storage times in the range of 0–3 ms the distribution of primary and secondary ions closely resembled those obtained with conventional ion sources operating at around 130 Pa pressure. This led to the demonstration that analysis by chemical ionization could easily be performed with the QUISTOR [65], and detailed investigations of the reactions of CH_5^+ ions with water and with ammonia were carried out [66]. The rate constant values obtained in this latter work made use of data [67] for the reaction

$$CO_2^{+\cdot} + H_2 \rightarrow CO_2H^+ + H^\cdot \tag{1.7}$$

for calibration purposes and suggested average ion kinetic energies in the 1–3-eV range which were consistent with calculations based on the pseudopotential well model proposed by Dehmelt [24] and considered in more detail in Chapter 3. The reaction represented by Eq. (1.7) is also employed in the measurement of isotope ratios using the ion trap as part of the Rosetta mission to characterize a comet (see Chapter 9).

The investigation of charge transfer effects has led to important increases in our understanding of the processes occurring within the ion trap. The first system studied was the reaction $Ar^{+\cdot} + CH_4$, and excellent data for Ar/CH_4 ratios in the range 1.0 : 0.1 to 1.0 : 3.4 were obtained [54]. Subsequently, in a study of the variation in the ratios of detected currents for $Ar^{+\cdot}$ and Ar^{2+} ions formed in the QUISTOR from argon under a variety of conditions, March and co-workers [56] invoked charge transfer as a process which leads to the selective migration of ions toward the center of the trap from which ions are extracted more efficiently. An alternative explanation advanced by Todd et al. [68] was based on the premise that space charge–induced shifts in the $\beta_z = 0$ boundary of the stability diagram would affect the stability of the two ionic species differently; however, in an extremely elegant subsequent experimental and theoretical study of mixtures of neon and carbon dioxide, the March group [69] demonstrated unequivocally the importance of ion migration, both for ions formed from these species and from the compound d_6-dimethyl sulfoxide.

Mention has already been made of the nonideal behavior of the ion trap that can be caused by space charge–induced defocusing of the ion cloud. Todd and co-workers [70] examined these effects in some detail using a technique in which the application of accurately measured DC and RF potentials to the ring electrode of a low-pressure trap allowed them to determine the precise working points at which the trapping of specific types of ions became possible. The results of these experiments indicated that distortion from the theoretical (single-ion) diagram becomes more pronounced as the value of m/z is reduced. When the data were replotted in terms of U–V coordinates, it was found that the lower apex (i.e., the operating region employed in the mass-selective storage mode) for each of the ions did indeed lie on a scan line, but one that had a positive intercept (of ~8.0 V) along the a_z axis. This is equivalent to a net defocusing potential acting along the axis of the trap. Todd et al. [71] have discussed their observations of space charge–induced effects in terms of the pseudopotential well model [24] and in relation to theoretical treatments given by Fischer [16] and Schwebel et al. [72] (see Chapter 3 for a more extensive discussion of this topic).

Progress in the study of ion/molecule reactions occurring within the ion trap was aided significantly by two important developments in technique: the selective ion reactor [73] and QUISTOR resonance ejection, both of which were developed by the group at Trent University. In the first of these, which was based on an original idea published by Bonner [74], a single reactant ion species may be isolated by applying appropriate DC pulses to the end-cap electrodes during the ion creation period, thus rendering unstable the trajectories of all ions except those of the selected precursor. So far this approach has not been used extensively, although a variant of the method is now employed in certain operating modes associated with mass-selective ejection (see Chapters 3 and 7).

QUISTOR resonance ejection (QRE) [75] utilizes a technique originally employed by Paul et al. [76] in their use of a quadrupole ion trap for the separation of isotopes, namely, the application of supplementary electric fields to energize ions mass selectively and thereby render unstable the trajectories of specific ion species. In the case of the QUISTOR this is effected by applying a low-amplitude alternating-current (AC) potential between the end-cap electrodes (in fact only one end-cap electrode need be energized since the other is held at earth during the storage period). By this method it has been possible to undertake a number of different types of investigations. For example, the rate constant for partial charge transfer reactions such as $Ar^{2+} \rightarrow Ar^{+\cdot}$ may readily be found by first resonating out primary $Ar^{+\cdot}$ ions formed during the ionization period and then monitoring changes in the Ar^{2+} and $Ar^{+\cdot}$ signal levels as the reaction period is increased. In the same manner, the coupling of reactant and product species involved in ion/molecule reactions may be demonstrated [77–79] in a way similar to the ion cyclotron double-resonance (ICDR) technique [80], and it has proved possible to initiate endothermic ionic processes and demonstrate a rate dependence upon amplitude of the resonant energy supplied [75]. In practice, it is found that the precise frequency at which a given ion comes into resonance deviates from that predicted theoretically, and this is analogous to the space charge–induced shifts in the stability boundaries noted earlier. By studying this effect in detail, Fulford et al. [75] were able to deduce that under their operating conditions with 2-propanol as the sample the saturation ion density was $6.4 \times 10^{12}\,m^{-3}$, which compared well with other literature values. This technique of resonant excitation has been applied to the study of collision-induced dissociation processes with the ion trap operating in the mass-selective mode [81] (see Chapters 3 and 7).

References pp. 25–33.

1.5. EARLY STUDIES OF THE THEORETICAL ASPECTS OF LOW-PRESSURE ION TRAP OPERATION

Initial attempts to model the behavior of quadrupole devices relied on the calculation of individual trajectories, assuming specific initial conditions of displacement, velocity, and phase angle of the RF drive potential at which the ions were formed, and the earliest work by Paul et al. [15, 76] relied upon finding analytical solutions to the equations of motion. Subsequently Lever [82] and Dawson and Whetten [34] used numerical integration of the Mathieu equation with the help of the fourth-order Runge–Kutta equation.

Two important advances which led to a considerable simplification of the computations were the use of matrix techniques, first applied by Richards et al. to the case of a quadrupole analyzer driven by a square wave [83], and the treatment of the ion motion by the methods of phase space dynamics, pioneered by Baril and Septier [84]. The detailed derivation of the expressions used in these techniques is beyond the scope of this historical survey, and the reader is referred to other sources [10, 85].

The major advantage of the phase space representation is that it is easy to envisage the behavior of the ensemble of ions at a specified phase angle with respect to the motion in a given direction: for stability all the values of the u, $\dot{u}$ coordinates must lie on or within an appropriate ellipse, and all those ions with values outside the boundary will be rejected. In this case the ellipse is termed an *acceptance ellipse*, and its area, proportional to the acceptance ε, represents the range of initial conditions that the system will accept. An alternative interpretation is that a focusing system will only transmit a beam of particles whose values of u and $\dot{u}$ are bounded by the ellipse, in which case ε is representative of the emergent beam and is called the *emittance* of the *beam ellipse*. For the QUISTOR–quadrupole combination there should ideally be *phase space matching* between the two components so that there will be optimum transmission when the emittance of the beam ellipse of ions ejected from the QUISTOR exactly coincides with the acceptance of the ellipse for the quadrupole. Such phase space matching has not been investigated for the QUISTOR–quadrupole system, although it has been considered for normal ion source–quadrupole mass filter combinations [86] and for tandem quadrupole systems [87]. Obviously, for a full description of the behavior of the ion trap, one must consider ellipses corresponding to the components of motion in the orthogonal x, y, and z directions; whereas the first two are in fact identical, motion in the z direction will be out of phase by half a cycle, and the value of u_{max} will be given by z_0.

Matrix methods, based upon phase space dynamics, were employed by Dawson and Lambert, who performed a detailed study of the behavior of the ion trap operating in the mass-selective storage mode [40]. In particular, they extended Fischer's earlier work [16] on the efficiency of trapping under various conditions and showed that radially-directed electron beams should be more effective in this regard than axially-directed ionization. They also demonstrated how ions formed with significant kinetic energies would be confined less efficiently than those with zero velocity; this work suggested that light ions such as $H_2^{+\cdot}$ and $He^{+\cdot}$ would be very difficult to trap. The efficiency of trapping light ions was also investigated by Baril [88], and the ability of the trap to confine multiply charged ions formed by sequential ionization was studied by Schwebel and co-workers [72, 89] and Baril [88].

Two further important contributions by Dawson included a description of how the phase space approach could be utilized to determine velocity distributions of the stored ions [90] and an analysis of the effects of collisions in terms of ways in which the modified velocity distributions could be represented by changes in the ellipse parameters [91]. In the first of these papers, it was assumed that if all points within a given phase space ellipse are occupied uniformly, then the number of ions having a particular velocity $\dot{u}$ will be proportional to the width of the ellipse at that $\dot{u}$ value; the average velocity may then be found by suitable integration. This treatment was extended subsequently by Todd et al. [92] and the results compared with calculations of velocities derived from the pseudopotential well model and one using a smoothed general solution of the Mathieu equation. Todd and co-workers [93] also adapted Dawson's velocity distribution approach to the calculation of the spatial distribution of ions (neglecting space charge and migration effects) and showed how the cylindrical volumes containing the ions varied with the phase angle of the RF drive potential for different values of the parameters a_u and q_u. With this treatment it was possible to offer an explanation for the shape of the *total-pressure curve* which is traced out when the ejected ion signal is plotted against the value of q_u at the time at which the ions are created and stored ($a_u = 0$).

Other investigations included (1) an appraisal of the accuracy of the pseudopotential well model in terms of phase space dynamics (where it was shown that only at very low q_u values (with $a_u = 0$), did the majority of the u, $\dot{u}$ coordinates lie simultaneously within the relevant ellipses for both models [94]) and (2) a time-of-flight analysis of ions ejected from the trap by the application of narrow withdrawal pulses which were synchronized with a specific phase angle of the RF drive potential [95, 96]. In the latter experiments it was shown that the velocities of the fastest ions corresponded to those with the calculated maximum velocity traveling toward the detector at the moment of ejection. Perhaps the most significant result of the phase-synchronized pulse-ejection experiments was to show that the optimum phase angle for ejection of the ions coincides with the time at which the velocities of the bulk of the ions in the radial phase are directed toward the z axis [71].

A comparative survey of recent methods employed for the simulation of ion trajectories in the quadrupole ion trap in given in Chapter 4.

1.6. RESEARCH ACTIVITIES WITH THE QUADRUPOLE ION TRAP

Concurrently with the characterization of the quadrupole ion trap, modeling of ion trajectories both with and without collisions with neutrals, and the development of theoretical treatments of ion trajectory behavior, a miscellany of research activities with the ion trap was being pursued. The variety of these activities was due, in part, to the appreciation of three further aspects of ion trap behavior. First, the number of ion/neutral collisions can be controlled, and the effects of such collisions are the quenching of an ion's internal energy and modification of its trajectory, particularly with respect to the maximum excursion from the center of the ion trap. Collisional focusing of the ion ensemble in this manner permitted ready illumination of virtually the entire ion cloud by a laser beam. Second, laser beam irradiation of the ion cloud caused not only photodissociation of ions but also trapping of photoproduct ions. Thus ions created in the trap, either by photodissociation (predominantly by multiphoton absorption, thus far) or by unimolecular fragmentation of metastable ions, may be trapped and mass analyzed. Third, while the field at the center of the ion trap is vital for prolonged storage of ions, the shapes of the electrodes to which the field-forming potentials are applied are not so critical. Thus it is possible to create ion-confining fields with a cylindrical ion trap wherein a cylindrical "barrel" electrode and two planar end-cap electrodes are substituted for the hyperboloidal electrodes of the quadrupole ion trap.

The focusing effects of ion/neutral collisions upon a trapped ion ensemble were shown dramatically by Dehmelt and co-workers [97, 98], thus supporting the argument of Bonner et al. [56] for collisional migration of ions. This focusing effect on trapped ions in collision with a light buffer gas was also pursued in great detail by André and co-workers [99, 100] and is vital to the subsequent commercial development of the ion trap [101]. In addition to inelastic collisions, reactive ion/molecule collisions were studied with particular emphasis on ion/molecule equilibria for the determination of proton affinities [102] and the chemistry of gaseous ions [103].

Laser irradiation has been employed extensively for sideband cooling of trapped ions by the groups of Dehmelt [104] and Wineland et al. [30, 105]. Though laser cooling, RF spectroscopy, and frequency standards (wherein the frequency of laser-induced light from ions confined in an ion trap may be used as an ultraprecise time standard) are beyond the scope of this volume, the line of demarcation is not firm. We have attempted to acknowledge particularly those contributions from pure physics to our understanding of the ion trap as it is applied to chemical problems. Two examples will serve to illustrate the point: the visible focused ion cloud shown by Dehmelt and co-workers [97, 98] and the laser probing of confined ions by Desaintfuscien and co-workers [106]. The former confirmed the focusing action due to ion collisions with a light gas, while the latter mapped the ion density as a function of working point, both with and without collisions; knowledge of spatial distribution and ion density is vital to the interpretation of ion signals in chemical applications. Laser-induced fluorescence from trapped ions has been employed in the determination of radiative lifetimes of trapped molecular ions of $HCl^{+\cdot}$ and $HBr^{+\cdot}$ [107] and in the investigations by Mahan, O'Keefe, and co-workers [108–111] of excited fragment ions formed in excited states by electron impact. Infrared multiphoton dissociation of trapped proton-bound dimer ions derived from aliphatic alcohols has been pursued by March and co-workers [112, 113] using a CO_2 continuous-wave laser, and the technique has been extended to photodissociation of protonated species for isomer differentiation [114].

The development of cylindrical ion traps [115–117] and the characterization of such devices [118–121] were followed rapidly by the utilization of these ion traps for RF spectroscopy [122]. Microtraps of modified geometry [123–125] have been used also for RF spectroscopy. Lifshitz and co-workers, in their studies of metastable ions fragmenting in the millisecond time range, have employed an ion trap with cylindrical geometry combined with a quadrupole mass filter as a means of obtaining time-resolved photoionization mass spectra [126, 127].

In tracing this history of the ion trap, we have seen that the interplay of theory and experiment has led to a greater understanding of the chemical and physical processes which take place within the device. Indeed, these investigations have continued with extensive theoretical work on the means by which one may trap, in the QUISTOR, ions generated externally and injected into the QUISTOR [128–131], and experimental studies [132, 133] which include conversion of an ion to a Rydberg state so that as a neutral species the ion may enter the ion trap wherein it is reionized [134]. Wanczek and co-workers used a cylindrical QUISTOR to trap ions formed by electron ionization of carbon dioxide clusters present in a supersonic beam [135–137] and, using a mass-selective technique, investigated cluster ion decay in the millisecond time range. However, as noted above, it is perhaps significant that up to this point in time the ion trap had still not been developed commercially. A detailed account of recent research utilizing cylindrical ion traps and microtraps is given in Chapter 6.

Although there had been early thoughts about the viability of the ion trap as a commercial mass spectrometer [138] and there was evidence that the trap could be employed for the enhancement of weak mass spectral peaks [139] and as a successful detector for use with a gas chromatograph [140], there were several factors working against its adoption as a routine analytical instrument. Specifically, it did not appear to offer any marked advantages over the existing instrumentation in relation to applications which were current in the mid-1970s. In addition, the technology was relatively unknown and appeared to be complex, especially regarding the efficient ejection of ions from the trap and the associated control systems for optimizing the performance of the electronic circuits. However, in 1980 and 1981 the position changed dramatically, with the discovery at Finnigan MAT [101] of an alternative means of generating mass spectra with the trap (which both simplified the operation and overcame the problems of ion ejection) and the advent of relatively inexpensive microcomputers such as the IBM PC and its successors.

1.7. MASS-SELECTIVE AXIAL EJECTION

In its most basic form, the operation of the ion trap in the mass-selective ejection mode employs the remarkably simple idea that if the working point (a_z, q_z) lying within the stability diagram (i.e., corresponding to a trapped ion) is moved along a "scan line" which intercepts either the $\beta_z = 0$ or $\beta_z = 1$ boundary, then the trapped ion will rapidly develop axial instability and be ejected from the ion trap into a suitably positioned detector. In practice, the easiest way of achieving axial instability is to supply the ring electrode of the trap with RF power only, so that the scan line becomes coincident with the q_z axis [53]. The cycle of operation commences by

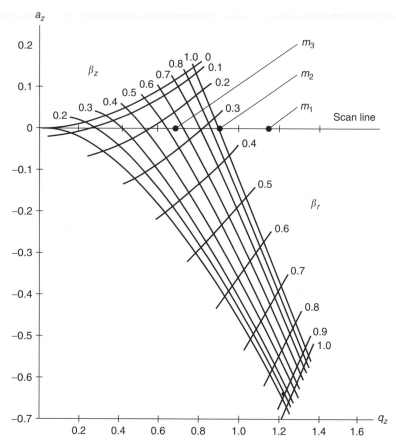

Figure 1.8. Stability diagram for the quadrupole ion trap plotted in (a_z, q_z) space (see text for definitions). The points marked m_1, m_2, and m_3 ($m_1 < m_2 < m_3$) refer to the coordinates of three ions: m_1 has already been ejected, m_2 is on the point of ejection, and the species m_3 is still trapped. Note that on this diagram, for ease of display, the q_z axis has been displaced downward from the $a_z = 0$ scan line. (From R. E. March and J. F. J. Todd (Eds.), *Practical Aspects of Ion Trap Mass Spectrometry*, Vol. III, page 11, Figure 1.7. CRC Press, Boca Raton, FL, 1995. Copyright 1995 by CRC LLC. Reproduced with permission of CRC Press LLC in the format Other Book via Copyright Clearance Center.)

creating ions with a pulse of electrons while the RF amplitude is held at a constant value; the RF amplitude is then ramped linearly so that the a_z, q_z values of the ions are moved to the $\beta_z = 1$ boundary, whereupon the ions are ejected in increasing order of m/z value, as indicated by the points marked m_1, m_2, and m_3 in Figure 1.8.

While the quality of the mass spectrum thus obtained is generally tolerable, it emerges that the presence of a light buffer gas, for example helium, within the trap has a considerably beneficial effect upon the mass spectrum in terms of resolution and sensitivity. Theoretical modeling [141, 142] suggests that this is because the moderating collisions cause the ions to migrate toward the center of the trap in a manner similar to the effects associated with charge transfer processes discussed earlier, so that on ejection they are better "bunched," both axially, which improves the resolution, and radially, which improves the ejection efficiency and hence the sensitivity. A further benefit of the migration effect is that the motion of the ions is less susceptible to field imperfections near the electrodes, so that the manufacturing tolerances are less stringent than those necessary for quadrupole mass analyzers. This, together with the small size and the use of sophisticated software, has led to a veritable explosion in the use of ion trap mass spectrometers, especially in gas chromatography/mass spectrometry (GC/MS) and liquid chromatography (LC)/MS applications, where multiple stages of MS/MS using the tandem-in-time methodology have provided uniquely useful information. Furthermore, new generations of commercial ion traps, based upon the electrode configuration of the linear quadrupole mass filter (see Chapter 5), have begun to appear, offering mass spectroscopists a bewildering choice of technology and variety of experimentation. These topics are considered in greater detail in later chapters of this book.

1.8. CONCLUSION

The quadrupole ion trap is an extraordinary device that functions both as an ion store, in which gaseous ions can be confined for a period of time, and as a mass spectrometer of considerable mass range and variable mass resolution. As a storage device, the ion trap acts as an "electric field test tube" for the confinement of gaseous ions, either positively charged or negatively charged, in the absence of solvent; the confining capacity arises from the formation of a trapping potential well when appropriate potentials are applied to the electrodes of the ion trap. In is simplest form, an ion trap permits the study of the spectroscopy and the chemistry of trapped ions; as a mass spectrometer, when combined with various ion selection and scanning techniques, the elucidation of ion structures by the use of repeated stages of mass analysis known as tandem mass spectrometry has added a new dimension to the armory of analytical techniques, especially in the biosciences. With the advent of new methods by which ions can be formed in the gas phase, from polar as well as from covalent molecules, and introduced subsequently into an ion trap, the range of applications of the quadrupole ion trap mass spectrometer, and from new generations of instruments evolving from it, is enormous.

REFERENCES

1. W. Paul, H. Steinwedel, Apparatus for separating charged particles of different specific charges, German Patent 944,900 (1956); U.S. Patent 2,939,952 (1960).
2. R. F. Post, L. Heinrich, University of California Radiation Laboratory Report (S. Shewchuck) UCRL 2209, Berkeley, CA (1953).
3. M. L. Good, University of California Radiation Laboratory Report UCRL 4146, Berkeley, CA (1953).

4. E. D. Courant, M. S. Livingston, H. S. Snyder, The strong-focusing synchrotron; a new high energy accelerator, *Phys. Rev.* **88** (1952) 1190–1196.

5. P. H. Dawson, N. R. Whetten, Quadrupoles, monopoles and ion traps. *Res. Dev.* **19**(2) (1969) 46–50.

6. W. Paul, H. Steinwedel, A new mass spectrometer without a magnetic field, *Z. Naturforsch.* **8a** (1953) 448–450.

7. K. Berkling aus Leipzig, *Der Entwurf eines Partialdruckmessers*, Physikalisches Institut der Universität, Bonn, Germany, 1956.

8. D. R. Denison, Operating parameters of a quadrupole in a grounded cylindrical housing, *J. Vac. Sci. Technol.* **8** (1971) 266–269.

9. P. E. Knewstubb, *Mass Spectrometry and Ion-Molecule Reactions*, Cambridge University Press, Cambridge, 1969.

10. P. H. Dawson (Ed.), *Quadrupole Mass Spectrometry and Its Applications*, Elsevier, Amsterdam, 1976. Reprinted as an American Vacuum Society classic by the American Institute of Physics (ISBN 1563964554).

11. P. H. Dawson, N. R. Whetten, The three-dimensional quadrupole ion trap, *Naturwissenschaften* **56** (1969) 109–112.

12. R. D. Knight, The general form of the quadrupole ion trap potential, *Int. J. Mass Spectrom. Ion Phys.* **51** (1983) 127–131.

13. E. Mathieu, Mémoire sur le mouvement vibratoire d'une membrane de forme elliptique, *J. Math. Pures Appl.* (*J. Liouville*) **13** (1868) 137.

14. N. W. McLachlan, *Theory and Applications of Mathieu Functions*, Clarendon, Oxford, 1947. See also R. Campbell, *Théorie Générale de l'Equation de Mathieu*, Masson, Paris, 1955.

15. W. Paul, O. Osberghaus, E. Fischer, *Forschungsberichte des Wirtschaft und Verkehrministeriums Nordrhein Westfalen*, No. 415, Westdeutscher Verlag, Köln and Opladen, 1958.

16. E. Fischer, Die dreidimensionale Stabilisirung von Ladungsträgern in einem Vierpolfeld, *Z. Phys.* **156**(1) (1959) 1–26.

17. V. von G. Rettinghaus, The detection of low partial pressures by means of the ion cage, *Z. Angew. Phys.* **22**(4) (1967) 321–326.

18. D. C. Burnham, D. Kleppner, Paper DCII. Practical limitations of the electrodynamic ion trap, *Bull. Am. Phys. Soc. Ser. II* **11** (1968) 70.

19. J. E. P. Syka, W. J. Fies, Jr., A Fourier transform quadrupole ion trap mass spectrometer. Proceedings of the 35th Annual ASMS Conference on Mass Spectrometry and Allied Topics, Denver, CO, May 24–29, 1987, pp. 767–768.

20. J. E. P. Syka, W. J. Fies, Jr., Quadrupole Fourier transform mass spectrometer and method, U.S. Patent 4,755,670 (1988).

21. D. E. Goeringer, R. I. Crutcher, S. A. McLuckey, Ion remeasurement in the radio frequency quadrupole ion trap, *Anal. Chem.* **67** (1995) 4164–4169.

22. V. E. Frankevich, M. H. Soni, M. Nappi, R. E. Santini, J. W. Amy, R. G. Cooks, Non-destructive ion trap mass spectrometer and method, US Patent 5,625,186 (1997).

23. M. Nappi, V. Frankevich, M. Soni, R. G. Cooks, Characteristics of a broad-band Fourier transform ion trap mass spectrometer, *Int. J. Mass Spectrom.* **177** (1998) 91–104.

24. H. G. Dehmelt, Radiofrequency spectroscopy of stored ions. I. Storage, *Adv. At. Mol. Phys.* **3** (1967) 53–72.

25. H. G. Dehmelt, Radiofrequency spectroscopy of stored ions. II. Spectroscopy. *Adv. At. Mol. Phys.* **5** (1969) 109–154.

26. H. A. Schuesslcr, Spin dependent charge transfer in low-energy collisions between helium ions and cesium atoms, *Metrologia* **13** (1977) 109–113.

27. D. J. Wineland, H. Dehmelt, Line shifts and widths of axial, cyclotron and G-2 resonances in tailored, stored electron (ion) cloud, *Int. J. Mass Spectrom. Ion Phys.* **16** (1975) 338–342; Erratum, **19** (1976) 251.

28. D. J. Wineland, W. M. Itano, J. C. Bergquist, J. J. Bollinger, J. D. Prestage, Optical pumping of stored atomic ions, *Ann. Phys. Fr.* **10** (1985) 737–748.

29. D. J. Wineland, W. M. Itano, J. C. Bergquist, J. J. Bollinger, J. D. Prestage, Spectroscopy of stored atomic ions, in R. S. Van Dyck, Jr. and E. N. Fortson (Eds.), *Atomic Physics*, Vol. 9, World Scientific, Singapore, 1985, p. 3.

30. D. J. Wineland, W. M. Itano, J. C. Bergquist, J. J. Bollinger, *Trapped Ions and Laser Cooling*, Selected Publications of the Ion Storage Group of the Time and Frequency Division, National Bureau of Standards, Boulder, CO, June 1985.

31. H. A. Schuessler, Stored ion spectroscopy, in W. Hanic and H. Kleinpoppen (Eds.), *Progress in Atomic Spectroscopy, Part B*, Plenum, New York, 1979, pp. 999–1029.

32. D. J. Wineland, W. M. Itano, R. S. Van Dyck, Jr., High-resolution spectroscopy of stored ions, *Adv. At. Mol. Phys.* **19** (1983) 135–185.

33. D. J. Wineland, W. M. Itano, J. J. Bollinger, J. C. Bergquist, H. Hemmati, Spectroscopy of stored ions using fluorescence techniques, *Soc. Photo-Opt. Instrum. Eng.* **426** (1983) 65–70.

34. P. H. Dawson, N. R. Whetten, Ion storage in three-dimensional, rotationally symmetric, quadrupole fields. I. Theoretical treatment, *J. Vac. Sci. Technol.* **5** (1968) 1–10.

35. P. H. Dawson, N. R. Whetten, Ion storage in three-dimensional, rotationally symmetric, quadrupole fields. II. A sensitive mass spectrometer, *J. Vac. Sci. Technol.* **5** (1968), 11–18.

36. P. H. Dawson, N. R. Whetten, Three-dimensional mass spectrometer and gauge, U.S. Patent 3,527,939 (1970).

37. P. H. Dawson, N. R. Whetten, *3-D Quadrupole Ion Trap Mass Spectrometer*, Air Force Cambridge Research Laboratory, Report No. AFCRL-69-0185 (1969).

38. C. S. Harden, P. E. Wagner, *Three-Dimensional Quadrupole Mass Analyzer. I. General Description*, Edgewood Arsenal Special Publications, Unclassified Report EASP 100-93 (1971).

39. C. S. Harden, P. E. Wagner, *A Three-Dimensional Quadrupolar Mass Analyzer. II. Operational Characteristics*, Edgewood Arsenal Technical Report EATR 4545 (1971), pp. 7–25.

40. P. H. Dawson, C. Lambert, A detailed study of the quadrupole ion trap, *Int. J. Mass Spectrom. Ion Phys.* **16** (1975) 269–280.

41. S. Mastoris, The three-dimensional quadrupole mass spectrometer, Ph.D. Thesis, University of Toronto, Institute of Aerospace Studies, Technical Note 172, Toronto, Canada, (1971).

42. E. P. Sheretov, V. A. Zenkin, Shape of the mass peak in a three-dimensional quadrupole mass spectrometer, *Sov. Phys. Tech. Phys.* **17** (1972) 160–162.

43. E. P. Sheretov, V. A. Zenkin, O. I. Boligatov, Three-dimensional quadrupole mass spectrometer with ion storage, *Gen. Exp. Tech.* **14** (1971) 195–197.

44. E. P. Sheretov, V. A. Zenkin, V. F. Samodurov, N. D. Veselkin, Three-dimensional quadrupole mass spectrometer with sweep of the mass spectrum by variation of the frequency of the supply signal, *Gen. Expt. Tech.* **16** (1973) 194–196.

45. E. P. Sheretov, A. F. Shalimov, G. A. Mogil'chenko, V. A. Zenkin, N. V. Veselkin, The control block of a three-dimensional quadrupole mass spectrometer with storage, *Prib. Tekh. Eksper.* **17** (1974) 482–484.

46. E. P. Sheretov, V. A. Zenkin, V. F. Samodurov, Three-dimensional accumulation quadrupole mass spectrometer, *Zh. Tekh. Fiz.* **43** (1973) 441.

47. E. P. Sheretov, V. A. Zenkin, G. A. Mogil'chenko, Charged-particle extraction from a three-dimensional quadrupole trap, *Sov. Phys. Tech. Phys.* **20** (1976) 1398–1400.

48. E. P. Sheretov, V. F. Samodurov, B. I. Kolotilin, H. K. Tuzhilkin, N. V. Veselkin, Three-dimensional quadrupole mass spectrometer with elliptical electrodes, *Prib. Tekh. Eksper.* **6** (1978) 115–117.

49. F. A. Londry, R. L. Alfred, R. E. March, Computer simulation of single-ion trajectories in Paul-type ion traps, *J. Am. Soc. Mass Spectrom.* **4** (1993) 687–705.

50. R. K. Julian, M. Nappi, C. Weil, R. G. Cooks, Multiparticle simulation of ion motion in the ion trap mass spectrometer: Resonant and direct current pulse excitation, *J. Am. Soc. Mass Spectrom.* **6** (1995) 57–70.

51. H. A. Bui, R. G. Cooks, Windows version of the ion trap simulation program ITSIM: A powerful heuristic and predictive tool in ion trap mass spectrometry, *J. Mass Spectrom.* **33** (1998) 297–304.

52. M. W. Forbes, M. Sharifi, T. R. Croley, Z. Lausevic, R. E. March, Simulation of ion trajectories in a quadrupole ion trap: A comparison of three simulation programs, *J. Mass Spectrom.* **34** (1999) 1219–1239.

53. G. C. Stafford, Jr., P. E. Kelley, J. E. P. Syka, W. E. Reynolds, J. F. J. Todd, Recent improvements in and analytical applications of advanced ion trap technology, *Int. J. Mass Spectrom. Ion Processes* **60** (1984) 85–98.

54. R. F. Bonner, G. Lawson, J. F. J. Todd, R. E. March, Ion storage mass spectrometry: Applications in the study of ionic processes and chemical ionization reactions, *Adv. Mass Spectrom.* **6** (1974) 377–384.

55. G. Lawson, J. F. J. Todd, R. F. Bonner, Theoretical and experimental studies with the quadrupole ion storage trap ('QUISTOR'), *Dyn. Mass Spectrom.* **4** (1975) 39–81.

56. R. F. Bonner, R. E. March, J. Durup, Effect of charge exchange reactions on the motion of ions in three-dimensional quadrupole electric fields, *Int. J. Mass Spectrom. Ion Phys.* **22** (1976) 17–34.

57. R. F. Bonner, R. E. March, The effects of charge exchange collisions on the motion of ions in three-dimensional quadrupole electric fields. Part II. Program improvements and fundamental results, *Int. J. Mass Spectrom. Ion Phys.* **25** (1977) 411–431.

58. G. Lawson, R. F. Bonner, J. F. J. Todd, The quadrupole ion store (QUISTOR) as a novel source for a mass spectrometer, *J. Phys. E. Sci. Instrum.* **6** (1973) 357–362.

59. J.-P. Schermann, F. G. Major, Heavy ion plasma confinement in an RF quadrupole trap, NASA Report X-524-71-343, Goddard Space Flight Center, Greenbelt, MD (1971). See also, Characteristics of electron-free plasma confinement in an RF quadrupole field, *Appl. Phys.* **16** (1978) 225–230.

60. J. F. J. Todd, G. Lawson, R. F. Bonner, Ion traps, in P. H. Dawson (Ed.), *Quadrupole Mass Spectrometry and Its Applications*, Elsevier, Amsterdam, 1976, Chapter VIII.

61. P. H. Dawson, I. Hedman, N. R. Whetten, A simple mass spectrometer, *Rev. Sci. Instrum.* **40** (1969) 1444–1450.

62. F. von Busch, W. Paul, Uber nichtlineare Resonanzen in elektrischen Massenfilter als folge Feldfehlern, *Z. Phys.* **164** (1961) 588–595.

63. N. R. Whetten, P. H. Dawson, Some causes of poor peak shapes in quadrupole field mass analysers, *J. Vac. Sci. Technol.* **6** (1969) 100–103.

64. G. Lawson, J. F. J. Todd, Some experiments with a three-dimensional quadrupole ion storage device. Abstract No. 44, Mass Spectrometry Group Meeting, Bristol, 1971.

65. J. F. J. Todd, R. F. Bonner, G. Lawson, A low-pressure chemical ionisation source: An application of a novel type of ion storage mass spectrometer, *J. Chem. Soc. Chem. Commun.* (1972) 1179–1180.

66. G. Lawson, R. F. Bonner, R. E. Mather, J. F. J. Todd, R. E. March, Quadrupole ion store (QUISTOR). Part I. Ion-molecule reactions in methane, water, and ammonia, *J. Chem. Soc. Faraday Trans. I* **72**(3) (1976) 545–557.

67. D. L. Smith, J. H. Futrell, Ion-molecule reactions in the carbon dioxide-hydrogen system by ion cyclotron resonance, *Int. J. Mass Spectrom. Ion Phys.* **10** (1973) 405–418.

68. J. F. J. Todd, R. M. Waldren, R. E. Mather, G. Lawson, On the relative efficiencies of confinement of Ar^+ and Ar^{2+} ions in a quadrupole ion storage trap (QUISTOR), *Int. J. Mass Spectrom. Ion Phys.* **28** (1978) 141–151.

69. M. C. Doran, J. E. Fulford, R. J. Hughes, Y. Morita, R. E. March, R. F. Bonner, Effects of charge-exchange reactions on the motion of ions in three-dimensional quadrupole electric fields. Part II. A two-ion model, *Int. J. Mass Spectrom. Ion Phys.* **33** (1980) 139–158.

70. R. F. Mather, R. M. Waldren, J. F. J. Todd, The characterization of a quadrupole ion storage mass spectrometer, *Dyn. Mass Spectrom.* **5** (1978) 71–85.

71. J. F. J. Todd, R. M. Waldren, R. E. Mather, The quadrupole ion store (QUISTOR). Part IX. Space-charge and ion stability. A. Theoretical background and experimental results, *Int. J. Mass Spectrom. Ion Phys.* **34** (1980) 325–349.

72. C. Schwebel, P. A. Moller, P.-T. Manh, Formation et confinement d'ions multichargés dans un champ quadrupolaire a haute fréquence, *Rev. Phys. Appl.* **10** (1975) 227–239.

73. J. E. Fulford, R. E. March, A new mode of operation for the three-dimensional quadrupole ion store (QUISTOR): The selective ion reactor, *Int. J. Mass Spectrom. Ion Phys.* **26** (1978) 155–162.

74. R. F. Bonner, Derivations of the field equations and stability parameters for three operating modes of the three-dimensional quadrupole, *Int. J. Mass Spectrom. Ion Phys.* **23**(4) (1977) 249–257.

75. J. E. Fulford, D.-N. Hoa, R. J. Hughes, R. E. March, R. F. Bonner, G. J. Wong, Radio-frequency mass selective excitation and resonant ejection of ions in a three-dimensional quadrupole ion trap, *J. Vac. Sci. Technol.* **17**(4) (1980) 829–835.

76. W. Paul, H. P. Reinhard, U. Von Zahn, Das elektrische massenfilter als massenspektrometer und isotopentrenner, *Z. Phys.* **152** (1958) 143–182.

77. M. A. Armitage, J. E. Fulford, D.-N. Hoa, R. J. Hughes, R. E. March, The application of resonant ion ejection to quadrupole ion storage mass spectrometry: A study of ion/molecule reactions in the QUISTOR, *Can. J. Chem.* **57** (1979) 2108–2113.

78. G. B. DeBrou, J. E. Fulford, E. G. Lewars, R. E. March, Ketene: Ion chemistry and proton affinity, *Int. J. Mass Spectrom. Ion Phys.* **26** (1978) 345–352.

79. M. A. Armitage, M. J. Higgins, E. G. Lewars, R. E. March, Methylketene. Ion chemistry and proton affinity, *J. Am. Chem. Soc.* **102** (1980) 5064–5068.

80. L. R. Anders, J. L. Beauchamp, R. C. Dunbar, J. D. Baldeschwieler, Ion cyclotron double resonance, *J. Chem. Phys.* **45** (1966) 1062–1063.

81. J. N. Louris, R. G. Cooks, J. E. P. Syka, P. E. Kelley, G. C. Stafford, J. F. J. Todd, Instrumentation, applications and energy disposition in quadrupole ion-trap tandem mass spectrometry, *Anal. Chem.* **59** (1987) 1677–1685.

82. R. F. Lever, Computation of ion trajectories in the monopole mass spectrometer by numerical integration of Mathieu's equation, *IBM J. Res. Develop.* **10**(1) (1966) 26–40.

83. J. A. Richards, R. M. Huey, J. Hiller, A new operating mode for the quadrupole mass filter, *Int. J. Mass Spectrom. Ion Phys.* **12** (1973) 317–339.

84. M. Baril, A. Septier, Ion storage in a three-dimensional high frequency, quadrupole field, *Rev. Phys. Appl.* **9** (1974) 525–531.

85. R. M. Waldren, J. F. J. Todd, The use of matrix methods and phase space dynamics for the modelling of r.f. quadrupole-type device performance, *Dyn. Mass Spectrom.* **5** (1978) 14–40.

86. P. H. Dawson, The acceptance of the quadrupole mass filter, *Int. J. Mass Spectrom. Ion Phys.* **17** (1975) 423–445.

87. P. H. Dawson, Ion optical properties of quadrupole mass filters, *Adv. Electron Elect. Phys.* **53** (1980) 153–208.

88. M. Baril, Phase space concepts and best operating point for the three-dimensional quadrupole ion trap, *Dyn. Mass Spectrom.* **6** (1981) 33–43.

89. P. A. Moller, P. T. Manh, C. Schwebel, A. Septier, Generation and storage of multiply-charged ions within a RF quadrupole trap, *IEEE Trans. Nucl. Sci.* **NS-23**(2) (1976) 991–993.

90. P. H. Dawson, Energetics of ions in quadrupole fields, *Int. J. Mass Spectrom. Ion Phys.* **20** (1976) 237–245.

91. P. H. Dawson, The influence of collisions on ion motion in quadrupole fields, *Int. J. Mass Spectrom. Ion Phys.* **24** (1977) 447–451.

92. J. F. J. Todd, R. M. Waldren, R. F. Bonner, The quadrupole ion store (QUISTOR). Part VIII. The theoretical estimation of ion kinetic energies: A comparative survey of the field, *Int. J. Mass Spectrom. Ion Phys.* **34** (1980) 17–36.

93. J. F. J. Todd, R. M. Waldren, D. A. Freer, R. B. Turner, The quadrupole ion store (QUISTOR). Part X. Space charge and ion stability. B. On the theoretical distribution and density of stored charge in RF quadrupole fields, *Int. J. Mass Spectrom. Ion Phys.* **35** (1980) 107–150.

94. J. F. J. Todd, D. A. Freer, R. M. Waldren, The quadrupole ion store (QUISTOR). Part XI. The model of ion motion in a pseudo-potential well: An appraisal in terms of phase-space dynamics, *Int. J. Mass Spectrom. Ion Phys.* **36** (1980) 185–203.

95. R. M. Waldren, J. F. J. Todd, The quadrupole ion store (QUISTOR). Part VI. Studies on phase-synchronised ion ejection: The effects of ejection pulse amplitude, *Int. J. Mass Spectrom. Ion Phys.* **31** (1979) 15–29.

96. R. E. Mosburg, Jr., M. Vedel, Y. Zerega, F. Vedel, J. André, A time-of-flight method for studying the properties of an ion cloud stored in an RF trap, *Int. J. Mass Spectrom. Ion Processes* **77** (1987) 1–12.

97. W. Neuhauser, M. Hohenstatt, P. E. Toschek, H. G. Dehmelt, Optical side-band cooling of visible atom cloud confined in parabolic well, *Phys. Rev. Lett.* **41**(4) (1978) 233–236.

98. W. Neuhauser, M. Hohenstatt, P. E. Toschek, H. G. Dehmelt, Visual observation and optical cooling of electrodynamically contained ions, *Appl. Phys.* **17** (1978) 123–129.

99. J. André, Etude théorique de l'influence des collisions élastiques sur un gaz dilué de particules chargées, confinées par un champ de radio-fréquence a symétrie quadrupolaire, *J. Phys.* **37** (1976) 719–730.

100. J. André, F. Vedel, M. Vedel, Invariance temporelle et propriétés statistiques énergétiques et spatiales d'ions confines dans une trappe quadrupolaire R.F., *J. Phys. Lett.* **40**(24) (1979) L633–L638.

101. P. E. Kelley, G. C. Stafford, Jr., D. R. Stephens, Method of mass analyzing a sample by use of a quadrupole ion trap, U. S. Patent 4,540,884 (1985); Canadian Patent 1,207,918 (1986).

102. A. Kamar, A. B. Young, R. E. March, Experimentally determined proton affinities of 4-methyl-3-penten-2-one, 2-propyl ethanoate, and 4-hydroxy-4-methyl-2-pentanone in the gas phase, *Can. J. Chem.* **64** (1986) 2368–2370.

103. J. E. Fulford, J. W. Dupuis, R. E. March, Gas phase ion/molecule reactions of dimethylsulfoxide, *Can. J. Chem.* **56** (1978) 2324–2330.

104. H. G. Dehmelt, in F. T. Arecchi, F. Strumia, and H. Walther (Eds.), *Advances in Laser Spectroscopy*, Plenum, New York, 1983, p. 153.

105. D. J. Wineland, Trapped ions, laser cooling, and better clocks, *Science* **226** (1984) 395–400.

106. F. Plumelle, M. Desaintfuscien, J. L. Duchene, C. Audoin, Laser probing of ions confined in a cylindrical radiofrequency trap, *Opt. Commun.* **34** (1980) 71–76.

107. C. C. Mariner, J. Pfaff, N. H. Rosenbaum, A. O'Keefe, R. J. Saykally, Radiative lifetimes of trapped molecular ions: $HC1^+$ and HBr^+, *J. Chem. Phys.* **78**(12) (1983) 7073–7076.

108. B. H. Mahan, A. O'Keefe, Electron impact dissociation of $CH_4(CD_4)$: Laser induced fluorescence of product $CH^+(CD^+)$, *Chem. Phys.* **69** (1982) 35–44.

109. F. J. Grieman, B. H. Mahan, A. O'Keefe, The laser-induced fluorescence spectrum of trapped methyliumylidene-d_1 ion, *J. Chem. Phys.* **72** (1980) 4246–4247.

110. F. J. Grieman, B. H. Mahan, A. O'Keefe, J. Winn, Laser-induced fluorescence of trapped molecular ions: The methylidyne cation $A^1\pi \leftarrow X^1\Sigma^+$ system, *Faraday Discuss. Chem. Soc.* **71** (1981) 191–203.

111. F. J. Grieman, B. H. Mahan, A. O'Keefe, The laser-induced fluorescence spectrum of trapped bromine cyanide cations, *J. Chem. Phys.* **74** (1981) 857–861.

112. R. J. Hughes, R. E. March, A. B. Young, Ion chemistry of *n*-butanol, Proc. 31st Ann. ASMS Conf. on Mass Spectrometry and Allied Topics, Boston, MA, May 8–13, 1983, pp. 747–748.

113. R. J. Hughes, R. E. March, A. B. Young, Multiphoton dissociation of ions derived from 2-Propanol in a QUISTOR with low-power cw infrared laser radiation, *Int. J. Mass Spectrom. Ion Phys.* **42** (1982) 255–263.

114. A. Kamar, A. B. Young, R. E. March, A comparative ion chemistry study of acetone, diacetone alcohol, and mesityl oxide, *Can. J. Chem.* **64** (1986) 1979–1988.

115. D. B. Langmuir, R. V. Langmuir, H. Shelton, R. F. Wuerker, Containment device, U.S. Patent 3,065,640 (1962).

116. M. Benilan, C. Audoin, Confinement d'ions par un champ électrique de radio-fréquence dans une cage cylindrique, *Int. J. Mass Spectrom. Ion Phys.* **11** (1973) 421–432.

117. G. Gabrielse, F. C. MacKintosh, Cylindrical penning traps with orthogonalized anharmonicity compensation, *Int. J. Mass Spectrom. Ion Phys.* **57** (1984) 1–18.

118. R. F. Bonner, J. E. Fulford, R. E. March, G. F. Hamilton, The cylindrical ion trap. Part I. General introduction, *Int. J. Mass Spectrom. Ion Phys.* **24** (1977) 255–269.

119. A. G. Nassiopoulos, P. A. Moller, A. Septier, Confinement d'ions dans une cage cylindrique à champ quadrupolaire HF. Application à la spectrometrie de masse. I. Partie théorique, *Rev. Phys. Appl.* **15** (1980) 1529–1541.

120. A. G. Nassiopoulos, P. A. Moller, A. Septier, Confinement d'ions dans une cage cylindrique à champ quadrupolaire HF. Application à la spectrometrie de masse. II. Partie expérimentale, *Rev. Phys. Appl.* **15** (1980) 1543–1552.

121. R. E. Mather, R. M. Waldren, J. F. J. Todd, R. E. March, Some operational characteristics of a quadrupole ion storage mass spectrometer having cylindrical geometry, *Int. J. Mass Spectrom. Ion Phys.* **33** (1980) 201–230.

122. M. Houssin, M. Jardino, M. Desaintfuscien, Experimental development to observe a two photon transition in stored Hg^+, poster session, EICOLS Conference, June 1987.

123. E. C. Beaty, Simple electrodes for quadrupole ion traps, *J. Appl. Phys.* **61**(6) (1987) 2118–2122.

124. J. C. Bergquist, D. J. Wineland, W. M. Itano, H. Hemmati, H. U. Daniel, G. Leuchs, Energy and radiative lifetime of the $5d^9\,6s^2\,{}^2D_{5/2}$ state in Hg II by doppler-free two-photon laser spectroscopy, *Phys. Rev. Lett.* **55**(15) (1985) 1567.

125. D. J. Wineland, J. C. Bergquist, W. M. Itano, J. J. Bollinger, C. H. Manney, Atomic-ion coulomb clusters in an ion trap, *Phys. Rev. Lett.* **59** (1987) 2935–2938.

126. C. Lifshitz, M. Goldenberg, Y. Malinovich, M. Peres, Photoionization mass spectrometry in the millisecond range, *Org. Mass Spectrom.* **17** (1982) 453–455.

127. C. Lifshitz, Y. Malinovich, Time-resolved photoionization mass spectrometry in the millisecond range, *Int. J. Mass Spectrom. Ion Processes* **60** (1984) 99–105.

128. M. Nand Kishore, P. K. Ghosh, Trapping of ions injected from an external source into a three-dimensional RF quadrupole field, *Int. J. Mass Spectrom. Ion Phys.* **29** (1979) 345–350.

129. J. F. J. Todd, D. A. Freer, R. M. Waldren, The quadrupole ion store (QUISTOR). Part XII. The trapping of ions injected from an external source: A description in terms of phase-space dynamics, *Int. J. Mass Spectrom. Ion Phys.* **36** (1980) 371–386.

130. C.-S. O, H. A. Schuessler, Confinement of pulse-injected external ions in a radiofrequency quadrupole ion trap, *Int. J. Mass Spectrom. Ion Phys.* **40** (1981) 53–66.

131. C.-S. O, H. A. Schuessler, Trapping of pulse injected Ions in a radiofrequency quadrupole trap, *Rev. Phys. Appl.* **17** (1982) 83.

132. M. Ho, R. J. Hughes, E. M. Kazdan, P. J. Matthews, A. B. Young, R. E. March, Isotropic collision induced dissociation studies with a novel hybrid instrument. Proc. 32nd Ann. ASMS Conf. on Mass Spectrometry and Allied Topics, San Antonio, TX, May 27–June 1, 1984, pp. 513–514.

133. J. E. Curtis, A. Kamar, R. E. March, U. P. Schlunegger, An improved hybrid mass spectrometer for collisionally activated dissociation studies. Proc. 35th Ann. ASMS Conf. on Mass Spectrometry and Allied Topics, Denver, CO, May 24–29, 1987, pp. 237–238.

134. M. Vedel, J. André, G. Brincourt, Y. Zerega, G. Werth, J. P. Schermann, Study of the SF_6^- ion lifetime in a r.f. quadrupole trap, *Appl. Phys.* **B34** (1984) 229–235.

135. R. H. Gabling, G. Romanowski, K.-P. Wanczek, A tandem cluster-beam-QUISTOR-quadrupole instrument for the study of cluster ions, *Int. J. Mass Spectrom. Ion Processes* **69** (1986) 153–162.

136. G. Romanowski, R. H. Gabling, K.-P. Wanczek, Bimolecular reactions of carbon dioxide microcluster ions with water, investigated with a tandem molecular beam QUISTOR-quadrupole instrument. Proc. 35th Ann. ASMS Conf. on Mass Spectrometry and Allied Topics, Denver, CO, May 24–29, 1987, pp. 777–778.

137. G. Romanowski, R. H. Gabling, K.-P. Wanczek, Study of CO_2 cluster ion decay kinetics with a supersonic molecular beam QUISTOR quadrupole instrument, *Int. J. Mass Spectrom. Ion Processes* **71** (1986) 119–127.

138. R. E. Finnigan, J. F. J. Todd, Private communication, 1970.

139. G. Lawson, J. F. J. Todd, Weak peak enhancement by selective ion trapping in a quadrupole storage source, *Anal. Chem.* **49**(11) (1977) 1619–1622.

140. M. A. Armitage, Applications of quadrupole ion storage mass spectrometry, M.Sc. Thesis, Trent University, Peterborough, ON, Canada, 1979.

141. J. F. J. Todd, J. J. Bexon, R. D. Smith, Stability diagram determination and ion trajectory modelling for the ion trap mass spectrometer. Proc. 35th Ann. ASMS Conf. on Mass Spectrometry and Allied Topics, Denver, CO, May 24–29, 1987, pp. 787–788.

142. J. F. J. Todd, J. J. Bexon, R. D. Smith, Ion trajectory modelling for the ion trap mass spectrometer, paper presented at the Sixteenth Meeting British Mass Spectrometry Society, York, U.K., 1987, pp. 128–131.

2

THEORY OF QUADRUPOLE INSTRUMENTS

Quadrupole Ion Trap Mass Spectrometry, Second Edition, By Raymond E. March and John F. J. Todd
Copyright © 2005 John Wiley & Sons, Inc.

2.1. PRELUDE

A gentle introduction is presented to the theory and application of the quadrupole mass filter (QMF) and the quadrupole ion trap (QIT) instruments. This presentation of the theoretical treatment or derivation flows through three stages. First, the general expression for the potential is modified by the Laplace condition. Second, it is recognized that the structure of each quadrupolar device is constrained by the requirement that the hyperbolic electrodes share common asymptotes. Third, the derivation is based on a demonstration of the equivalence of the force acting on an ion in a quadrupole field and the force derived from the Mathieu equation; this equivalence permits the application of the solutions of Mathieu's equation to the confinement of gaseous ions. For the quadrupole ion trap, no assumption is made a priori concerning the ratio r_0/z_0.

2.2. INTRODUCTION

In electric- and magnetic-sector mass spectrometers, static fields are applied so as to produce a constant force on the ions under study and to transmit a given ion species through the instrument. Under these circumstances of constant acceleration, the ion trajectories are well defined and easily calculated. Quadrupole devices, however, are dynamic instruments in which ion trajectories are influenced by a set of time-dependent or dynamic forces that render their trajectories somewhat more difficult to predict. Ions in such quadrupole fields experience strong focusing in which the restoring force, which drives the ions back toward the center of the device, increases linearly with displacement from the origin. Hybrid mass spectrometers are instruments in which two or more analyzers of differing types are combined in any constructive order.

As noted in Chapter 1, the motion of ions in quadrupole fields is described mathematically by the solutions to a second-order linear differential equation described by Mathieu in 1868 [1]. The equations governing the motion of ions in quadrupole fields have been treated in an introductory fashion [2–4] and in detail [5–9]. The relevant mathematics have been examined by McLachlan [10]. From Mathieu's investigation of the mathematics of vibrating stretched skins, he was able to describe solutions in terms of regions of stability and instability. We can apply these solutions and the concepts of stability and instability to describe the trajectories of ions confined in quadrupole devices and to define the limits to ion trajectory stability: for stable solutions the displacement of the ion periodically passes through zero, whereas for unstable solutions the displacement increases without limit to infinity. To adopt the solutions to the Mathieu equation, we must verify that the equation of motion of an ion confined in a quadrupole device can be described by the Mathieu equation.

The path that we shall follow is to equate an expression for a force (mass × acceleration) in Mathieu's equation with an expression that gives the force on an ion in a quadrupole field. This comparison, which is laid out below in simple mathematical terms, allows us to express the magnitudes and frequencies of the potentials applied to the electrodes of quadrupolar devices, the sizes of these quadrupolar devices, and the mass/charge ratio of ions confined therein in terms of Mathieu's trapping parameters a_u and q_u, where u represents the coordinate axes x, y, and z. On this basis, we shall adopt the idea of stability regions in a_u, q_u space in order to discuss the confinement, and limits thereto, of gaseous ions in quadrupole devices.

The following treatment is intended to provide a common base for the understanding of the mathematics and physical theory of quadrupole devices. While such an understanding is not a prerequisite for the operation of quadrupolar devices, it is essential for an appreciation of the capabilities and performance of the two-dimensional QMF and linear ion trap (LIT), and the three-dimensional QIT. Furthermore, the effort required for comprehension of the theory is well within the capability of the average chemistry graduate. It is hoped that this treatment will serve as a guide to the novitiate and as an aide memoire to the researcher.

We commence by considering an expression for the electric potential within a quadrupolar device, that is, *between* the electrodes. When this expression is subjected to the Laplace condition, the quadrupole field in each direction can be derived, in turn, for each of the two-dimensional QMF and the three-dimensional QIT. An expression for the force within a two-dimensional quadrupole field is derived and is compared with the corresponding expression from the Mathieu equation. Similarly, an expression for the force within a three-dimensional quadrupole field is derived and is compared with the corresponding expression from the Mathieu equation. In this manner, the stability parameters for each of the QMF and the QIT are derived. Stable solutions to the Mathieu equation are examined. By examination of stability criteria, we proceed to a discussion of regions of stability and instability in which we develop stability diagrams. Particular attention is devoted to the QIT stability diagram closest to the origin in stability parameter space. Finally, a rigorous treatment of the general expression for the potential is presented so as to complete the theory of quadrupole instruments. This approach permits an introduction to the contributions of fields of higher order than quadrupole, that is, hexapole and octopole, and acts as a preparation for Chapter 3.

In a logical mathematical approach to the theory of quadrupolar devices, one should begin with an examination of the field in one dimension and proceed to examine two- and three-dimensional devices. Previously, the lowly two-dimensional QMF has been given relatively scant treatment [5] because of late the central interest of readers was the QIT. However, the QMF and quadrupole rod arrays play important roles in modern mass spectrometers both in pure quadrupole devices, such as the triple-stage quadrupole, and in hybrid instruments with, for example, a time-of-flight analyser. In addition, the emergence of commercial LITs using one or more quadrupole rod arrays requires that a more detailed examination be presented here so as to supplement the discussion of LITs in Chapter 5.

2.3. THEORY OF QUADRUPOLAR DEVICES

The theory presented here is based on the behavior of a single ion in an infinite, ideal quadrupole field in the total absence of any background gas. As a single ion is rarely of great value and background gases cannot be excluded totally, in reality the theory of ion containment must be considered along with the dynamical aspects of ion behavior presented in the following chapter.

The term *quadrupolar* refers to the fact that the potential at a point within such a device depends upon the square of the distance from the origin of reference and not the fact that the QMF comprises an array of four rod-shaped electrodes!

In a quadrupolar device described with reference to rectangular coordinates, the potential $\phi_{x, y, z}$ at any given point within the device can be expressed in its most general form as

$$\phi_{x, y, z} = A(\lambda x^2 + \sigma y^2 + \gamma z^2) + C \tag{2.1}$$

where A is a term independent of x, y, and z that includes the electric potential applied *between* the electrodes of opposing polarity (an RF potential either alone or in combination with a DC potential), C is a "fixed" potential (which also may be an RF potential either alone or in combination with a DC potential) applied effectively to all the electrodes so as to "float" the device, and λ, σ, and γ are weighting constants for the x, y, and z coordinates, respectively. It can be seen from Eq. (2.1) that in each coordinate direction the potential increases quadratically with x, y, and z, respectively, and that there are no "cross terms" of the type xy, and so on. This property of Eq. (2.1) has important implications for the treatment of the motion of ions within the field, in that we can consider the components of motion in the x, y, and z directions to be independent of each other.

In an electric field, it is essential that the Laplace condition

$$\nabla^2 \phi_{x, y, z} = 0 \tag{2.2}$$

be satisfied; that is, the second differential of the potential at a point be equal to zero, where

$$\nabla^2 = \frac{\partial^2}{\partial x^2} + \frac{\partial^2}{\partial y^2} + \frac{\partial^2}{\partial z^2}. \tag{2.3}$$

Once the quadrupole potential given in Eq. (2.1) is substituted into the Laplace equation (2.2), we obtain

$$\nabla^2 \phi = \frac{\partial^2 \phi}{\partial x^2} + \frac{\partial^2 \phi}{\partial y^2} + \frac{\partial^2 \phi}{\partial z^2} = 0. \tag{2.4}$$

The partial derivatives of the field are found as

$$\frac{\partial \phi}{\partial x} = \frac{\partial}{\partial x}(A\lambda x^2) = 2A\lambda x \tag{2.5}$$

and

$$\frac{\partial^2 \phi}{\partial x^2} = 2\lambda A. \tag{2.6}$$

Likewise

$$\frac{\partial^2 \phi}{\partial y^2} = 2\sigma A \qquad \text{and} \qquad \frac{\partial^2 \phi}{\partial z^2} = 2\gamma A. \tag{2.7}$$

Substitution of Eqs. (2.6 and 2.7) into Eq. (2.4) yields

$$\nabla^2 \phi = A(2\lambda + 2\sigma + 2\gamma) = 0. \tag{2.8}$$

Clearly, A is nonzero; therefore we obtain

$$\lambda + \sigma + \gamma = 0. \tag{2.9}$$

An infinite number of combinations of λ, σ, and γ exist which satisfy Eq. (2.9); however, the simplest that have generally been chosen in practice are, for the two-dimensional QMF and the LIT,

$$\lambda = -\sigma = 1 \qquad \gamma = 0 \tag{2.10}$$

and for the cylindrically symmetric three-dimensional QIT,

$$\lambda = \sigma = 1 \qquad \gamma = -2. \tag{2.11}$$

2.3.1. The Quadrupole Mass Filter (QMF)

Substituting the values in Eq. (2.10) into Eq. (2.1) gives

$$\phi_{x,y} = A(x^2 - y^2) + C. \tag{2.12}$$

To establish a quadrupolar potential of the form described by Eq. (2.12), we have to consider a configuration comprising two pairs of electrodes having hyperbolic cross sections formed according to equations of the general type

$$\frac{x^2}{x_0^2} - \frac{y^2}{a^2} = 1 \tag{2.13}$$

for the x pair of rod electrodes and

$$\frac{x^2}{b^2} - \frac{y^2}{y_0^2} = -1 \tag{2.14}$$

for the y pair of rod electrodes corresponding, respectively, to the conditions $x = \pm x_0$ when $y = 0$ and $y = \pm y_0$ when $x = 0$. Equations (2.13) and (2.14) describe two complementary rectangular hyperbolas and, in order to establish a quadrupolar potential, it is a condition that the hyperbolas share common asymptotes even though it is not a requirement that $x_0 = y_0$, as was demonstrated by Knight [11] also to be the case for the analogous derivation of the potential within a QIT (see later). Using standard mathematical procedures, it is easy to show that the slopes of the asymptotes of Eqs. (2.13) and (2.14) are $m = \pm a/x_0$ and $m' = \pm y_0/b$, respectively. Since, as we have noted already, Eqs. (2.13) and (2.14) are rectangular hyperbolas, the asymptotes will therefore have slopes of $\pm 45°$, so that $m = m' = \pm\tan 45° = \pm1$, whence $a = \pm x$ and $b = \pm y$.

In practice, all the commercial mass filters described to date and using hyperbolic electrodes have been constructed symmetrically according to the condition

$$x_0 = y_0 = r_0 \tag{2.15}$$

where r_0 is the radius of the inscribed circle tangential to the inner surface of the electrodes. It is easy to demonstrate mathematically that, under these conditions, the asymptotes bisect the gaps between adjacent electrodes; as a result, the quality of the quadrupolar potential will be maintained at greater distances from the origin, even though the electrode surfaces are truncated at some point, rather than extending (ideally) to infinity. Thus from Eq. (2.15) the equations for the electrode surfaces become

$$x^2 - y^2 = r_0^2 \tag{2.16}$$

for the x pair of electrodes and

$$x^2 - y^2 = -r_0^2 \tag{2.17}$$

for the y pair of electrodes.

2.3.1.1. QMF with Round Rods

On a further practical aspect of the construction of QMFs, it should be noted that most modern instruments use arrays of round rods to reduce costs and to simplify construction. When circular rods of radius r are used, a good approximation to a quadrupole field can be obtained when the radius of each rod is made equal to 1.148 times the desired r_0 value [12]. Thus, for example, when using round rods of $\frac{1}{4}$ in. diameter, they must be mounted so that $r_0 = 2.766$ cm. Recently, Gibson and Taylor [13] have questioned this assertion and have claimed that it is not possible to give a single figure for r/r_0 because the results are influenced to a small extent by the form of the ion beam entering the QMF. They found that a value in the range $r = 1.12 \times r_0$ to $r = 1.13 \times r_0$ produces the best performance.

2.3.1.2. The Structure of the QMF

A QMF comprised of circular rods is shown in Figure 2.1. Each pair of opposite rods is electrically connected, thereby establishing a two-dimensional quadrupole field in the x–y plane. The ions enter and travel in the z direction. While traveling in the z direction, the ions also oscillate in the x–y plane,

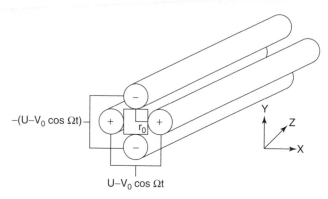

$-(U-V_0 \cos \Omega t)$

r_0

$U-V_0 \cos \Omega t$

Figure 2.1. Quadrupole mass filter. The ions enter and travel in the z direction, while oscillating in the x–y plane. The oscillation is controlled by the DC (U) and RF (V) potentials applied to each pair of rods. Only those ions with stable trajectories at the selected U and V values will travel the length of the QMF and be detected.

due to the potentials applied to the rods. This oscillation is a property of the mass/charge ratio of a given ion species. Therefore, ions of a specific mass/charge ratio will all react equally to the electric potentials imposed by the quadrupole assembly. Under appropriate electrical conditions, ions of a single mass/charge ratio will have a stable trajectory for the entire length of the quadrupole. A QMF can be operated so as to transmit either all ions or only a specific range of mass/charge ratios and to focus them at the exit aperture. The ions transmitted impinge subsequently onto the detector.

An electric potential, $\phi_{y\,pair}$ is applied to the vertical rod pair in Figure 2.1 and $\phi_{x\,pair}$ is applied to the horizontal rod pair such that the pairs of rods are out of phase with each other. Two points should be noted here: first, each rod (or electrode) in a rod array is either in phase or out of phase with the remaining three rods and, second, the potential along the z axis of the rod array is zero.

2.3.1.3. *Quadrupolar Potential*   As noted earlier, the actual quadrupolar potential to which an ion is subjected, ϕ_0, is given by the *difference* between the potentials applied to the x pair and the y pair of electrodes. Thus

$$\phi_0 = \phi_{x\,pair} - \phi_{y\,pair}. \tag{2.18}$$

Considering now, say, the x pair of electrodes, since the potential must be the same across the whole of the electrode surface, we can write that when $y = 0$, $x_0^2 = r_0^2$, so that substituting in Eq. (2.12) we have

$$\phi_{x\,pair} = A(r_0^2) + C. \tag{2.19}$$

Likewise we have

$$\phi_{y\,pair} = A(-r_0^2) + C \tag{2.20}$$

so that from Eqs. (2.18), (2.19), and (2.20) we find

$$\phi_0 = 2Ar_0^2 \tag{2.21}$$

$$\therefore A = \frac{\phi_0}{2r_0^2}. \tag{2.22}$$

Hence Eq. (2.12) becomes

$$\phi_{x,y} = \frac{\phi_0}{2r_0^2}(x^2 - y^2) + C \tag{2.23}$$

whence we note that at the origin ($x = 0$, $y = 0$) we have

$$\phi_{0,0} = C. \tag{2.24}$$

If the electrode structure is floated at zero (ground) potential, then $C = 0$, so that Eq. (2.23) becomes

$$\phi_{x,y} = \frac{\phi_0}{2r_0^2}(x^2 - y^2) \tag{2.25}$$

which is the expression for the potential often found in standard texts.

We proceed now to examine the motion of an ion when subjected to the potential given by Eq. (2.25). If we consider first the component of motion in the x direction, then putting $y = 0$ in Eq. (2.25) gives

$$\phi_{x,0} = \frac{\phi_0 x^2}{2r_0^2} \tag{2.26}$$

so that the electric field at the point (x, 0) is

$$\left(\frac{d\phi}{dx}\right)_y = \frac{\phi_0 x}{r_0^2}. \tag{2.27}$$

As a result, the force acting on an ion, F_x, at a point (x, 0) is given by

$$F_x = -e\left(\frac{d\phi}{dx}\right)_y = -e\frac{\phi_0 x}{r_0^2} \tag{2.28}$$

where the negative sign indicates that the force acts in the opposite direction to increasing x. Since force = mass × acceleration, from Eq. (2.28) we can write

$$m\left(\frac{d^2x}{dt^2}\right) = -e\frac{\phi_0 x}{r_0^2}. \tag{2.29}$$

Let us now consider a real system in which

$$\phi_0 = 2(U + V \cos \Omega t) \tag{2.30}$$

where V is the zero-to-peak amplitude of a RF potential oscillating with angular fre-

References pp. 71–72.

quency Ω (expressed in radians per second) and $+U$ is a DC voltage applied to the x pair of electrodes while a DC voltage of $-U$ volts is applied to the y pair of electrodes. Thus from Eqs. (2.29) and (2.30)

$$m\left(\frac{d^2x}{dt^2}\right) = -2e\frac{(U+V\cos\Omega t)x}{r_0^2} \tag{2.31}$$

which may be expanded to

$$\frac{d^2x}{dt^2} = -\left(\frac{2eU}{mr_0^2} + \frac{2eV\cos\Omega t}{mr_0^2}\right)x. \tag{2.32}$$

2.3.1.4. The Mathieu Equation   The canonical or commonly accepted form of the Mathieu equation is

$$\frac{d^2u}{d\xi^2} + (a_u - 2q_u\cos 2\xi)u = 0 \tag{2.33}$$

where u is a displacement, ξ is a dimensionless parameter equal to $\Omega t/2$ such that Ω must be a frequency as t is time, and a_u and q_u are additional dimensionless stability parameters which, in the present context of quadrupole devices, are in fact "trapping" parameters. It can be shown by substituting $\xi = \Omega t/2$ and using operator notation to find

$$\frac{d}{dt} = \frac{d\xi}{dt}\frac{d}{d\xi} = \frac{\Omega}{2}\frac{d}{d\xi} \tag{2.34}$$

so that

$$\frac{d^2}{dt^2} = \frac{d\xi}{dt}\frac{d}{d\xi}\left(\frac{d}{dt}\right) = \frac{\Omega^2}{4}\frac{d^2}{d\xi^2} \tag{2.35}$$

that we can write

$$\frac{d^2u}{dt^2} = \frac{\Omega^2}{4}\frac{d^2u}{d\xi^2}. \tag{2.36}$$

Substitution of Eq. (2.36) into Eq. (2.33), substituting Ωt for 2ξ, and rearranging yield

$$\frac{d^2u}{dt^2} = -\left(\frac{\Omega^2}{4}a_u - 2\times\frac{\Omega^2}{4}q_u\cos\Omega t\right)u. \tag{2.37}$$

We can now compare directly the terms on the right-hand sides of Eqs. (2.32) and (2.37), recalling that u represents the displacement x, to obtain

$$-\left(\frac{2eU}{mr_0^2} + \frac{2eV\cos\Omega t}{mr_0^2}\right)x = -\left(\frac{\Omega^2}{4}a_x - 2\times\frac{\Omega^2}{4}q_x\cos\Omega t\right)x \tag{2.38}$$

whence one deduces the relationships

$$a_x = \frac{8eU}{mr_0^2\Omega^2} \tag{2.39}$$

and

$$q_x = \frac{-4\,eV}{mr_0^2\Omega_0^2}\,. \tag{2.40}$$

When this derivation is repeated to obtain the force on an ion in the y direction in a QMF, one finds that $a_x = -a_y$ and $q_x = -q_y$; this relationship is obtained because $\lambda = -\sigma = 1$. The a_u, q_u trapping parameters are particularly interesting because they are functions of the magnitude of either the DC voltage or the RF voltage applied to a quadrupolar device, the RF frequency (in radians per second), the mass/charge ratio of a given ion species, and the size (r_0) of the device; that is, the trapping parameters are functions of the instrumental parameters that govern the various operations of a quadrupole device. Other parameters of interest, β_u and ω_u, can be derived from the a_u, q_u trapping parameters. Both β_u and ω_u describe the nature of the ion trajectories and are considered later in more detail: β_u is a complex function of a_u and q_u, and ω_u is the so-called secular frequency of the ion motion in the u direction.

2.3.1.5. Regions of Stability of the QMF

The solutions to the Mathieu equation are now accessible to us and can be interpreted in terms of ion trajectory stability (and instability) in each of the x and y directions, of confinement within the totality of the quadrupole field (when conditions correspond simultaneously to ion trajectory stability in each of the x and y directions), and of the characteristic fundamental secular frequencies of ion motion in the x and y directions. Because the fields in the QMF are uncoupled, one need examine the solutions to the Mathieu equation in one dimension only. The solutions obtained for one dimension only can then be combined using the appropriate values of λ and σ.

The primary concern in our utilization of a QMF lies with the criteria that govern the stability (and instability) of the trajectory of an ion within the field, that is, the experimental conditions that determine whether an ion moves in a stable trajectory within the device or is ejected from the device and either lost or detected externally (as in a LIT, discussed in Chapter 5). The boundaries between stable and unstable regions of the stability diagram correspond to those values of a_u and q_u for which the parameter β_u is an integer, that is, 0, 1, 2, 3, ... These limits have been shown to correspond to combinations of cosine and sine elliptic series (see later).

For these values, the solutions to the Mathieu equation are periodic but unbound, and they represent, in practical terms, the point at which the trajectory of an ion becomes unbound. Stable, bound solutions for the Mathieu equation in the x direction are obtained when the a, q parameters lie within the region composed of multiple lines shown in Figure 2.2. This region, which is located in the vicinity of the origin, is bounded by the characteristic curves a_0 and b_1 for which $\beta_u = 0$ and $\beta_u = 1$, respectively. The values of these characteristic curves [10] are given by

$$a_0 = -\frac{q^2}{2} + \frac{7q^4}{128} - \frac{29q^6}{2304} + \cdots \tag{2.41}$$

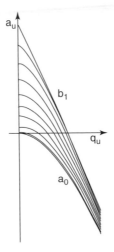

Figure 2.2. Mathieu stability region for u direction closest to origin, plotted in (a_u, q_u) space.

and

$$b_1 = 1 - q - \frac{q^2}{8} + \frac{q^3}{64} + \frac{q^4}{1536} + \cdots . \tag{2.42}$$

Within this stable region are shown the iso-β_u lines that correspond to the set of a_u, q_u points that have the same value of β_u.

In Figure 2.3 are shown additional regions of stability and instability in a_u, q_u space. In Figure 2.4, these additional stability regions are identified with each of the x and y directions. The x stable regions shown in Figure 2.4a are obtained by multiplying the regions shown in Figure 2.2 by unity and the y stable regions shown in Figure 2.4b are obtained by multiplying those regions by -1; this difference between the two sets of curves reflects the values of the weighting constants λ and σ ($\lambda = 1$, $\sigma = -1$, $\gamma = 0$) for the two-dimensional quadrupole device. Figures 2.3 and 2.4 are shown in a simplified fashion in that only the positive values are shown along the q_u axis because symmetry exists about the a_u axis. Regions in which the values of a_u and q_u represent stable solutions to the Mathieu equation are shaded; for those regions that are not shaded, the solutions to the Mathieu equation are unstable.

Presented in Figure 2.5 is the Mathieu stability diagram [1] in one dimension of (a_u, q_u) space showing the regions delineated by characteristic numbers of a cosine-type function (a_m) of order m and a sine-type function (b_m) of order m. This diagram is labeled in the terminology used by McLachan [10]. The boundaries of even order are symmetric about the a_u axis, but the boundaries of odd order are not; however, the diagrams themselves appear to be symmetric about the a_u axis.

A diagram that represents ion trajectory stability (and instability) regions in both the x and y directions can be constructed by overlapping parts (a) and (b) of Figure 2.4, as shown in Figure 2.6. Here, four stability regions labeled A, B, C, and D are identified; regions A and D lie on the q_u axis and are symmetric about this axis while regions

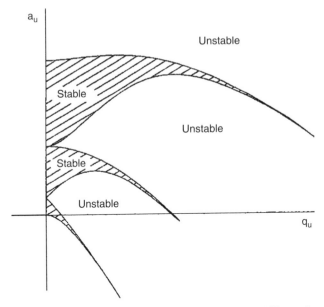

Figure 2.3. Three Mathieu stability regions (with alternating instability regions) in u direction closest to origin, plotted in (a_u, q_u) space.

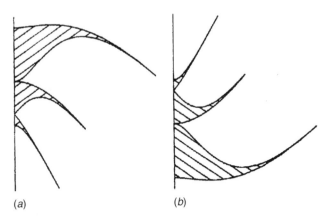

Figure 2.4. Graphical representation of three Mathieu stability regions: (a) x stable, as in Figure 2.3; (b) y stable. The y-stable region is obtained as -1 times the x-stable region.

B and C are disposed symmetrically away from the q_u axis. When one envisages the overlap of Figure 2.5 with the inverse of Figure 2.5, it is clear that there are many regions of stability, though most of them are not accessible at this time [5, 14–17].

References pp. 71–72.

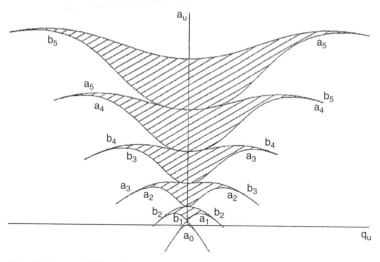

Figure 2.5. Mathieu stability diagram in one dimension of (a_u, q_u) space. The characteristic curves $a_0, b_1, a_1, b_2, \ldots$ divide the plane into regions of stability and instability. The even-order curves are symmetric about the a_u axis, but the odd-order curves are not. The diagram itself, however, appears to be symmetric about the a_u axis.

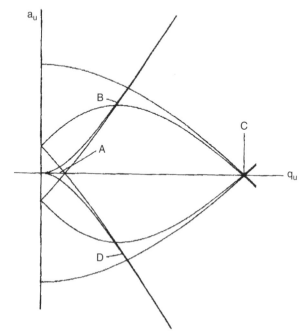

Figure 2.6. Mathieu stability diagram in two dimensions (x and y). Regions of simultaneous overlap are labeled A, B, C, and D.

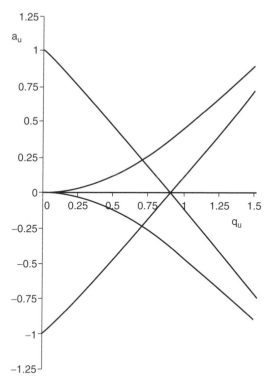

Figure 2.7. Boundaries of stability region A in a_u and q_u for QMF. The boundaries represent the limits in a_u and q_u for stable ion trajectories and satisfy the Mathieu equation. The stability diagram is symmetric about the q_u axis. Conventionally, the stability diagram in a_x, q_x space is presented. Usually, only the upper part of this stability diagram is depicted.

We shall focus upon stability region A. Region A, as is shown in Figure 2.7, is bounded by the characteristic curves for which $\beta_x = 0$, 1 and $\beta_y = 0$, 1; thus within this region of stability $0 < \beta_x < 1$ and $0 < \beta_y < 1$. Often, only the upper part of this diagram is shown because stability region A is symmetric about the q_u axis. For an ion to have a stable trajectory in a quadrupole rod array, its a_x, q_x coordinates must lie within the diamond-shaped region shown in Figure 2.7. When an ion's a_x, q_x coordinates lie within stability region A, its a_y, q_y coordinates must necessarily lie also within stability region A.

2.3.1.6. Mass Selectivity of the QMF   When a quadrupole rod array is operated with an RF potential only, all a_x, q_x coordinates must lie on the q_x axis because all values of a_x are zero, that is, $U = 0$ in Eq. (2.39). In this RF-only mode of operation, the rod array acts in a "total-ion" mode as all ions above a fixed mass/charge ratio are transmitted through the device provided that their properties upon entry to the

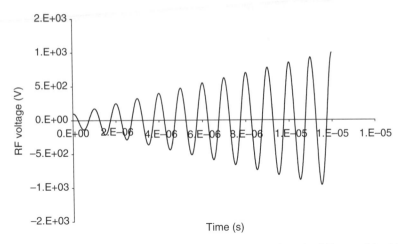

Figure 2.8. RF amplitude ramp. To scan a selected mass region, the RF potential which is applied to each pair of rods is ramped (increased at a constant rate), changing the amplitude but not the frequency.

device are appropriate for transmission. Normally, the RF potential is either fixed or increased incrementally (as shown in Figure 2.8) so as to enhance the efficiency of transmission of ions of higher mass/charge ratio.

The mass selectivity of the QMF is established by varying the magnitudes of the DC (U) and AC or RF (V) voltages (Figure 2.9) applied at a constant ratio to each pair of rods. One pair of rods (x–z plane) is connected to a positive DC voltage, U, along with an RF voltage, $V \cos \Omega t$. The second pair of rods (y–z plane) is connected to a negative DC voltage, $-U$, and a RF voltage, $-V \cos \Omega t$, which is equal in amplitude but of opposite phase to that applied to the other pair of rods (x–z plane). The difference between these applied voltages is given by

$$\phi_0 = 2(U + V \cos \Omega t). \tag{2.30}$$

The use of two potentials affects ions of various mass/charge ratios in different fashions. In the x–z plane, the RF potential will greatly affect light ions. These ions will oscillate in phase with the RF drive potential and the amplitude of oscillation will increase until the ion is lost either via contact with the rods or ejection from the quadrupole rod assembly. Heavier atoms are less affected by the RF drive, remaining near the center of the quadrupole rod assembly, and are emitted through the exit aperture at the downstream end of the quadrupole rod array. Thus the field in the x–z plane acts as a high-mass pass filter. In contrast, in the y–z plane, both heavy and light atoms are drawn toward the negative DC potential of the rods; however, light atoms are refocused toward the center by the RF drive and maintain stable trajectories. Thus the field in the y–z plane acts as a low-mass pass filter. Therefore, by selection of an appropriate DC/RF ratio, the two filters can be overlapped such that only ions within a small range of mass/charge ratios maintain stable trajectories and reach the detector. By ramping both the DC and RF potentials with a constant ratio of U/V,

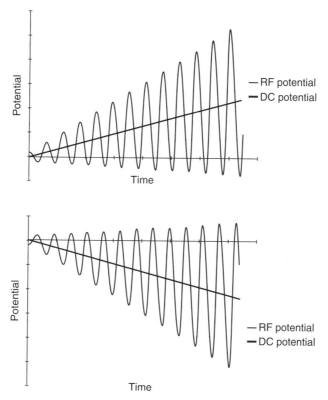

Figure 2.9. Upper: ramping of both DC and RF potentials applied to rods in the *x–z* plane, which act as the high-mass pass filter. Lower: ramping of both negative DC and RF potentials applied to rods in *y–z* plane, which act as the low-mass pass filter. The upper and lower diagrams are mirror images and are therefore 180° out of phase from each other.

an entire mass spectrum of mass/charge ratios can be scanned. A mass spectrum is defined by the American Society for Mass Spectrometry (`www.asms.org`) as a spectrum obtained when a beam of ions is separated according to the mass/charge ratios of the ionic species contained within it.

Alternately, the upper part of the stability diagram shown in Figure 2.7 can be displayed in *U, V* space as shown in Figure 2.10. In *U, V* space, there is a stability diagram for each ion species; diagrams for ions of mass/charge ratio M_1, M_4, and M_6 are shown in this figure, where $M_1 < M_4 < M_6$. Thus the *U, V* coordinates at the apices for each of the ions in Figure 2.10 correspond to the a_x, q_x coordinates at the upper apex in Figure 2.7. Hence during operation of a QMF where the *U* and *V* amplitudes are close to those corresponding to the apex for M_4 in Figure 2.10, only M_4 ions will be transmitted and detected. The angled line in Figure 2.10 is the *operating line* or *scan line*. When *U* and *V* are increased at a constant ratio, the scan line passes close

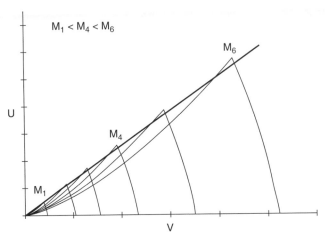

Figure 2.10. Superimposed stability diagrams in U, V space for ions in order of increasing mass/charge ratio. By ramping the DC and RF potentials appropriately, only the peak of each individual stability diagram will be intersected. Therefore, the ions will be transmitted selectively from the QMF, with the ions of lowest mass/charge ratio being transmitted first.

to the apex for each ion in order of increasing mass/charge ratio. In practice, relatively few ions will be transmitted when the scan line is close to the apex but the mass resolution will be high in that few, if any, other types of ions will be transmitted. As the slope of the scan line is reduced, ion signal intensity increases but at the expense of mass resolution.

Discussion of the secular frequencies of ions in a QMF is presented in the following treatment of the QIT.

2.3.2. The Quadrupole Ion Trap (QIT)

The QIT is an extraordinary device that functions both as an ion store in which gaseous ions can be confined for a period of time and as a mass spectrometer of large mass range, variable mass resolution, and high sensitivity. As a storage device, the QIT confines gaseous ions, either positively charged or negatively charged, in the absence of solvent. The confining capacity of the QIT arises from the formation of a trapping potential well when appropriate potentials are applied to the electrodes of the ion trap.

That the basic theory of operation of quadrupole devices was enunciated almost 100 years before the QIT and the related QMF were invented by Paul and Steinwedel [18] is a shining example of the inherent value of sound basic research. The pioneering work of the inventors was recognized by the award of the 1989 Nobel Prize in Physics to Wolfgang Paul [19], together with Norman Ramsay and Hans Dehmelt.

2.3.2.1. The Structure of the QIT As noted in Chapter 1, the QIT mass spectrometer consists essentially of three shaped electrodes that are shown in open array in Figure 2.11. Two of the three electrodes are virtually identical and, while having hyperboloidal geometry, resemble small inverted saucers; these saucers are the so-called end-cap electrodes and each has one or more holes in the center. One end-cap electrode contains the "entrance" aperture through which electrons and/or ions can be gated periodically while the other is the "exit" electrode through which ions pass to a detector. The third "ring" electrode has an internal hyperboloidal surface: in some early designs of ion trap systems, a beam of electrons was gated through a hole in this electrode rather than an end-cap electrode (see Chapter 1). The ring electrode is positioned symmetrically between two end-cap electrodes, as shown in Figures 1.2 and 2.12; Figure 2.12a shows a photograph of an ion trap cut in half along the axis of cylindrical symmetry while Figure 2.12b is a cross section of an ideal ion trap showing the asymptotes and the dimensions r_0 and z_0, where r_0 is the radius of the ring electrode in the central horizontal plane and $2z_0$ is the separation of the two end-cap electrodes measured along the axis of the ion trap.

The electrodes in Figure 2.12 are truncated for practical purposes, but in theory they extend to infinity and meet the asymptotes shown in the figure. The asymptotes arise from the hyperboloidal geometries of the three electrodes. The geometries of the electrodes are defined so as to produce an ideal quadrupole potential distribution that, in turn, will produce the necessary trapping field for the confinement of ions.

Figure 2.11. Three electrodes of QIT shown in open array.

References pp. 71–72.

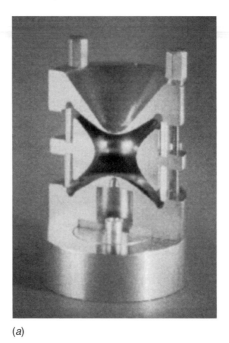

(a)

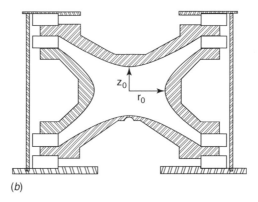

(b)

Figure 2.12. Quadrupole ion trap: (a) photograph of ion trap cut in half along axis of cylindrical symmetry; (b) schematic diagram of three-dimensional ideal ion trap showing asymptotes and dimensions r_0 and z_0.

2.3.2.2. Electrode Surfaces   We have already seen that for the cylindrically symmetric QIT the values of λ, σ, and γ given in Eq. (2.11) satisfy the Laplace condition [Eq. (2.2)] when it is applied to Eq. (2.1). Thus we have

$$\phi_{x, y, z} = A(x^2 + y^2 - 2z^2) + C. \tag{2.43}$$

To proceed, we must convert Eq. (2.43) into cylindrical polar coordinates employing the standard transformations $x = r \cos \theta$, $y = r \sin \theta$, $z = z$. Thus Eq. (2.43) becomes

$$\phi_{r,z} = A(r^2 - 2z^2) + C. \tag{2.44}$$

It should be noted that in making this transformation the angular orientation of the x–y coordinate plane is lost. The effect of this is that when the equations of motion of the ions are developed in a manner analogous to that presented earlier for the QMF there is an implicit assumption that each ion possesses zero angular velocity around the z axis.

We can now follow the procedure adopted by Knight [11] and write the equations for the electrode surfaces in generalized forms as

$$\frac{r^2}{r_0^2} - \frac{z^2}{a^2} = 1 \quad \text{(ring electrode)} \tag{2.45}$$

and

$$\frac{r^2}{b^2} - \frac{z^2}{z_0^2} = -1 \quad \text{(end-cap electrode)}. \tag{2.46}$$

It should be remembered that when plotted out these equations represent cross sections through the electrodes which, of course, possess cylindrical symmetry around the z axis. In terms of their respective geometric forms, the ring electrode is a single-sheet hyperboloid and the pair of end-cap electrodes comprises a double-sheet hyperboloid. As with the QMF, we note the conditions $r = \pm r_0$ when $z = 0$ and $z = \pm z_0$ when $r = 0$. From Figure 2.12b we recall that $2r_0$ is the innermost diameter of the ring electrode and $2z_0$ is the closest distance between the innermost surfaces of the end-cap electrodes; a and b are geometric quantities which will be evaluated shortly.

Following standard mathematical procedures we can deduce that the slopes of the asymptotes of the ring electrode are

$$m = \pm \frac{a}{r_0} \tag{2.47}$$

and those of the end-cap electrodes are

$$m' = \pm \frac{z_0}{b}. \tag{2.48}$$

As in the previous treatment of the mass filter, we note that in order to establish a quadrupolar field the ring and the end-cap electrodes must share common asymptotes, so that $m = m'$, and from Eqs. (2.47) and (2.48) we can write

$$a^2 b^2 = r_0^2 z_0^2. \tag{2.49}$$

References pp. 71–72.

Hence substituting for a^2 in Eq. (2.45) for the ring electrode we obtain

$$\frac{r^2}{r_0^2} - \frac{z^2 b^2}{r_0^2 z_0^2} = 1 \tag{2.50}$$

thus

$$r^2 = r_0^2 + \frac{z^2 b^2}{z_0^2}. \tag{2.51}$$

Since the value of the potential given by Eq. (2.44) must be a constant across the electrode surfaces, we can establish conditions under which $\phi_{r,z}$ is independent of r and z by the following procedure, in which we replace r^2 in the term $r^2 - 2z^2$ to obtain, for the ring electrode,

$$r^2 - 2z^2 = r_0^2 + \frac{z^2 b^2}{z_0^2} - 2z^2$$

$$= r_0^2 + \frac{z^2(b^2 - 2z_0^2)}{z_0^2}. \tag{2.52}$$

Similarly for the end-cap electrodes we have, from Eq. (2.46),

$$r^2 = \frac{z^2 b^2}{z_0^2} - b^2 \tag{2.53}$$

$$r^2 - 2z^2 = \frac{z^2(b^2 - 2z_0^2)}{z_0^2} - b^2. \tag{2.54}$$

Thus $\phi_{r,z}$ becomes constant when $b^2 - 2z_0^2 = 0$

$$\therefore b^2 = 2z_0^2 \tag{2.55}$$

and therefore from Eq. (2.49) we have

$$a^2 = \tfrac{1}{2} r_0^2. \tag{2.56}$$

Hence the equations for the electrode surfaces now become

$$\frac{r^2}{r_0^2} - \frac{2z^2}{r_0^2} = 1 \quad \text{(ring electrode)} \tag{2.57}$$

and

$$\frac{r^2}{2z_0^2} - \frac{z^2}{z_0^2} = -1 \quad \text{(end-cap electrode)}. \tag{2.58}$$

Also from Eqs. (2.47) and (2.48), equating the gradients of the asymptotes, we obtain

$$m = \pm \frac{a}{r_0} = \pm \frac{r_0}{\sqrt{2} r_0} = m' = \pm \frac{z_0}{b} = \pm \frac{z_0}{\sqrt{2} z_0}$$

$$= \pm \frac{1}{\sqrt{2}}. \tag{2.59}$$

This relationship corresponds to the asymptotes having an angle of 35.264° with respect to the radial plane of the ion trap.

2.3.2.3. Quadrupolar Potential   As in the case of the QMF presented earlier, we must now proceed to evaluate the constants A and C in the general expression for the potential within the ion trap, Eq. (2.44). By analogy with Eq. (2.18) we define a quadrupolar potential ϕ_0 in terms of the difference between the potentials applied to the ring and the pair of end-cap electrodes,

$$\phi_0 = \phi_{ring} - \phi_{endcaps}. \tag{2.60}$$

Recalling that $r = \pm r_0$ when $z = 0$ and $z = \pm z_0$ when $r = 0$, substitution into Eq. (2.44) gives

$$\phi_{ring} = A(r_0^2) + C \tag{2.61}$$

and

$$\phi_{endcaps} = A(-2z_0^2) + C \tag{2.62}$$

so that from Eq. (2.60)

$$\phi_0 = A(r_0^2 + 2z_0^2) \tag{2.63}$$

whence

$$A = \frac{\phi_0}{r_0^2 + 2z_0^2} \tag{2.64}$$

and therefore from Eq. (2.44)

$$\phi_{r,z} = \frac{\phi_0(r^2 - 2z^2)}{r_0^2 + 2z_0^2} + C. \tag{2.65}$$

As with the QMF, we proceed to evaluate the constant C by taking account of the potentials actually connected to the electrodes of opposing polarity. However, whereas in the mass filter equal and opposite potentials are applied to the x pair and the y pair of electrodes, respectively, in the QIT the end-cap electrodes are normally held at ground (i.e., zero) potential while a unipolar RF potential and any DC voltage are applied to the ring electrode only. Thus Eq. (2.65) becomes

$$\phi_{0,z_0} = \phi_{endcaps} = \frac{\phi_0(0 - 2z_0^2)}{r_0^2 + 2z_0^2} + C = 0$$

and

$$\therefore \ C = \frac{2\phi_0 z_0^2}{r_0^2 + 2z_0^2} \tag{2.66}$$

whence

$$\phi_{r,z} = \frac{\phi_0(r^2 - 2z^2)}{r_0^2 + 2z_0^2} + \frac{2\phi_0 z_0^2}{r_0^2 + 2z_0^2}. \tag{2.67}$$

From Eq. (2.67) we see that the potential at the center of the ion trap $(0, 0)$ is no longer zero but is at a potential equal to a fraction $[2z_0^2/(r_0^2 + 2z_0^2)]$ of that applied to the ring electrode.

While this mode of connecting the potentials does not affect the motion of the ions within the ion trap, which still "see" an electric field developed by a quadrupolar potential, it does have the effect of halving the maximum value of mass/charge ratio for ions that may be stored in (and ejected from) the ion trap compared to the situation where equal and opposite potentials are applied to the ring and the pair of end-cap electrodes [see also Eq. (2.75) below]. Having set the pair of end-cap electrodes at ground potential, we must now define the value of ϕ_0 in terms of the real-system potentials applied to the ring electrode. Thus we have

$$\phi_0 = (U + V \cos \Omega t) \tag{2.68}$$

where the quantities are defined in the same manner as those in Eq. (2.30). In doing this, it should be noted that the factor 2 is absent from the right-hand side of Eq. (2.68) when compared to Eq. (2.30); this situation arises from the "asymmetric" connection of the potentials to the electrodes of the ion trap, in contrast to the method of connection employed in the mass filter. Substituting from Eq. (2.68) into Eq. (2.67) therefore gives

$$\phi_{r,z} = \frac{(U + V \cos \Omega t)(r^2 - 2z^2)}{r_0^2 + 2z_0^2} + \frac{2(U + V \cos \Omega t)z_0^2}{r_0^2 + 2z_0^2}. \tag{2.69}$$

As with the QMF, the components of the ion motion in the radial (r) and the axial (z) directions may be considered independently, so for the axial direction we can write, by analogy with Eqs. (2.28) and (2.29),

$$\mathbf{F}_z = -e\left(\frac{d\phi}{dz}\right)_r = e\frac{4\phi_0 z}{r_0^2 + 2z_0^2}$$

$$= m\left(\frac{d^2z}{dt^2}\right). \tag{2.70}$$

Hence from Eq. (2.68) we have

$$m\left(\frac{d^2z}{dt^2}\right) = \frac{4e(U + V \cos \Omega t)z}{r_0^2 + 2z_0^2} \tag{2.71}$$

which may be expanded to give

$$\frac{d^2z}{dt^2} = \left(\frac{4eU}{m(r_0^2 + 2z_0^2)} + \frac{4eV \cos \Omega t}{m(r_0^2 + 2z_0^2)}\right)z. \tag{2.72}$$

As with the mass filter, we recognize the similarity of Eq. (2.72) with the Mathieu equation (2.33):

$$\frac{d^2u}{d\xi^2} + (a_u - 2q_u \cos 2\xi)u = 0$$

so that using the transformations given in Eqs. (2.34), (2.35), and (2.36) and replacing u by z, we can write

$$\left[\frac{4eU}{m(r_0^2 + 2z_0^2)} + \frac{4eV \cos \Omega t}{m(r_0^2 + 2z_0^2)}\right]z = -\left(\frac{\Omega^2}{4}a_z - 2 \times \frac{\Omega^2}{4}q_z \cos \Omega t\right)z \qquad (2.73)$$

whence one deduces the relationships

$$a_z = -\frac{16eU}{m(r_0^2 + 2z_0^2)\Omega^2} \qquad (2.74)$$

and

$$q_z = \frac{8eV}{m(r_0^2 + 2z_0^2)\Omega^2} . \qquad (2.75)$$

When this derivation is repeated for the radial component of motion at a fixed value of z, one finds that

$$a_z = -2a_r \qquad \text{and} \qquad q_z = -2q_r \qquad (2.76)$$

again arising from the values of λ, σ, and γ inserted into the general equation (2.1) when it is applied to the QIT.

So far in this formulation of the motion occurring within the ion trap we have made no assumption concerning the relationship between the dimensions r_0 and z_0. Historically we see that ever since the early descriptions of the ion trap [20, 21] the relationship

$$r_0^2 = 2z_0^2 \qquad (2.77)$$

has been selected as a requirement for forming the ideal quadrupolar potential distribution. Furthermore, we note that with the identity given in Eq. (2.77) the asymptotes with the gradients $\pm 1/\sqrt{2}$ [Eq. (2.59)] will pass through the coordinates $\pm r_0$, $\pm z_0$. Knight [22] has shown that in practical ion trap systems with truncated electrodes, under the conditions of Eq. (2.77), the asymptotes bisect the gaps between the ring electrode and the end-cap electrodes at high values of r and z, thereby minimizing the contributions to the potential of higher order terms (see also the Appendix to this chapter and Chapter 3). It should be noted that inserting the relationship given in Eq. (2.77) into Eq. (2.67) shows that under the conditions where the end-cap electrodes are held at earth potential the potential at the center of the ion trap is equal to half that applied to the ring electrode.

2.3.2.4. An Alternative Approach to QIT Theory

An alternative approach to the derivation of the potential in a QIT is given in Ref. 9, Vol. 1, Chapter 2; we present a shortened version of this approach here. This approach has the advantage of introducing components of the trapping potential of order higher than quadrupolar. A solution of Laplace's equation in spherical polar coordinates (ρ, θ, ϕ) for a system with axial symmetry (such as is the case for the QIT) is obtained from the theory of differential equations and has the general form

$$\phi(\rho, \theta, \varphi) = \phi_0 \sum_{n=0}^{\infty} A_n \frac{\rho^n}{r_0^2} P_n(\cos \theta) \tag{2.78}$$

where A_n are arbitrary coefficients and $P_n(\cos \theta)$ denotes a Legendre polynomial. When $\rho^n P_n(\cos \theta)$ is expressed in cylindrical polar coordinates, Eq. (2.79) is obtained as

$$\phi_{r,z} = \phi_0 \left(A_2 \frac{r^2 - 2z^2}{2r_0^2} + A_3 \frac{3r^2 z - 2z^3}{2r_0^3} + A_4 \frac{3r^4 - 24r^2 z^2 + 8z^4}{8r_0^2} \right.$$

$$\left. + A_5 \frac{15r^4 z - 40r^2 z^3 + 8z^5}{8r_0^5} + A_6 \frac{5r^6 - 90r^4 z^2 + 120r^2 z^4 - 16z^6}{16r_0^6} \right). \tag{2.79}$$

The values of $n = 0, 1, 2, 3, 4, 5, 6$ correspond to the monopole, dipole, quadrupole, hexapole, octopole, decapole, and docecapole components, respectively, of the potential field ϕ. Higher-order field components such as hexapole and octopole can play important roles in the operation of modern ion trap mass spectrometers (see Chapters 3 and 8).

2.3.2.5. Regions of Ion Trajectory Stability

Quadrupole ion trap operation is concerned with the criteria that govern the stability (and instability) of the trajectory of an ion within the field, that is, the experimental conditions that determine whether an ion is stored within the device or is ejected from the device and either lost or detected externally.

The solutions to Mathieu's equation are of two types: (i) periodic but unstable and (ii) periodic and stable. Solutions of type (i) are called Mathieu functions of integral order and form the boundaries of unstable regions on the stability diagram. The boundaries, which are referred to as characteristic curves or characteristic values, correspond to those values of the new trapping parameter β_z that are integers, that is, $0, 1, 2, 3, \ldots$; β_z is a complex function of a_z and q_z to which we shall return. The boundaries represent, in practical terms, the point at which the trajectory of an ion becomes unbounded.

Solutions of type (ii) determine the motion of ions in an ion trap. The stability regions corresponding to stable solutions of the Mathieu equation in the z direction are shaded and labeled z stable in Figure 2.13a. The stability regions corresponding to stable solutions of the Mathieu equation in the r direction are shaded and labelled r stable in Figure 2.13b; it can be seen that they are doubled in magnitude along the

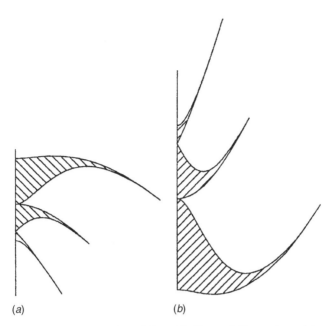

Figure 2.13. Graphical representation of three Mathieu stability regions: (*a*) z stable, as in Figure 2.3 for x stable regions; (*b*) r stable. The r-stable region is obtained as -2 times the z-stable region.

ordinate and inverted, that is, multiplied by -2. It is seen from Eq. (2.76) that $a_z = -2a_r$ and $q_z = -2q_r$, that is, the stability parameters for the r and z directions differ by a factor of -2.

Ions can be stored in the ion trap provided that their trajectories are stable in the r and z directions simultaneously; such trajectory stability is obtained in the region closest to the origin, that is, region A in Figure 2.14 analogous to Figure 2.6. Regions A and B are referred to as stability regions; region A is of the greatest importance at this time (region B remains to be explored) and is shown in greater detail in Figure 2.15. The coordinates of the stability region in Figure 2.15 are the Mathieu parameters a_z and q_z. Here, we plot a_z versus q_z rather than using the general parameters a_u versus q_u in order to avoid confusion. In Figure 2.15, the $\beta_z = 1$ stability boundary intersects with the q_z axis at $q_z = 0.908$; this working point is that of the ion of lowest mass/charge ratio [i.e., low-mass cutoff (LMCO), as discussed below] that can be stored in the ion trap for given values of r_0, z_0, V, and Ω.

2.3.3. Secular Frequencies

A three-dimensional representation of an ion trajectory in the ion trap, as shown in Figure 2.16, has the general appearance of a Lissajous curve or figure-of-eight

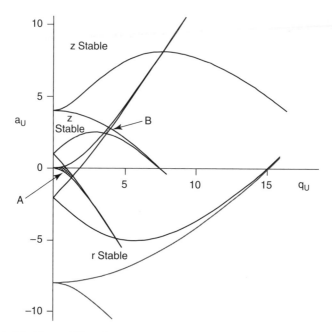

Figure 2.14. Mathieu stability diagram in (a_z, q_z) space for QIT in both r and z directions. Regions of simultaneous overlap are labeled A and B. While the axes are labeled a_u and q_u, the diagrammatic representation shown here shows the ordinate and abscissa scales in units of a_z and q_z, respectively.

composed of two fundamental frequency components, $\omega_{r,0}$ and $\omega_{z,0}$ of the secular motion [23]. The description "fundamental" implies that there exist other higher-order (n) frequencies and the entire family of frequencies is thus described by $\omega_{r,n}$ and $\omega_{z,n}$. These secular frequencies are given by

$$\omega_{u,n} = (n + \tfrac{1}{2}\beta_u)\Omega \qquad 0 \leq n < \infty \tag{2.80}$$

and

$$\omega_{u,n} = -(n + \tfrac{1}{2}\beta_u)\Omega \qquad -\infty < n < 0 \tag{2.81}$$

where

$$\beta_u \approx \sqrt{(a_u + \tfrac{1}{2}q_u^2)} \tag{2.82}$$

for $q_u < 0.4$. It should be noted that while the fundamental axial secular frequency $\omega_{z,0}$ is usually given in units of hertz in the literature and referred to simply as ω_z it should be expressed in radians per second. At this time, the higher-order frequencies are of little practical significance.

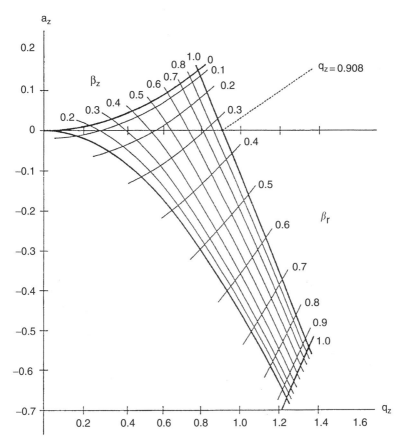

Figure 2.15. Stability diagram in (a_z, q_z) space for region of simultaneous stability A in both r and z directions near origin for three-dimensional QIT; the iso-β_r and iso-β_z lines are shown in the diagram. The q_z axis intersects the $\beta_z = 1$ boundary at $q_z = 0.908$, which corresponds to q_{max} in the mass-selective instability mode. Conventionally, the stability diagram in (a_z, q_z) space is presented.

   It should be noted further that the definition of β_u given in Eq. (2.82) above is only an approximation, known as the *Dehmelt approximation* (see also Chapter 3). A precise value of β_u is obtained from a continued fraction expression in terms of a_u and q_u, as shown later in the complete solution to the Mathieu equation.

   The resemblance of the simulated ion trajectory shown in Figure 2.16 to a roller coaster ride is due to the motion of an ion on the potential surface shown in Figure 2.17. The oscillatory motion of the ion results from the undulations of the potential surface that can be envisaged as a rotation of the potential surface. The simulation of the ion trajectory was carried out using the ITSIM simulation program [24], while the potential surface was generated [25] from Eq. (2.79) by calculating $\phi_{r, \phi, z}$ for

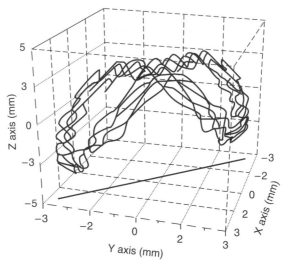

Figure 2.16. Trajectory of a trapped ion of m/z 105. The initial position was selected randomly from a population with an initial Gaussian distribution [full width at half maximum (FWHM) of 1 mm]; $q_z = 0.3$; zero initial velocity. The projection onto the x–y plane illustrates planar motion in three-dimensional space. The trajectory develops a shape that resembles a flattened boomerang. (Reprinted from the *International Journal of Mass Spectrometry and Ion Processes,* vol. 161, M. Nappi, C. Weil, C. D. Cleven, L. A. Horn, H. Wollnik, R. G. Cooks, "Visual representations of simulated three-dimensional ion trajectories in an ion trap mass spectrometer," Fig. 1, 77–85 (1997), with permission from Elsevier.)

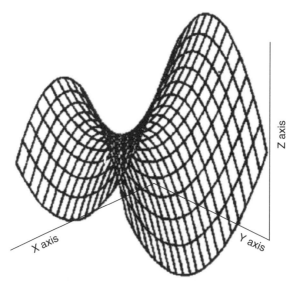

Figure 2.17. Pure quadrupole field, or potential surface, for QIT. Note the four poles of the surface and the similarity of the field shape to the trajectory in Figure 2.16.

$A_2^0 = 1$ and all of the other coefficients equal to zero for increment steps of 1 mm in both radial and axial directions.

Some of the secular frequencies defined by Eqs. (2.80) and (2.81) are illustrated in Figure 2.18. In this figure are shown the results of a power spectral Fourier analysis of the trajectory of an ion of m/z 100 calculated using ITSIM [24]. The essential trapping parameters were $r_0 = 10$ mm, $z_0 = 7.071$ mm, $\Omega/2\pi = 1.1$ MHz, $q_z = 0.40$, and the background pressure was zero. The ion's fundamental axial secular frequency, $\omega_{z,0}$, is of interest because it is the axial motion of an ion that is excited during axial modulation (see Chapter 4); this frequency, $\beta_z\Omega/2$, was observed

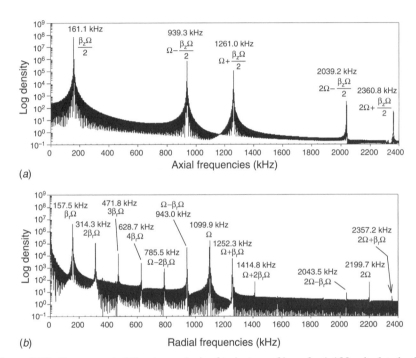

(a)

(b)

Figure 2.18. Power spectral Fourier analysis of trajectory of ion of m/z 100 calculated using ITSIM. The essential trapping parameters were $r_0 = 10$ mm, $z_0 = 7.071$ mm, $\Omega/2\pi = 1.1$ MHz, $q_z = 0.40$, and the background pressure was zero. The data were collected at intervals of 100 ns for 1 ms: (a) axial frequencies; (b) radial frequencies. The magnitude of the frequency band is plotted on a logarithmic scale and shows the intensity of each harmonic lower by several orders of magnitude from the fundamental secular frequency. Note that the radial position data were obtained from $r = \sqrt{x^2 + y^2}$, and not from either x or y; thus the radial fundamental secular frequency is observed at $\beta_r\Omega$ rather than $\beta_r\Omega/2$. (Reprinted from the *Journal of Mass Spectrometry*, vol. 34, M. W. Forbes, M. Sharifi, T. R. Croley, Z. Lausevic, R. E. March, "Simulation of ion trajectories in a quadrupole ion trap: A comparison of three simulation programs," Fig. 13, 1219–1239 (1999). © John Wiley & Sons Limited. Reproduced with permission.)

References pp. 71–72.

at 160.91142 kHz and is in close agreement with that shown in the left-hand side (LHS) of Figure 2.18a. Also shown here are two sets of complementary frequencies corresponding to $\Omega \pm \beta_z\Omega/2$ and $2\Omega \pm \beta_z\Omega/2$. In Figure 2.18$b$, where the radial frequencies are shown, the central feature is the RF drive frequency Ω; also shown are two sets of complementary frequencies corresponding to $\Omega \pm \beta_r\Omega$ and $2\Omega \pm 2\beta_r\Omega$, the harmonic frequencies $\beta_r\Omega$, $2\beta_r\Omega$, and $3\beta_r\Omega$, and the set of frequencies corresponding to 2Ω and $2\Omega \pm \beta_r\Omega$.

2.3.4. Calculations

On many occasions, while working with a QIT, it becomes necessary to calculate some of the ion-trapping parameters, such as q_z, the LMCO value (see below), β_z, and the secular frequency ω_z. In modern ion trap instruments, these calculations can be carried out using the accompanying software, but it is instructive to examine the manner in which each of the following parameters is calculated.

Let us consider an ion of butylbenzene (m/z 134) in a normal stretched ion trap that has a ring electrode of radius $r_0 = 1.00$ cm and with $z_0 = 0.783$ cm (corresponding to an electrode spacing, $2z_0$, of 15.66 mm) and under the following conditions:

$$U = 0$$
$$V = 757 \text{ V}_{(0-p)} \text{ at } 1.05 \text{ MHz}$$
$$\Omega = 2\pi f = 2\pi \times 1.05 \times 10^6 \text{ rad s}^{-1}$$
$$m = 134 \text{ Da} = 134 \times 10^{-3} \text{ kg mol}^{-1}$$
$$\text{Avogadro's number} = 6.022 \times 10^{23} \text{ mol}^{-1}$$

2.3.4.1. q_z and LMCO From Eq. (2.75), we recall that $q_z = 8eV/m(r_0^2 + 2z_0^2)\Omega^2$. Thus

$$q_z = \frac{8(1.602 \times 10^{-19}\text{C})(757 \text{ kg m}^2 \text{ s}^{-2} \text{ C}^{-1})(6.022 \times 10^{23} \text{ mol}^{-1})}{(134 \times 10^{-3} \text{ kg mol}^{-1})[(1.000 + 1.226) \times 10^{-4} \text{ m}^2](2\pi \times 1.05 \times 10^6 \text{ s}^{-1})^2}$$

$$= 0.450$$

We have now calculated that m/z 134 has a q_z value of 0.450 under these conditions, but what is the LMCO value at q_z slightly less than 0.908? Because $m \times q_z = \text{const}$ at constant V from Eq. (2.75), the LMCO value can be calculated as

$$(\text{LMCO})(0.908) = (m/z \ 134)(0.450)$$

Rearranging,

$$\text{LMCO} = (m/z \ 134)(0.450)/(0.908)$$
$$= m/z \ 66.4$$

That is, with a potential of 757 $V_{(0\text{-}p)}$ applied to the ring electrode, only those ions of $m/z > 66.4$ will be stored. The potential V to be applied to the ring electrode to effect a given LMCO is given as

$$V = \frac{\text{LMCO} \times 757 V_{(0\text{-}p)}}{m/z\ 66.4} = (11.40 \times \text{LMCO}) V_{(0\text{-}p)}$$

This calculation is particularly useful when an ion is to be fragmented and one wishes to know the low mass/charge limit for fragment ions stored, that is, the LMCO.

2.3.4.2. β_z From Eq. (2.82), we see that β_z is given approximately by $(q_z^2/2)^{1/2}$; thus, when $q_z = 0.450$, $\beta_z = 0.318$. However, we have exceeded the limit of the Dehmelt approximation relating q_z and β_z and so the calculated value of β_z is high by about 5%. For m/z 1340, where $q_z = 0.0450$ such that the above approximation is valid, $\beta_z = 0.0318$.

2.3.4.3. ω_z From Eq. (2.80), the fundamental axial secular frequency ω_z (or, more properly, $\omega_{z,0}$) is given by $\beta_z \Omega/2$; thus, when $\beta_z = 0.318$ and $\Omega = 2\pi \times 1.05 \times 10^6$ rad s^{-1}, $\omega_z = 1.049 \times 10^6$ rad s^{-1} or, more conventionally, $\omega_z = 167$ kHz; ω_z is correspondingly high by about 5%. However, for m/z 1340, $\omega_z = 16.7$ kHz.

2.3.4.4. Mass Range The upper limit of the mass range is given by the mass/charge ratio having a q_z value of, let us say, exactly 0.900 when the maximum RF amplitude is applied to the ring electrode. From Eq. (2.75) it is seen that $m \times q_z/V = \text{const}$; this constant can be evaluated from the above expression for q_z as 0.0797. With $q_z = 0.900$ and $V = 7340$ $V_{(0\text{-}p)}$, the mass range is found as 650 Da.

2.3.5. The Complete Solution to the Mathieu Equation

The complete solution to the Mathieu equation (2.33) is composed of two linear independent solutions, $u_1(\xi)$ and $u_2(\xi)$, such that

$$u = \Gamma u_1(\xi) + \Gamma' u_2(\xi) \tag{2.83}$$

where Γ and Γ' are constants of integration that depend upon the initial conditions of position u_0, velocity u_0, and RF phase ξ_0. A corollary of Floquet's theorem states that there will always exist a solution to Eq. (2.83) of the form

$$u(\xi) = e^{\mu n} \varphi(\xi) \tag{2.84}$$

where μ is a constant and φ has period π. The functions u_1 and u_2 are chosen to be even and odd, respectively, such that

$$u_1(\xi) = u_1(-\xi); \quad u_2(\xi) = -u_2(\xi). \tag{2.85}$$

Thus we can write

$$u(\xi) = \Gamma e^{\mu\xi}\varphi(\xi) + \Gamma' e^{-\mu\xi}\varphi(-\xi). \tag{2.86}$$

From Fourier's theorem, a periodic function may be expressed as an infinite sum of exponential terms so that we may write

$$\varphi(\xi) = \sum_{n=-\infty}^{\infty} C_{2n}\exp(2ni\xi) \qquad \text{and} \qquad \varphi(-\xi) = \sum_{n=-\infty}^{\infty} C_{2n}\exp(-2ni\xi) \tag{2.87}$$

so that Eq. (2.86) becomes

$$u(\xi) = \Gamma e^{\mu\xi}\sum_{n=-\infty}^{\infty} C_{2n,\,u}\exp(2ni\xi) + \Gamma' e^{-\mu\xi}\sum_{n=-\infty}^{\infty} C_{2n,\,u}\exp(-2ni\xi). \tag{2.88}$$

The $C_{2n,\,u}$ coefficients are factors that describe the amplitudes of ion motion and depend only on a_u and q_u.

The term μ is referred to as the characteristic exponent and may be real, imaginary, or complex; its value determines the type of solution to the Mathieu equation, and it may be expressed as $\mu = \alpha + i\beta$.

The solutions are of two types:

(i) stable where μ remains finite as ξ increases, and
(ii) unstable where μ increases without limit as ξ increases.

Only solutions where $\alpha = 0$ are possibly stable; if $\alpha \neq 0$, then one of the terms $e^{\mu\xi}$ or $e^{-\mu\xi}$ will tend to infinity as ξ increases; thus such solutions must be unstable.

These four possibilities and their consequences have been summarized by Dawson [5] as follows:

1. μ is real and not zero; here, one of the terms $e^{\mu\xi}$ or $e^{-\mu\xi}$ will increase without limit, and the solution is not stable.
2. μ is complex; with this condition, the solutions are not stable.
3. $\mu = im$, where m is an integer; here, the solutions are periodic but unstable. These solutions are called Mathieu functions of integral order and form the boundaries between stable and unstable regions on the stability diagram. The boundaries are referred to as characteristic curves or characteristic values.
4. $\mu = i\beta$, that is, imaginary, and β is not a whole number. These solutions are periodic and stable.

From the constraint that α must be zero, the solution of the Mathieu equation becomes

$$u(\xi) = \Gamma \sum_{n=-\infty}^{\infty} C_{2n}\exp(2n+\beta)i\xi + \Gamma' \sum_{n=-\infty}^{\infty} C_{2n}\exp-(2n+\beta)i\xi \tag{2.89}$$

and with substitution of the trigonometric identity

$$\exp i\theta = \cos \theta + i \sin \theta \qquad (2.90)$$

the expression for stable solutions becomes

$$u(\xi) = A \sum_{n=-\infty}^{\infty} C_{2n} \cos(2n + \beta)\xi + B \sum_{n=-\infty}^{\infty} C_{2n} \sin(2n + \beta)\xi \qquad (2.91)$$

where

$$A = (\Gamma + \Gamma') \qquad \text{and} \qquad B = i(\Gamma - \Gamma'). \qquad (2.92)$$

The differential of Eq. (2.91) gives an expression for the ion velocity in the quadrupole field

$$\frac{du}{dt} = A \sum_{n=-\infty}^{\infty} C_{2n}(2n + \beta) \sin(2n + \beta)\xi - B \sum_{n=-\infty}^{\infty} C_{2n}(2n + \beta) \cos(2n + \beta)\xi \qquad (2.93)$$

that is useful for simulation studies.

2.3.6. Secular Frequencies

Two series of frequencies are described by $\omega_{r,n}$ and $\omega_{z,n}$, as in Eqs. (2.80) and (2.81). These secular frequencies are obtained readily from Eq. (2.93) when we recall that $\xi = \frac{1}{2}\Omega t$ and $u = r, z$. As the magnitudes of the $C_{2n,u}$ coefficients fall off rapidly as n increases, the higher-order frequencies are of little practical significance.

As discussed previously, it should be noted that the definition of β_u given in Eq. (2.82) is only an approximation; β_u is defined precisely by a continued-fraction expression in terms of a_u and q_u, as shown in the equation

$$\beta_u^2 = a_u + \cfrac{q_u^2}{(\beta_u + 2)^2 - a_u - \cfrac{q_u^2}{(\beta_u + 4)^2 - a_u - \cfrac{q_u^2}{(\beta_u + 6)^2 - a_u - \cdots}}}$$
$$+ \cfrac{q_u^2}{(\beta_u - 2)^2 - a_u - \cfrac{q_u^2}{(\beta_u - 4)^2 - a_u - \cfrac{q_u^2}{(\beta_u - 6)^2 - a_u - \cdots}}}. \qquad (2.94)$$

2.4. CONCLUSIONS

The theory of ion trap operation differs from those of other mass spectrometers and presents an exciting challenge to the mass spectrometry community; it is hoped that this introduction to quadrupole ion trap mass spectrometry will be useful to those wishing to overcome this barrier and will enable them to enjoy the delights of this nascent branch of mass spectrometry.

APPENDIX

One of the most original and valuable contributions to the ion trap literature was the 1983 paper by Knight [11]. Up to that time, it was generally accepted that the ring electrode and end-cap electrode geometries are constrained by the relation between r_0, the ring electrode intercept, and z_0, the end-cap electrode intercept, according to $r_0 = \sqrt{2} z_0$. Knight showed that there are no constraints upon r_0 and z_0 and that RF ion traps of all geometries are described by a single stability diagram if the Mathieu equation parameters are defined properly. In his development of the most general treatment of the QIT potential, Knight introduced a constraint that the two hyperboloids (end-cap electrodes as one hyperboloid and ring electrode as an other) have the same asymptote. The slope of the common asymptote was shown to be $1/\sqrt{2}$, independent of r_0 and z_0, and thus the angle of the asymptote above the r axis is 35.26° for all QITs. Knight wrote, "A simple geometric calculation shows that the asymptote of the $r_0 = \sqrt{2} z_0$ trap bisects, at high values of r and z, the ring electrode–end-cap electrode gap."

In preparing this chapter, the authors decided to carry out this simple geometric calculation for their own satisfaction. The task consumed an inordinate amount of time and the objective was not achieved. Eventually, it was decided to approach Dr. Knight and to request his geometric calculation. To his credit, Dr. Knight responded readily to our request, so readily that we suspect he spent much of his President's Day holiday weekend at the task. The calculation shown below may or may not be the same as that carried out some 20 years previously, but it is similar. The calculation is certainly not simple and, while it is not difficult, it is somewhat devious!

Although when Knight's paper first appeared it tended to be cited as possibly being a sort of perfectionist refinement, in fact, now that most practical commercial traps have "stretched" geometries where $r_0^2 < 2z_0^2$, the use of his extended equations to evaluate a_u and q_u is essential, even as an approximation when the equations for the electrode surfaces are still of the form specified for the "unstretched" version.

Proof That the Asymptote of a QIT Bisects the Ring Electrode to End-Cap Electrode Gap at High Values of r and z if $r_0 = \sqrt{2} z_0$ First consider the ring electrode, as shown in Figure 2.19. A point on the ring electrode with coordinates (r, z) is distance d from the asymptote, measured perpendicular to the asymptote. From the equation for the ring electrode,

$$\frac{r^2}{r_0^2} - \frac{2z^2}{r_0^2} = 1 \tag{2.95}$$

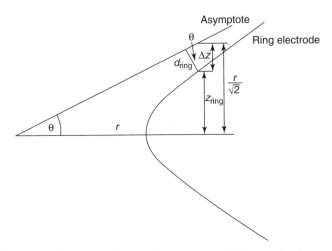

Figure 2.19. Geometric construction for determination of distance d_{ring} between ring electrode and asymptote in QIT.

the height of the ring electrode above the radial axis, z_{ring}, is

$$z_{ring} = \sqrt{\frac{r^2 - r_0^2}{2}} = \frac{r}{\sqrt{2}}\sqrt{1 - \frac{r_0^2}{r^2}}. \tag{2.96}$$

As Ref. 11 shows, the asymptote has the equation $z = (1/\sqrt{2})r$, independent of r_0 and z_0. Therefore the height of the asymptote above the radial axis is $r/\sqrt{2}$. Thus, at radial distance r, the vertical distance between the electrode and the asymptote, Δz, is

$$\Delta z = \frac{r}{\sqrt{2}} - z_{ring} = \left(\frac{r}{\sqrt{2}} - \frac{r}{\sqrt{2}}\sqrt{1 - \frac{r_0^2}{r^2}}\right) = \frac{r}{\sqrt{2}}\left(1 - \sqrt{1 - \frac{r_0^2}{r^2}}\right). \tag{2.97}$$

Figure 2.19 shows that $d = \Delta z \cos\theta$, where θ is the angle of the asymptote. Because the slope of the asymptote is $1/\sqrt{2}$, meaning that $\tan\theta = 1/\sqrt{2}$, it follows that $\cos\theta = \sqrt{\frac{2}{3}}$. Thus at radial distance r, the spacing between the ring electrode and the asymptote, d_{ring}, is

$$d_{ring} = \Delta z \cos\theta = \sqrt{\frac{2}{3}}\Delta z = \frac{r}{\sqrt{3}}\left(1 - \sqrt{1 - \frac{r_0^2}{r^2}}\right). \tag{2.98}$$

This is an exact expression.

The treatment of the end-cap electrode is similar (see Figure 2.20), but this time $\Delta z = z_{endcap} - r/\sqrt{2}$. The end-cap electrode equation is

$$\frac{r^2}{2z_0^2} - \frac{z^2}{z_0^2} = -1 \tag{2.99}$$

References pp. 71–72.

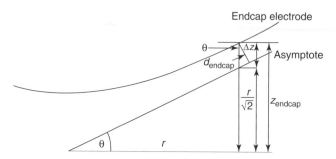

Figure 2.20. Geometric construction for determination of distance d_{endcap} between end-cap electrode and asymptote.

from which we can write

$$z_{endcap} = \sqrt{\frac{r^2 + 2z_0^2}{2}} = \frac{r}{\sqrt{2}}\sqrt{1 + \frac{2z_0^2}{r^2}}. \tag{2.100}$$

Thus at radial distance r, the spacing between the end-cap electrode and the asymptote, d_{endcap}, is

$$d_{endcap} = \Delta z \cos\theta = \sqrt{\frac{2}{3}}\Delta z = \frac{r}{\sqrt{3}}\left(\sqrt{1 + \frac{2z_0^2}{r^2}} - 1\right). \tag{2.101}$$

This is also an exact expression.

Now consider the situation in which r and z are "large," meaning that $r_0/r \ll 1$ and $z_0/r \ll 1$. We can use the binomial approximation $(1 \pm x)^{1/2} \approx 1 \pm \frac{1}{2}x$ if $x \ll 1$ to write

$$d_{ring} = \frac{r}{\sqrt{3}}\left(1 - \sqrt{1 - \frac{r_0^2}{r^2}}\right) \approx \frac{r}{\sqrt{3}}\left[1 - \left(1 - \frac{1}{2}\frac{r_0^2}{r^2}\right)\right] = \frac{r_0^2}{2\sqrt{3}r} \tag{2.102}$$

and

$$d_{endcap} = \frac{r}{\sqrt{3}}\left(\sqrt{1 + \frac{2z_0^2}{r^2}} - 1\right) \approx \frac{r}{\sqrt{3}}\left[\left(1 + \frac{1}{2}\frac{2z_0^2}{r^2}\right) - 1\right] = \frac{z_0^2}{\sqrt{3}r}. \tag{2.103}$$

In general, $d_{ring} \neq d_{endcap}$. However, if $r_0 = \sqrt{2}z_0$, then $r_0^2 = 2z_0^2$ and we find that

$$d_{ring} = \frac{r_0^2}{2\sqrt{3}r} = \frac{2z_0^2}{2\sqrt{3}r} = \frac{z_0^2}{\sqrt{3}r} = d_{endcap}. \tag{2.104}$$

Thus the asymptote bisects the electrode gap if $r_0 = \sqrt{2}z_0$.

The degree to which this is true can be tested by using the exact expressions for d_{ring} and d_{endcap}. Table 2.1 shows the difference $\Delta d = d_{ring} - d_{endcap}$ as a fraction of the gap spacing $d_{gap} = d_{ring} + d_{endcap}$. The end-cap electrode is always closer to the asymptote, but the difference has dropped to less than 3% by $r = 3r_0$ and to 1% at $r = 5r_0$.

TABLE 2.1. Calculated Values for d_{ring} and d_{endcap} as
Radial Distance from Center of Ion Trap is Increased

r	d_{ring}	d_{endcap}	$\Delta d/d_{gap}$
$2r_0$	$0.1547r_0$	$0.1363r_0$	$0.063 = 6.3\%$
$3r_0$	$0.09906r_0$	$0.09369r_0$	$0.028 = 2.8\%$
$4r_0$	$0.07333r_0$	$0.07106r_0$	$0.016 = 1.6\%$
$5r_0$	$0.05832r_0$	$0.05717r_0$	$0.010 = 1.0\%$

REFERENCES

1. E. Mathieu, *Math. Pure Appl. (J. Liouville)* **13** (1868) 137.
2. P. H. Dawson, N. R. Whetten, *J. Vac. Sci. Technol.* **5** (1968) 1.
3. P. H. Dawson, N. R. Whetten, *J. Vac. Sci. Technol.* **5** (1968) 11.
4. J. E. Campana, *Int. J. Mass Spectrom. Ion Phys.* **33** (1980) 101.
5. P. H. Dawson, *Quadrupole Mass Spectrometry and Its Applications*, Elsevier, Amsterdam, 1976. Reprinted as an "American Vacuum Society Classic" by the American Institute of Physics. (ISBN 1563964554)
6. R. E. March, R. J. Hughes, J. F. J. Todd, *Quadrupole Storage Mass Spectrometry*, Wiley Interscience, New York, 1989.
7. G. Lawson, J. F. J. Todd, R. F. Bonner, *Dyn. Mass Spectrom.* **4** (1975) 39.
8. R. E. March, *J. Mass Spectrom.* **32** (1997) 351.
9. R. E. March, J. F. J. Todd (Eds.), *Practical Aspects of Ion Trap Mass Spectrometry*, Modern Mass Spectrometry Series, Vols. I–III, CRC Press, Boca Raton, FL, 1995.
10. N. W. McLachan, *Theory and Applications of Mathieu Functions*, Clarendon, Oxford, 1947. See also R. Campbell, *Théorie Générale de l'Equation de Mathieu*, Masson, Paris, 1955.
11. R. D. Knight, The general form of the quadrupole ion trap potential, *Int. J. Mass Spectrom. Ion Processes* **51** (1983) 127–131.
12. D. R. Denison, *J. Vac. Sci. Technol.* **8** (1971) 266.
13. J. R. Gibson, S. Taylor, *Rapid Commun. Mass Spectrom.* **15** (2001) 1960.
14. P. H. Dawson, Higher zones of stability for the quadrupole mass filter, *J. Vac. Sci. Technol.* **11**(6) (1974) 1151.
15. P. H. Dawson, Y. Bingqi, Quadrupoles operating in the second stability region, paper presented at the Thirty-Second Annual Conference of the American Society for Mass Spectrometry and Allied Topics, San Antonio, TX, May/June 1984, pp. 505–506.
16. P. H. Dawson, Y. Bingqi, The second stability region of the quadrupole mass filter. II. Experimental results, *Int. J. Mass Spectrom. Ion Phys.* **56** (1984) 41.
17. P. H. Dawson, Y. Bingqi, The second stability region of the quadrupole mass filter. I. Ion optical properties, *Int. J. Mass Spectrom. Ion Phys.* **56** (1984) 25.
18. W. Paul, H. Steinwedel, Apparatus for separating charged particles of different specific charges, German Patent 944,900 (1956); U.S. Patent 2,939,952 (June 7, 1960).

19. W. Paul, *Angew. Chem.* **29** (1990) 739.

20. W. Paul, O. Osberghaus, E. Fischer, *Forschungsberichte des Wirtschaft and Verkehrministeriums Nordrhein Westfalen*, No. 415, Westdeutschter Verlag, Koln and Opladen, 1958, pp. 6–7.

21. R. F. Wuerker, H. Shelton, R. V. Langmuir, *J. Appl. Phys.* **30** (1959) 342.

22. R. D. Knight, personal communication, 2004.

23. M. Nappi, C. Weil, C. D. Cleven, L. A. Horn, H. Wollnik, R. G. Cooks, *Int. J. Mass Spectrom. Ion Processes* **161** (1997) 77.

24. H.-P. Reiser, R. E. Kaiser, Jr., R. G. Cooks, *Int. J. Mass Spectrom. Ion Processes* **121** (1992) 49.

25. M. Splendore, personal communication, 1999.

3

DYNAMICS OF ION TRAPPING

3.1. INTRODUCTION

In Chapter 2 we developed a generalized approach to the theory of ion motion occurring within quadrupole devices, and this treatment was subsequently applied to the quadrupole mass filler (QMF) and linear ion trap (LIT), each of which utilizes a two-dimensional field, and to the quadrupole ion trap (QIT), in which ions are confined three dimensionally. This general approach to the theory provided a vehicle for the introduction of the Mathieu equation, the trapping parameters a_u and q_u, and the concept of *stable* and *unstable* ion trajectories as well as a discussion of the parameter β_u and the associated secular frequency ω_u for each independent component of the ion motion; supporting specimen calculations were given for q_z, the low mass cut-off value (LMCO), β_z, ω_z, and the upper limit of the mass range of stored ions. Finally, a more rigorous alternative approach to the derivation of the potential within an ion trap was considered together with an account of the general solution of the Mathieu equation.

However, as was clearly stated (Section 2.3), the whole of the treatment given in Chapter 2 is based on the assumption that we are examining the behavior of a single ion in an infinite, ideal quadrupole field in the total absence of any background gas. Clearly, in most applications a single ion is rarely of great value and background gases cannot be excluded totally; in reality, therefore, the theory of ion containment must be considered along with the dynamical aspects of ion behavior, including the effects arising from the presence of other trapped ions (i.e., space charge), the existence of higher order (nonquadrupole) field components and collisional effects, both with the background and with deliberately added neutral gases. Furthermore, in practice, it is obviously necessary to create ions within the quadrupole field or to inject them from an external source and, in most applications, to eject them for detection. Other possible experiments may make use of the ability to excite resonantly the motion of ions mass-selectively through the application of additional potentials oscillating with frequencies that match one or more of the secular frequencies $\omega_{u,n}$.

All these topics are considered in this chapter under the general heading "dynamics of ion trapping." First we consider a simplifying approximation to the theory of containment that assumes that the motion of the ions can be described in terms of their being trapped within a *pseudopotential well*. This approximation allows us, in turn, to develop models that predict the maximum number of ions that may be stored within an ion trap and to make estimates of the kinetic energies of ion motion. From here we examine the influence of higher order field components and the effects of *nonlinear resonances* on the motion of the ions, together with such phenomena as *black holes* and *mass shifts*.

3.2. THE PSEUDOPOTENTIAL WELL MODEL

Examination of the ion trajectory modeled in Figure 2.16 shows that the motion can be regarded as being a combination of a low-frequency ("secular") oscillation superimposed upon which is a high-frequency ripple. It was this characteristic that led Wuerker et al. [1] to formulate an approach that was later extended by Dehmelt [2] to become what is now known as the pseudopotential well model. Essentially, at low values of β_u, it is assumed that, by neglecting the high-frequency ripple, the motion of ions along the u coordinate can be approximated to that of a charged particle undergoing simple harmonic motion in a parabolic potential well. By relating the depth of this well to the trapping parameters a_u and q_u, and thus to the experimental operating conditions of the device, it is possible to derive estimates of the density of trapped ions and to evaluate their kinetic energies (in the absence of collisional effects). The treatment that follows is based on the derivation reported by Todd et al. [3].

Let us consider initially the axial (z) motion of the ion; we may rewrite the Mathieu equation (2.33) as

$$\frac{d^2z}{d\xi^2} = -(a_z - 2q_z \cos 2\xi)z \tag{3.1}$$

where

$$a_z = -\frac{16eU}{m(r_0^2 + 2z_0^2)\Omega^2} \qquad \text{[Eq. (2.74)]}$$

and

$$q_z = \frac{8eV}{m(r_0^2 + 2z_0^2)\Omega^2}. \qquad \text{[Eq. (2.75)]}$$

If we now substitute for the value of the displacement z a combination of the secular displacement, Z, and the micromotion associated with the ripple as δ, then we can write

$$z = Z + \delta. \tag{3.2}$$

Building in two further assumptions, that $\delta \ll Z$ and $d\delta/dt \gg dZ/dt$, Eq. (3.1) can now be approximated to

$$\frac{d^2\delta}{d\xi^2} = -(a_z - 2q_z \cos 2\xi)Z. \tag{3.3}$$

Making the further assumptions that $a_z \ll q_z$ and Z is constant over a period of the RF oscillation, Eq. (3.3) integrates to

$$\delta = \frac{-q_z Z}{2} \cos 2\xi \tag{3.4}$$

so that substitution for δ into Eq. (3.2) gives

$$z = Z - \frac{q_z Z}{2} \cos 2\xi \tag{3.5}$$

whence Eq. (3.1) becomes

$$\frac{d^2z}{d\xi^2} = -a_z Z + \frac{a_z q_z Z}{2} \cos 2\xi + 2q_z Z \cos 2\xi - q_z^2 Z \cos^2 2\xi. \tag{3.6}$$

Since the acceleration $d^2\delta/d\xi^2$ resulting from the application of the RF "drive" potential when averaged over a period of the RF drive is zero, the acceleration of the secular motion averaged over the same period is

$$\left\langle \frac{d^2Z}{d\xi^2} \right\rangle_{av} = \frac{1}{\pi} \int_0^\pi \frac{d^2z}{d\xi^2} \, d\xi. \tag{3.7}$$

Integration of Eq. (3.7) between these limits, and using the relationship given in Eq. (2.36) to convert into units of time gives, therefore,

$$\frac{d^2Z}{dt^2} = -\left(a_z + \frac{q_z^2}{2}\right)\frac{\Omega^2}{4} Z. \tag{3.8}$$

Equation (3.8) corresponds to a simple harmonic motion with the general form

$$\frac{d^2Z}{dt^2} = -\frac{\beta_z^2\Omega^2 Z}{4} = -\omega_z^2 Z \tag{3.9}$$

where

$$\beta_z = \left(a_z + \frac{q_z^2}{2}\right)^{1/2} \tag{3.10}$$

and the secular frequency ω_z is given by

$$\omega_z = \tfrac{1}{2}\beta_z\Omega. \tag{3.11}$$

It should be noted that Eq. (3.10) is identical to the expression for the Dehmelt approximation given in Eq. (2.82).

Taking Eq. (3.8) under the special conditions of $a_z = 0$ (i.e., no DC potential applied between the ring and the end-cap electrodes) and substituting for q_z from Eq. (2.75), we have

$$\frac{d^2Z}{dt^2} = -\frac{8e^2V^2}{m^2(r_0^2+2z_0^2)^2\Omega^2} Z \tag{3.12}$$

which may be written as

$$m\frac{d^2Z}{dt^2} = -e\frac{d\bar{D}_z}{dZ}. \tag{3.13}$$

Here $d\bar{D}_z/dZ$ is the electric field due to a "pseudopotential" $\bar{D}_z$ acting in the z direction such that

$$\frac{d\bar{D}_z}{dZ} = \frac{8eV^2}{m(r_0^2+2z_0^2)^2\Omega^2} Z \tag{3.14}$$

which on integrating between $Z = 0$ and $Z = z_0$ gives

$$\bar{D}_z = \frac{4eV^2z_0^2}{m(r_0^2 + 2z_0^2)^2\Omega^2} = \frac{mz_0^2\Omega^2}{16e} q_z^2 \tag{3.15}$$

by substitution for V from Eq. (2.75). An exactly analogous treatment of the motion in the r direction gives

$$\overline{D}_r = \frac{eV^2 r_0^2}{m(r_0^2 + 2z_0^2)^2 \Omega^2} = \frac{mr_0^2 \Omega^2}{16e} q_r^2. \tag{3.16}$$

Partial substitution from Eq. (2.75) for q_z in Eq. (3.15) gives

$$\overline{D}_z = \frac{z_0^2}{2(r_0^2 + 2z_0^2)} V q_z \tag{3.17}$$

whence if $r_0^2 = 2z_0^2$ [Eq. (2.77)]

$$\overline{D}_z = \tfrac{1}{8} V q_z. \tag{3.18}$$

Similarly, it may be shown that under these conditions

$$\overline{D}_r = \tfrac{1}{8} V q_r. \tag{3.19}$$

From Eqs. (3.15) and (3.16) we can now immediately compare the relative values of $\overline{D}_z$ and $\overline{D}_r$ and find that

$$\frac{\overline{D}_z}{\overline{D}_r} = \frac{4z_0^2}{r_0^2} \tag{3.20}$$

whence we see that for the "standard" configuration of $r_0^2 = 2z_0^2$ [Eq. (2.77)]

$$\overline{D}_z = 2\overline{D}_r. \tag{3.21}$$

Thus we deduce that under the conditions of the Dehmelt approximation, that is, for low values of q_z ($\ll 0.4$, see below), the ions can be treated as though they were trapped in a pseudopotential well of the form illustrated in Figure 3.1, for which the profiles in the r and z directions are shown in Figure 3.2.

It is instructive to consider a test of this derivation of the pseudopotential well model by asking the following question: if we apply a positive DC potential U to the ring electrode, under what conditions might the trajectories become unstable so that the ions will be ejected from the trap along the z direction? Since we are working with an ion trap in which the end-cap electrodes are held at ground potential, we must employ Eq. (2.67) in order to determine the effective defocusing potential $\phi_{0,0}$, which is equal to the depth of the pseudopotential well $\overline{D}_z$, that is, $\overline{D}_z = \phi_{0,0}$. Putting $r = z = 0$ and $\phi_0 = U$ in Eq. (2.67) gives

$$\phi_{0,0} = \frac{2Uz_0^2}{r_0^2 + 2z_0^2} \tag{3.22}$$

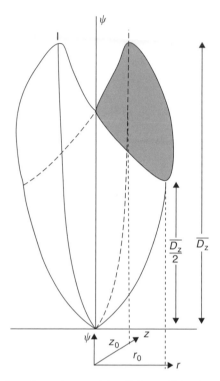

Figure 3.1. Representation of pseudopotential well for confining ions within an ion trap having $r_0^2 = 2z_0^2$ when operated in RF-only mode ($a_z = 0$). Sections through the well in the r–z plane are elliptical. (Reprinted from *Mass Spectrometry Reviews*, Vol. 10, J. F. J. Todd, "The ion trap mass spectometer—Past, present and future (?)," Fig. 14, pages 3–52 (1991). © John Wiley & Sons Limited. Reproduced with permission.)

so that on rearranging Eq. (2.74) in order to obtain an expression for U

$$U = -\frac{m(r_0^2 + 2z_0^2)\Omega^2}{16e}a_z \tag{3.23}$$

we can deduce that

$$\phi_{0,0} = -\frac{mz_0^2\Omega^2}{8e}a_z. \tag{3.24}$$

Consequently, on equating this expression for $\phi_{0,0}$ with that for $\overline{D}_z$ given in Eq. (3.15) and simplifying, we find that

$$a_z = -\tfrac{1}{2}q_z^2 \tag{3.25}$$

which, when substituted into Eq. (3.10), gives $\beta_z = 0$. In other words, the addition of a positive, defocusing, potential to the ring electrode sufficient to oppose the trapping effect of the pseudopotential well in the axial direction when the ion trap is operating

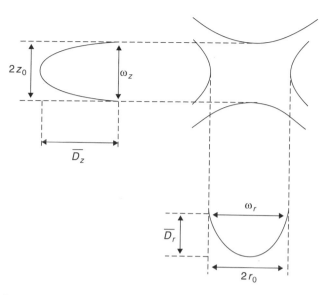

Figure 3.2. Parabolic pseudopotential well in the r- and z-directions for ion trap having $r_0^2 = 2z_0^2$ when operated in RF-only mode ($a_z = 0$). Symbols $\bar{D}_r$ and $\bar{D}_z$ indicate the depths of the wells and ω_r and ω_z represent the fundamental frequencies of secular oscillations in the radial and axial direction, respectively. (Reprinted from *Dynamic Mass Spectrometry*, Vol. 4, D. Price and J. F. J. Todd (Eds.), G. Lawson, J. F. J. Todd, R. F. Bonner, "Theoretical and experimental studies on the quadrupole ion storage source," Heyden and Son, Fig. 22, pages 39–81 (1976). © John Wiley & Sons Limited. Reproduced with permission.)

at a given value of the trapping parameter q_z will cause ion ejection at the corresponding point a_z on the $\beta_z = 0$ boundary of the stability diagram shown in Figure 2.15.

3.2.1. Specimen Calculation of the Pseudopotential Well Depth $\bar{D}_z$

The depth of the pseudopotential well in the z direction for a trap in which $r_0^2 = 2z_0^2$ and with $U = 0$ may be estimated from Eq. (3.18):

$$\bar{D}_z = \tfrac{1}{8}Vq_z$$

Thus for $Ar^{+\cdot}$ ions trapped according the conditions listed in Table 1.2, that is, m/z 40 but at $q_z = 0.38$ with a drive potential $V = 90$ $V_{0\text{-p}}$, we have

$$\bar{D}_z = \tfrac{1}{8}(0.38)(90) = 4.3 \text{ V}$$

3.2.2. Some Applications of the Pseudopotential Well Model

Apart from the well model offering a fairly simple mental picture of the manner in which ions are held within the QIT, there have been two particular applications that have afforded more quantified answers concerning the properties of the QIT: what is the maximum number of ions that can be stored under a given set of operating conditions and what is the range of the kinetic energies of the trapped ions? While the latter question is now more effectively answered by the various simulation methods described in Chapter 4, historically it was the pseudopotential well approach that provided the first clues about ion "temperatures." These two applications are now considered below.

3.2.2.1. Estimation of the Effects of Space Charge

Introduction   We have already noted above, and especially in Chapter 2, that one of the underlying assumptions when deriving the equations of motion of ions within quadrupole devices is that there is only a single ion moving under the influence of the electric field. In most practical applications there are, of course, very many ions to consider, and because of coulombic interactions, each of these will influence the behavior of the others. Indeed, it is probably self-evident that there must be a maximum *space charge* limit to the number of ions that can be confined within the quadrupole field. For the QMF operating as an analytical instrument, the instantaneous density of ions injected into the field from the ion source is always very low so that effects of space charge on the performance of the analyzer can usually be ignored, although this was not the case in the development of a large-scale QMF as an isotope separator [4]. For the LIT and the QIT very different conditions apply as we create ions "internally" or inject them from an external source, and in Chapter 1, Eq. (1.4), we have already met this concept in terms of the "saturation conditions" where $N = N_\infty$. The ensuing discussion concentrates predominantly on the QIT; a comparison of the capacities of the QIT and the LIT is given in Chapter 5.

As with many aspects of ion trap behavior, the influence of space charge on ion trajectory stability was first considered by the Bonn group in 1959 in a paper under the authorship of Fischer [5]. He noted that if it is assumed that the space charge is distributed uniformly over the whole of the trapping field, then this acts as an effective DC voltage increment, ΔU, which causes shifts in the a values, Δa_r and Δa_z. These shifts, in turn, lead to displacements of the boundaries of the stability diagram as well as to predicted shifts in the working points (a_u, q_u) and consequently of the secular frequencies. A similar effect has been discussed by Schwebel et al. [6] and Sheretov et al. [7]. Todd et al. [3] carried out an extensive series of experimental determinations of the shifts of the boundaries of the stability diagrams for different ions, under the influence of varying amounts of space charge, in order to estimate the numbers of trapped ions, and Schuessler [8] used the shift in secular frequencies to estimate the ion number, as was also done by Fulford et al. [9].

Dehmelt [2] was the first person to formulate a detailed model for the treatment of space charge and the equilibrium temperature of the stored ion cloud, based upon his pseudopotential well approach; direct simulations by Ghosh and co-workers [10, 11] showed the influence of the initial injection density on the initial periods of the motion and the stability of ions in the trap. Studies of charge exchange reactions [12–14] and investigations of the collisional cooling with a buffer gas [2, 15, 16] took account of momentum transfer to calculate ion trajectories and showed that the distributions of the ions depended strongly on the environment of the ionic population. More recently, two detailed publications by Marshall and co-workers [17, 18] have provided detailed analyses of the space charge distributions in traps, the latter paper being of especial interest because it compares the behavior of the Paul trap with that of the Penning (ion cyclotron resonance (ICR) trap) and the combined trap (which comprises a Paul trap with a superimposed axial magnetic field).

In the account which follows, we first derive an expression for the space charge–limited ion density based upon the pseudopotential well model, as originally formulated by Dehmelt [2], but with the refinement utilized in Chapter 2, namely that no assumption is made a priori concerning the r_0:z_0 ratio; we then compare the results of this treatment with the computations presented in Ref. 18.

Estimation of Space Charge by Means of the Pseudopotential Well Model   The model of ion trapping in a pseudopotential well is somewhat artificial in that the well does not actually exist until an ion is held in a stable trajectory within the ion trap! Every ion that is placed subsequently in the ion trap must modify the fields experienced by those ions already present: eventually the space charge limit is reached when the trapping potential ψ is balanced by the repulsive electrostatic potential ϕ_i arising within the ion ensemble. If we assume that this trapping potential is in fact provided by the pseudopotential derived above, then from the Poisson relation we can write

$$[\text{Electrostatic potential, } \phi_i] + [\text{Pseudopotential, } \psi] = \text{const} \qquad (3.26a)$$

such that

$$-\nabla^2 \phi_i = \nabla^2 \psi = 4\pi \rho_{max} \qquad (3.26b)$$

where ρ_{max} is the theoretical space charge–limited ion density.

Let us assume that the ion trap is being operated with zero DC bias between the electrodes, that is, $U = 0$, and that the ions are confined within an inscribed cylindrical volume within the ion trap where the radius of the cylinder is r_0 and the height of the axis is $2z_0$. If we assume that the pseudopotential acting in each of the x and y directions is $\bar{D}_r$ and that acting in the z direction is $\bar{D}_{zr}$, then we can express the pseudopotential $\psi_{x, y, z}$ acting at a point (x, y, z) as

$$\psi_{x, y, z} = \frac{\bar{D}_r}{r_0^2}(x^2 + y^2) + \frac{\bar{D}_z}{z_0^2}z^2. \qquad (3.27)$$

Thus

$$\nabla \psi = \frac{\bar{D}_r}{r_0^2}(2x + 2y) + \frac{2\bar{D}_z z}{z_0^2} \qquad (3.28)$$

and

$$\nabla^2\psi = \frac{4\overline{D}_r}{r_0^2} + \frac{2\overline{D}_z}{z_0^2} \tag{3.29}$$

which, on substituting for $\overline{D}_r$ from Eq. (3.21), gives

$$\nabla^2\psi = \frac{3\overline{D}_z}{z_0^2}. \tag{3.30}$$

Equating this with Eq. (3.26b) gives

$$\rho_{max} = \frac{3\overline{D}_z}{4\pi z_0^2} \tag{3.31}$$

or

$$N_{max} = \frac{3\overline{D}_z}{4\pi \, ez_0^2}. \tag{3.32}$$

Substitution for $\overline{D}_z$ from Eq. (3.15) gives

$$N_{max} = \frac{3}{\pi} \frac{V^2}{m(r_0^2 + 2z_0^2)^2\Omega^2} \tag{3.33}$$

or, on making the substitution from Eq. (2.75),

$$N_{max} = \frac{3V}{8\pi e(r_0^2 + 2z_0^2)} q_z \tag{3.34}$$

or, repeating the substitution, we have

$$N_{max} = \frac{3}{64\pi} \frac{m\Omega^2}{e^2} q_z^2. \tag{3.35}$$

Hence, we see from Eq. (3.33) that when operating the ion trap at a fixed drive potential V and fixed frequency Ω the maximum number of stored ions, N_{max}, of a given species is inversely proportional to their mass and independent of their charge; conversely, when working at a fixed value of q_z the value of N_{max} is directly proportional to the mass of the ion and inversely proportional to the square of its charge.

Specimen Calculation of the Maximum Ion Density N_{max}   The maximum ion density is given by Eq. (3.35):

$$N_{max} = \frac{3}{64\pi} \frac{m\Omega^2}{e^2} q_z^2$$

so that, again taking some of the conditions given in Table 1.2 for the trapping of Ar$^+$· ions such that m/z is 40 at $q_z = 0.38$ with a drive frequency of 0.762 MHz, we have

$$N_{max} = \frac{3}{64\pi} \frac{(40\,g\,mol^{-1})(2\pi \times 0.762 \times 10^6\,s^{-1})^2(0.38)^2}{4.803 \times 10^{-10}\,g^{1/2}cm^{3/2}s^{-1})^2(6.022 \times 10^{23}\,mol^{-1})}$$

$$= 0.45 \times 10^7\,cm^{-3}$$

which is, somewhat fortuitously, in good agreement with the experimental value of N_∞ ($=1.04 \times 10^7 \text{ cm}^{-3}$ at $q_z = 0.59$) reported in Table 1.2 [19].

Recent Discussions on Space Charge In their comparative study of the equilibrium ion density distribution in the Penning, Paul, and combined ion traps, Li et al. [18] considered the case where the standard dimensional relationship

$$r_0^2 = 2z_0^2 \qquad \text{[Eq. (2.77)]}$$

applied and, for the Paul trap, the operating pressure was in the range 10^{-6}–10^{-3} Torr with the end-cap electrodes held at zero potential. Under these conditions it was argued that, because of the frequent ion-neutral collisions, the ions reached thermal equilibrium quite rapidly. Two extreme sets of conditions were therefore examined: the *high temperature–low density limit* and the *low temperature–high density limit*. In the former it was argued that the ion–ion Coulomb interaction energy is much less than the ion kinetic energy and that the ion density distribution is therefore Gaussian and determined by the ion-trapping field and the ion temperature. Calculated density distributions for singly charged ions of mass 2000 amu under conditions of $q_z = 0.38$ are shown in Figure 3.3 for the radial (upper) and axial (lower) directions for three different temperatures. When the equivalent calculations applied to the Penning (ICR) and the Paul traps are compared, it is found that in the former the radial ion density distribution lies closer to the z axis and is less sensitive to changes in temperature than in the latter, whereas the reverse situation applies in respect of the axial distributions.

With regard to the conditions of low temperature–high density, the ion kinetic energy is much less than the ion–ion Coulomb interaction energy and the ion density distribution determines the maximum possible ion density in the trap. In exploring this high-ion-density case, the authors drew upon the pseudopotential well model derived above, in which the maximum ion density distribution represents a balance between the Coulombic forces and the force from the trapping potential, as formulated in Eq. (3.25). In doing this they invoked the Dehmelt approximation in which $q_z < 0.4$ and determined the value of N_{max} using the expression

$$N_{max} = \frac{3V}{4\pi m r_0^4 \Omega^2} \tag{3.36}$$

which is derived from Eq. (3.33) by substituting for z_0^2 from Eq. (2.77). Computed maximum ion densities expressed as a function of the drive potential V for a range of high-mass ions are shown in Figure 3.4. As noted above, under this formulation the maximum ion densities depend upon V, Ω, and m but not on the ion charge. This independence of the ion charge was also found to apply to the Penning trap, where it was shown that the value of N_{max} is given by

$$N_{max} = \frac{B^2}{8\pi m c^2} \tag{3.37}$$

in which B is the magnetic field intensity and c is the speed of light.

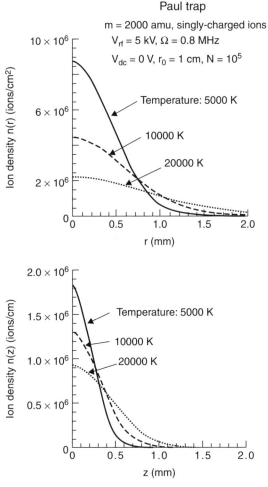

Figure 3.3. Plots showing radial (top) and axial (bottom) ion density spatial distributions at thermal equilibrium in a quadruple ion trap ($r_0^2 = 2z_0^2$) in limits of high temperature and low ion density. (Reprinted by permission of Elsevier from "Comparison of equilibrium ion density distribution and trapping force in Penning, Paul, and combined ion traps," by G.-Z. Li, S. Guan, and A. G. Marshall, *Journal of American Society for Mass Spectrometry,* Vol. 9, pages 473–481, Copyright 1998, by the American Society for Mass Spectrometry.)

Practical Consequences of Space Charge   In terms of the applications of the ion trap as an analytical mass spectrometer, control of the trapped ion density is extremely important. This is because space charge can cause a significant departure from ideal behavior, an effect that becomes especially important when there is a requirement for the system to be able to provide routine, reproducible, and quantitative analytical data. As far as mass-selective operation of the trap is concerned, space charge may cause shifts in the peak positions, and hence mass assignments, as well

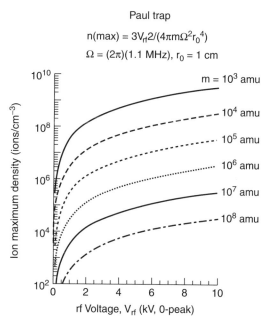

Figure 3.4. Plots showing the maximum ion density in a quadrupole ion trap ($r_0^2 = 2z_0^2$) as a function of the RF drive potential at thermal equilibrium and in the limits of low temperature and high ion density. (Reprinted by permission of Elsevier from "Comparison of equilibrium ion density distribution and trapping force in Penning, Paul, and combined ion traps," by G.-Z. Li, S. Guan, and A. G. Marshall, *Journal of American Society for Mass Spectrometry*, Vol. 9, pages 473–481, Copyright 1998, by the American Society for Mass Spectrometry.)

as peak broadening and consequent loss of resolution. In addition, at high sample pressures, reactions between the ions and neutral molecules may cause significant spectral changes (e.g., via "self-chemical ionization") so that the spectra could change substantially during the elution of a chromatographic peak. These problems were especially apparent with the early commercial instruments, where it was found that there could be a significant mismatch between the electron ionization (EI) mass spectra recorded with the ion trap compared to those obtained with "standard" instruments and held in mass spectral library collections.

The problems are essentially twofold. First, high sample concentrations combined with significant trapping times of, say, several milliseconds lead to the occurrence of ion/molecule reactions, thus changing the identities of the ions being analyzed and also causing a loss of quantitative response. Second, the buildup of ion density within the ion trap can lead to space charge effects, substantially modifying the electric fields to which the ions are being subjected (thereby causing, e.g., shifts in the secular frequencies of the ions and hence the apparent positions of the iso-β lines and the boundaries of the stability diagram), resulting in changes in the mass/charge ratio assignments of the ions.

The situation may be viewed rather more quantitatively as follows [20]. The number of ions (N) formed in the trap by electron ionization is given by

$$N = k[S]it \qquad (3.38)$$

where k is a constant from which we see that there should be a linear dependence of N upon the sample concentration $[S]$, the electron beam current i, and the duration of the ionization period t. Thus in a GC/MS application, for example, as the value of $[S]$ changes during the elution of a component of the analyte, there will be linearity of response between the ion trap signal and the sample concentration only when the ideal behavior described by Eq. (3.38) is maintained. The limits of ideality are governed by the considerations represented by Figure 3.5, where it is evident that outside the clear rectangle (the "ideal" region) there are shaded areas where space charge effects and ion/molecule reactions will become appreciable. The easiest way in which to remain within the ideal region as the value of $[S]$ changes is to alter the ionization time in a controlled manner using the method of *automatic gain control* (AGC) [20–22]. The idea is to incorporate two ionization stages into the scan function, as

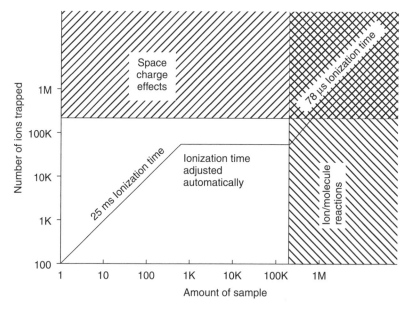

Figure 3.5. Ion trap response with automatic gain control (AGC) showing the region (unshaded) where ions created from the sample should not experience space charge effects and/or ion/molecule reactions, which would otherwise distort the mass spectral performance. (Reprinted from *Mass Spectrometry Reviews*, Vol. 10, J. F. J. Todd, "The ion trap mass spectrometer – Past, present and future (?)," Fig. 22, pages 3–52 (1991). © John Wiley & Sons Limited. Reproduced with permission.)

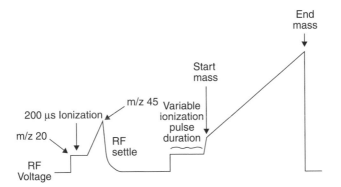

Figure 3.6. The AGC scan function and representations of the associated ion signals. (Reprinted from *Mass Spectrometry Reviews*, Vol. 10, J. F. J. Todd, "The ion trap mass spectrometer – Past, present and future (?)," Fig. 23, pages 3–52 (1991). © John Wiley & Sons Limited. Reproduced with permission.)

indicated in Figure 3.6. A scan function is a diagram showing the temporal variations of the various potentials applied during the course of an ion trap experiment. The duration of application of each of the potentials is indicated normally along with the duration of each "cooling" period (see also below). The first "test" ionization time is of fixed duration, for example, 200 μs, after which ions formed from the background gases (typically up to m/z 45) are removed by rapid mass-selective ejection and discarded, and the remaining analyte ions are detected without further mass analysis by, for example, reducing the value of V to zero or rapidly stepping V to a very high value so as to eject the ions without any resolution. The "total ion" signal measured in this "prescan" is then used to calculate the optimum ionization time, which is then set automatically before effecting the second, "analytical," stage of the function in order to avoid the effects noted above. This procedure occurs each time the scan function is repeated, and the resulting ionization times are recorded along with spectral intensities in order to normalize the data before retrieval.

This approach is clearly an extremely elegant method, in which a degree of "machine intelligence" is employed, and has enabled ion trap mass spectrometry to be established as a standard quantitative analytical method. Similar control over the ion population is, of course, necessary when the ions have been created externally and injected into the ion trap or where the analyte ions have been created internally by means of an ion/molecule reaction, that is, chemical ionization (see below and Chapter 7). In the Finnigan ion trap documentation, this technique has been termed automatic reaction control ('ARC'); essentially the same kind of approach has been employed by Bruker in its ESQUIRE range of instruments under the name *ion charge control* (ICC).

For the LIT (see Chapter 5), there is clearly a need to control the space charge also, and a recent patent by Hager describes, according to Claim 1 of the patent (Ref. 23, Col. 6):

A method of setting a fill time for a mass spectrometer including a linear ion trap the method comprising:

(a) operating the mass spectrometer in a transmission mode; (b) supplying ions to the mass spectrometer; (c) detecting ions passing through at least part of the mass spectrometer in a preset time period to determine the ion current; (d) from a desired maximum charge density for the ion trap and the ion current determining a fill time for the ion trap; (e) operating the mass spectrometer in a trapping mode to trap ions in the ion trap, and filling the ion trap for the fill time determined in step (d); and (f) obtaining an analytical spectrum from ions trapped in the ion trap.

3.2.2.2. Ion Kinetic Energies

One of the advantages of the pseudopotential well model is the ease with which it provides an estimate of the kinetic energies of the trapped ions. Recalling that Z represents the secular displacement of an ion along the z axis of the ion trap and that the secular frequency for simple harmonic oscillation within the well is ω_z [see Eq. (3.9)], then in the absence of collisions an ion with the maximum allowable velocity and the maximum displacement, z_0 would approximately follow the path

$$Z(t) = z_0 \sin \omega_z t. \tag{3.39}$$

Hence

$$\dot{Z}(t) = z_0 \omega_z \cos \omega_z t \tag{3.40}$$

so that combining Eqs. (3.10) and (3.11) in order to derive an expression for ω_z and assuming that $a_z = 0$, we obtain

$$\dot{Z}(t) = \frac{z_0 q_z \Omega}{2\sqrt{2}} \cos \omega_z t \tag{3.41}$$

which gives a maximum value of the velocity in the z direction as

$$\dot{Z}(t_{\max}) = \dot{u}_{\max}(z) = \frac{z_0 q_z \Omega}{2\sqrt{2}} \tag{3.42}$$

corresponding to the condition of $\cos \omega_z t = 1$.

The average value of the velocity, $\langle \dot{u}_z \rangle$, is obtained from

$$\langle \dot{u} \rangle = \frac{\int_0^\tau \dot{u}_z(t)\, dt}{\int_0^\tau dt} \tag{3.43}$$

where τ is the half period of the secular oscillation π/ω_z, to give

$$\langle \dot{u}_z \rangle = \frac{z_0 q_z \Omega}{\sqrt{2}\pi} \tag{3.44}$$

which is equivalent to a mean kinetic energy of

$$\tfrac{1}{2}m\langle \dot{u}_z\rangle^2 = \langle E(z)\rangle = \frac{mz_0^2 q_z^2 \Omega^2}{4\pi^2}. \tag{3.45}$$

Combining Eq. (3.45) with Eq. (3.15) gives

$$\langle E(z)\rangle = \frac{4}{\pi^2}e\overline{D}_z. \tag{3.46}$$

Similarly, since $E_{max} = \tfrac{1}{2}m\langle \dot{u}_{max}\rangle^2$, combining Eq. (3.42) with Eq. (3.15) gives the maximum kinetic energy for motion along the z axis as

$$E_{max}(z) = e\overline{D}_z \tag{3.47}$$

To calculate the total ion energy, we must take account of both the radial and axial components of motion. If we follow the assumption made in Chapter 2, in which we state that any given ion possesses zero angular momentum around the z axis, then we can consider that any individual ion has a two-dimensional trajectory, so that the effective total kinetic energy is derived by adding the radial and axial components together, that is,

$$E(\text{total}) = E(r) + E(z). \tag{3.48}$$

The values of the average radial kinetic energy, $\langle E(r)\rangle$, and the maximum radial kinetic energy, $E_{max}(r)$ may be found by applying the foregoing derivations but using the equations relevant to the radial component of motion to obtain, respectively,

$$\langle E(r)\rangle = \frac{4}{\pi^2}e\overline{D}_r \tag{3.49}$$

and

$$E_{max}(r) = e\overline{D}_r \tag{3.50}$$

thus giving

$$\langle E(\text{total})\rangle = \frac{4}{\pi^2}e(\overline{D}_r + \overline{D}_z) \tag{3.51}$$

and

$$E(\text{total})_{max} = e(\overline{D}_r + \overline{D}_z). \tag{3.52}$$

Recalling the relationship between $\overline{D}_r$ and $\overline{D}_z$ given in Eq. (3.20), Eqs. (3.51) and (3.52) become

$$\langle E(\text{total})\rangle = \frac{4}{\pi^2}e\overline{D}_z\left(\frac{r_0^2}{4z_0^2} + 1\right) \tag{3.53}$$

and

$$E(\text{total})_{\text{max}} = e\bar{D}_z\left(\frac{r_0^2}{4z_0^2} + 1\right). \tag{3.54}$$

Hence for the standard configuration, in which $r_0^2 = 2z_0^2$ [Eq. (2.77)], we have

$$\langle E(\text{total})\rangle = \frac{6}{\pi^2}e\bar{D}_z \tag{3.55}$$

and

$$E(\text{total})_{\text{max}} = \tfrac{3}{2}e\bar{D}_z. \tag{3.56}$$

Specimen Calculations of Mean and Maximum Total Ion Kinetic Energies Taking the sample calculation given previously, in which the value of $\bar{D}_z$ was found to be 4.3 V using Eq. (3.18), then substitution of this quantity into Eqs. (3.55) and (5.56) gives values for $\langle E(\text{total})\rangle$ and $E(\text{total})_{\text{max}}$ as 2.6 and 6.5 eV, respectively.

This model for calculating ion kinetic energies is clearly somewhat crude: it considers only the fundamental component of the secular motion of the ions [compare Eq. (2.80)] and also really applies only within the limits of the Dehmelt approximation ($q_z < 0.4$). Invoking this approximation is equivalent to saying that we should ignore any possible contribution at any instant arising from the high-frequency component of the ion's motion due to the applied RF drive potential. Nevertheless, the approach does provide an easy means of estimating the ion kinetic energy from the operating parameters of the ion trap, as opposed to using more complex simulation techniques (see Chapter 4). In an early detailed study of the methods for estimating the kinetic energies of trapped ions, Todd et al. [24] compared the above calculation based upon the pseudopotential well model with results derived from computing the individual ion trajectories and from the use of phase space dynamics: generally, the resulting ion kinetic energies agreed to within a margin of approximately ±50%. The method of phase space dynamics affords a means of considering the entire ensemble of trapped ions (in a collision-free environment and in the absence of space charge) and provides an excellent way of visualizing the spatial and velocity distributions of the trapped ions at different phase angles of the applied RF drive potential [25]. Essentially, at any RF phase angle, all ions with stable trajectories must lie on or within an elliptical boundary plotted in (u, $\dot{u}$) space: The instantaneous orientation of an ellipse relative to the coordinate axes is determined by the value of the phase angle, but the area (or *emittance*) remains constant, corresponding to conservation of the total number of ions within the quadrupole field. Over a complete RF cycle there will be an infinite number of ellipses, so that normally one considers a family of ellipses plotted at intervals of, say, one-quarter of a cycle. Of some interest to the present discussion is a comparison of the overlap between a family of phase space ellipses and the corresponding time-invariant ellipse for the simple harmonic motion described by the pseudopotential well model for different values of the trapping parameter q_z (with $a_z = 0$) [26].

3.3. HIGHER ORDER FIELD COMPONENTS AND NONLINEAR RESONANCES

In the introduction to this chapter we noted that the underlying theory of quadupole devices as derived in Chapter 2 was predicated on the basis that we were considering an ideal system, containing a single ion. In the previous section we have considered the effects that ion interactions may have on the practical behavior of the ion trap mass spectrometer and have seen how the pseudopotential well model affords a simple means of estimating both the maximum number of ions that may be confined and approximate values for the kinetic energy of the trapped ions.

In this section we shall examine how the practical behavior of real systems is dependent upon certain electrical and geometrical limitations, in particular that the RF drive potential may not be strictly sinusoidal, the fact that the electrode surfaces do not extend to infinity (i.e. they are truncated), that there may be inaccuracies in the spacing between the electrodes, and that there may be imperfections in the shapes and precision of the electrode surfaces. This last-mentioned limitation is clearly exacerbated by the need to have one or more holes through the electrodes in order to admit electrons or ions and/or to eject the ions for detection, a situation that may also lead to field penetration arising from potentials applied to electrodes external to the ion trap. We shall note that there may be local regions of instability embedded within the stability diagram where ions are not trapped, the so-called "black holes" or "black canyons." These effects arise from the presence of superimposed higher order field components, resulting in the occurrence of *nonlinear resonances*. We shall see also that these nonidealities may lead to so-called "mass shifts" whereby the mass spectrum of the ejected ions exhibits peaks that do not appear to have the correct assignment of *m/z*. However, in anticipation of Chapters 7 and 8, we shall find also that, as with many discoveries in science, once the origin of this undesirable behavior was determined and characterized, the relevant theory has been used to advantage in improving the performance of ion trap mass spectrometers.

In discussing this particular topic, the material has been deliberately divided into separate parts that are contained within different chapters of the book. The following account looks at the historical background and underlying theory of nonlinear resonances and describes how their influences on the motion of the trapped ions and on the phenomenon of mass shifts have been explored, while, as indicated above, Chapters 7 and 8 concentrate upon the beneficial aspects of these resonance effects, demonstrating how they have been deliberately designed into modern ion trap instruments.

3.3.1. Historical Background

As with many fundamental aspects of ion trap behavior, the discovery and basic understanding of the existence of nonlinear resonance effects, initially with the QMF, can be traced back to the work of Wolfgang Paul and his research group at Bonn. One of

Paul's earliest applications of the mass filter was as a separator for stable isotopes and, in particular, a report by von Busch and Paul [27] describes a system comprising 2-cm-diameter rod electrodes of length 3 m and $r_0 = 1.5$ cm driven with a potential oscillating at 2.6 MHz that was used to enrich ^{25}Mg from a naturally occurring mixture of ^{24}Mg, ^{25}Mg, and ^{26}Mg. Apart from describing how this process could be effected by applying supplementary AC voltages of the appropriate frequencies between the x pair of electrodes, to remove selectively ^{24}Mg and ^{26}Mg (i.e., the first example of the application of resonant ejection), the authors also reported that there were additional fields (referred to by them as "an additional quadrupole field") that yielded "interference fields" at certain zones within the stability envelope. These were considered to be mathematically anologous to the betatron oscillations of protons observed in a synchrotron [28]. When ions moved into resonance with these fields, they were ejected with exponentially increasing amplitudes, as opposed to the linear increase of amplitudes observed with the application of supplementary voltages to the rod electrodes, an effect that was seen to be advantageous in improving the resolution.

These effects were explored in greater detail by von Busch and Paul in an accompanying paper [29] which showed that plots of ion current versus q exhibited resonance "dips." By applying a supplementary AC potential of variable frequency between the x pair of electrodes, it was possible to demonstrate that, when $a = 0$, the minima occurred at points corresponding to $\beta_x = \beta_y$ values of $\frac{1}{3}$, $\frac{1}{2}$, and $\frac{2}{3}$ and that other resonances were also observed when $a \neq 0$. Drawing upon the work of Hagedorn [30], Hagedorn and Schoch [31], Schoch [32], and Barbier and Schoch [33], von Busch and Paul [29] were able to characterize this behavior mathematically in terms of "interference potentials" arising from superimposed multipole fields. It was noted that, although deviations from the ideal field might not normally be detectable, they can become apparent under conditions when the resonance effects arise from the presence of these small interference potentials: this behavior was considered to be similar to that observed in circular accelerators with alternating field gradients. A further effect which was also noted [29] is that interferences may arise when the RF drive potential contains minor traces of subharmonics, as might occur, for example, if the output frequency is derived by a frequency-doubling process within the signal generator.

The first attempt to extend the discussion of nonlinear resonance effects to the three-dimensional QIT was by Dawson and Whetten [34], who had observed peak splitting in mass spectral peaks obtained using the mass-selective storage method of operation [35]. While this paper offers an early insight into this aspect of ion trap behavior, Wang and Franzen [36, 37] have argued that, in part, the conclusions are erroneous, arising from an incorrect extension of the two-dimensional model based upon the synchrotron to the three-dimensional ion trap with cylindrical symmetry about the z axis. Dawson and Whetten also examined nonlinear resonance effects in the QMF and in the monopole mass spectrometer [38]. The treatment of ion motion in two-dimensional multipole fields was examined in depth by Friedman et al. [39], and a detailed study of these systems was carried out by Szabo [40] and Hägg and Szabo [41–43]; Davis and Wright have reported upon the computer modeling of fragmentation processes in multipole collision cells [44].

3.3.2. Fundamental Aspects of Nonlinear Resonances

The aim of this section is to provide the reader with a "feel" for the underlying physics of nonlinear resonances, without attempting to provide mathematically rigorous derivations. This approach is based mainly upon the seminal work of Jochen Franzen and Yang Wang and their colleagues, and reference should be made to their publications for a full and detailed account of this somewhat complex material [36, 37, 45–47].

3.3.2.1. Multipole Fields We begin by taking as our "reference" system an ideal quadrupole field that has been developed by applying a potential

$$\phi_0 = (U + V \cos \Omega t) \qquad \text{[Eq. (2.68)]}$$

whose oscillating component is a pure sinusoidal waveform, to an ion trap in which the geometry is defined by

$$r_0^2 = 2z_0^2. \qquad \text{[Eq. (2.77)]}$$

Expressed in spherical coordinates (where ρ is the distance from the origin and θ and φ are angles), the generalized potential $\phi(\rho, \theta, \varphi)$ at a point whose coordinates are (ρ, θ, φ) is given by [48]

$$\phi(\rho, \theta, \varphi) = \phi_0 \sum_{n=0}^{\infty} A_n \frac{\rho^n}{r_0^n} P_n(\cos \theta) \qquad \text{[Eq. (2.78)]}$$

in which r_0 is a scaling factor (i.e., the internal radius of the ring electrode), the values of A_n are weighting factors, and the expressions P_n are Legendre polynomials of order n. Due to the assumed rotational symmetry about the z axis of the trap, the potential does not in fact depend upon the angle φ.

Considering only the quadrupole term, that is, $n = 2$, with all values $n \neq 2$ equal to zero, and converting to cylindrical polar coordinates in which we replace ρ^2 by $r^2 + z^2$, the value of $P_2(\cos \theta)$ takes the form

$$P_2(\cos \theta) = \frac{r^2 - 2z^2}{2\rho^2} \qquad (3.57)$$

so that Eq. (2.78) becomes

$$\phi_{r, z} = \phi_0 A_2 \frac{r^2 - 2z^2}{2r_0^2}. \qquad (3.58)$$

This is essentially the same form as Eq. (2.65) written with $r_0^2 = 2z_0^2$ and without the constant potential at the center of the ion trap [which would be equated with the term in which $n = 0$ in Eq. (2.78)]. If we replace r^2 by $x^2 + y^2$ in Eq. (3.58), it

References pp. 125–132.

should be noted that, as with Eq. (2.65), Eq. (3.58) also satisfies the Laplace condition

$$\nabla^2 \phi_{x, y, z} = 0. \qquad \text{[Eq. (2.2)]}$$

The full expansion of Eq. (2.78) in cylindrical coordinates, taking values from $n = 2$ to $n = 6$, is

$$\phi_{r, z} = \phi_0 \left(A_2 \frac{r^2 - 2z^2}{2r_0^2} + A_3 \frac{3r^2z - 2z^3}{2r_0^3} + A_4 \frac{3r^4 - 24r^2z^2 + 8z^4}{8r_0^2} \right.$$

$$\left. + A_5 \frac{15r^4z - 40r^2z^3 + 8z^5}{8r_0^5} + A_6 \frac{5r^6 - 90r^4z^2 + 120r^2z^4 - 16z^6}{16r_0^6} \right).$$

$$\text{[Eq. (2.79)]}$$

Again it is a necessary condition that each term in Eq. (2.79) satisfies the Laplace condition, Eq. (2.2), and as noted above, this can be verified by replacing r^2 with $x^2 + y^2$. For example the term for $n = 3$ becomes

$$\phi_{x, y, z} = \phi_0 A_3 \frac{3x^2z + 3y^2z - 2z^3}{2r_0^3} \qquad (3.59)$$

whence, following Eq. (2.4),

$$\nabla^2 \phi_{x, y, z} = \phi_0 A_3 \frac{6z + 6z - 12z}{2r_0^3} = 0. \qquad (3.60)$$

The reader may care to verify this condition for the remaining terms in Eq. (2.79) and thus discover the reason for the apparently random set of coefficients for the different terms in r and z.

As noted in Chapter 2, the successive terms in Eq. (2.79) represent the presence of quadrupole ($n = 2$), hexapole ($n = 3$), octopole ($n = 4$), decapole ($n = 5$), and dodecapole ($n = 6$) components to the potential, in proportions according to the respective weighting factors A_n. Thus the number of poles is equal to $2n$, and the multipoles are classified as being either "even" for even values of n or "odd" for odd values of n. It will also be seen that in Eq. (2.79) the sum of the exponents to which r and z are raised equals the value of n in each respective term. A further point of some significance is that whereas for the even multipoles the exponents of the terms in r and z are symmetrical and all the exponents are even [e.g., for the octopole expression ($n = 4$) there are terms in r^4, r^2z^2, and z^4], for the odd multipoles the exponents of the terms in r and z are nonsymmetrical, the exponents of z are all odd, and, in particular, there are no terms of the form r^n.

The effect of this difference can be seen from Figures 3.7a and 3.7b. These show the potential surfaces within the ion trap cavity for the pure hexapole and pure octopole terms, respectively, and were obtained by plotting the appropriate terms in Eq. (2.79) with the value of ϕ_0 held constant; similar types of plots may be obtained for the decapole ($n = 5$) and dodecapole ($n = 6$) terms [49]. These potential surfaces

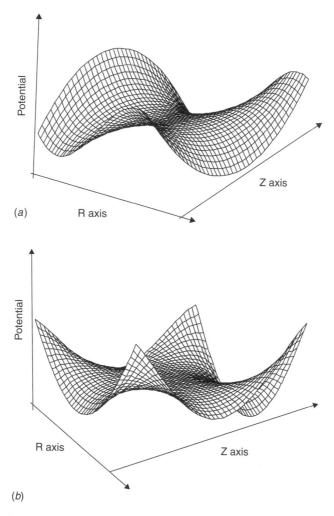

Figure 3.7. (*a*) Graph showing surface for a hexapole potential plotted according to Eq. (3.59). (*b*) Graph showing surface for an octopole potential plotted according to third term on right-hand side of Eq. (2.79). (Reprinted from *Rapid Communications in Mass Spectrometry*, Vol. 9, W. Mo, M. L. Langford, J. F. J. Todd, "Investigation of 'ghost' peaks caused by nonlinear fields in the ion trap mass spectrometer," Figs. 3 and 4, pages 107–113 (1995). © John Wiley & Sons Limited. Reproduced with permission.)

should be compared with that for the pure quadrupole potential given in Figure 2.17. Figure 3.7*b* reveals both the rotational and planar symmetry of the octopole potential, whereas the hexapole potential (Figure 3.7*a*) is clearly asymmetric, a feature that Wang et al. [37] have argued was ignored by Dawson and Whetten [34] in their early discussion of nonlinear resonances in the ion trap, as noted previously. It should be emphasized at this point that Figures 3.7*a* and 3.7*b* show the pure hexapole and octopole potentials, respectively, and two-dimensional configurations utilizing these higher order potentials are employed as ion guides in certain commercial ion sources and in tandem mass spectrometers; however, in an ion trap possessing the field imperfections outlined earlier in this section, the resulting contributions from these higher order multipoles to the main quadrupole trapping potential within the QIT may be of the order of only a few percent (see, e.g., Chapter 8). Furthermore, depending upon the precise electrode configuration, the contributions of the different higher order terms may be "positive" or "negative," as reflected in the values of the corresponding factors A_n in Eq. (2.79).

The impact of the presence of higher order multipole fields on the behavior of the ion trap has been well summarized in a concise form by Wang et al. [37] in the following comparison between the behavior of ions in a trap subjected to a pure quadrupole driving potential and one in which there are superimposed weak multipole electric fields.

Pure Quadrupole Potential

- The ion movement in the axial (z) direction is completely decoupled from that in the perpendicular radial (r) direction, as evidenced by the absence of "cross terms" of the form rz in the expression for the quadrupole potential, Eq. (3.58).

- The RF field amplitude varies in a linear manner with distance from the center in the r and z directions of the cylindrical coordinates, as determined by the first differential, $\nabla\phi$, of Eq. (3.58), and has only one parameter describing the periodicity. In this context, the pure quadupole device might be termed a "linear" ion trap, although, as discussed in Chapter 5, this expression is now applied in a totally different context to describe ion traps that are based on the linear electrode configuration associated with two-dimensional QMFs.

- The stability of ion trajectories of a given mass/charge ratio in an infinitely large quadrupole field does not depend on their initial starting conditions (position and velocity); it depends only on the field parameters. In practice, however, with fields confined by the dimensions of the electrode structure, this is not necessarily true.

- Only the two mass-related trapping parameters a and q of the DC and RF fields, respectively, determine whether the solution for the ion movement is stable or whether the oscillations of the ions increase their amplitude without limit to infinity. This condition is expressed by the well-known stability diagrams for QITs (see Chapter 2).

- If the parameters a and q are kept inside the stability region of the stability diagram, in the absence of any auxiliary AC potentials applied between the end-cap electrodes, the ions perform stable secular oscillations in the r and z directions, respectively, with frequencies lower than half that of the driving voltage applied to the ion trap, and can be described by the so-called iso-β_r and iso-β_z lines, which map onto the stability region.

- The frequency of the secular oscillation of an ion is independent of its displacement from the center.

If we now superimpose weak multipole fields (e.g., hexapole, octopole, decapole, dodecapole, and higher order fields), the resulting nonlinear field ion traps exhibit some effects which differ considerably from those of the linear field trap.

Addition of Multipole Potentials

- The components of the RF field amplitude arising from the higher order multipoles are nonlinear in the r and z directions of the cylindrical coordinates; this is seen from the expressions for $\nabla \phi$ in respect of the terms in Eq. (2.79) for $n > 2$, where now the dependence of the field varies with r and/or z raised to powers greater than 1.

- For multipoles higher than or equal to hexapoles, the secular frequencies of oscillation are no longer constant for constant field parameters; they now become amplitude dependent.

- The ion trajectories in the r and z directions become amplitude dependent and are now coupled because of the existence of cross terms of the form rz in the expressions for the higher order multipole potentials.

- Several types of nonlinear resonance conditions exist for each type of multipole superposition, forming resonance lines within the stability region of the stability diagram (see below). These are the nonlinear resonances that were first detected by von Busch and Paul [29].

- Contrary to many experimental observations (see below), the trajectories of ions with nonlinear resonances do not always exhibit instability. They take up energy from the RF drive field and thus increase their secular oscillation amplitude. Because of the amplitude dependence of the secular frequency, this frequency now drifts out of resonance, resulting in a kind of beat motion. The maximum amplitude of the secular oscillation, in this case, is dependent on the initial conditions (location and speed) of the ions at the beginning of the resonance (see also Chapter 8).

To understand the origin of the phenomenon of nonlinear resonances, Franzen and co-workers [37, 47], following the approach of Hagedorn [30], developed a general treatment of the trapping potential by making use of Hamilton's equations, in which they employed a Hamiltonian function h which was split into a function $h^{(2)}$ representing kinetic energy and potential energy arising from the basic quadupolar component and a perturbing part h' describing the potential energy due to the higher order multipole components:

$$h = h^{(2)} + h'. \tag{3.61}$$

From this, the equation of motion of the ions in Cartesian coordinates in, for example, the x direction is expressed as

$$\frac{\partial^2 x}{\partial \xi^2} + [a_r - 2q_r \cos(2\xi)]x = r_0^2(a_r - 2q_r \cos 2\xi)\left(\frac{\partial \phi'}{\partial x}\right) \tag{3.62}$$

where the term $\partial \phi'/\partial x$ refers to the electric field resulting from the superimposed multipole components and $\xi = \Omega t/2$. When, in Eq. (2.79), $A_n = 0$ for $n > 2$, the right-hand side of Eq. (3.62) becomes zero, and the normal Mathieu equation (2.33) associated with the basic quadupole potential results. Expressions similar to Eq. (3.62) are obtained for the motion in the y and z directions. Note that, as with the derivation of the expression for the pure quadrupole potential in Chapter 2 [Eq. (2.44)], the angular velocity, and hence the angular momentum, of ions around the z axis is assumed to be zero.

In the ensuing treatment, the Hamiltonian functions for the linear (quadrupole) and nonlinear (multipole) components are transformed such that the resulting functions, when expanded into a Taylor series and rearranged, lead to an expression in which the denominator has the form

$$2v - \{\beta_r[(j - k) + (l - m)] + \beta_z(o - p)\} \tag{3.63}$$

where v is an integer greater than or equal to 0 and the integers $j + k + l + m + o + p = n$, with $n > 2$, j, k, l, m, o, $p \geq 0$, and $0 \leq \beta_r$, $\beta_z \leq 1$ for the first region of the Mathieu stability diagram (see Figure 2.15). If the condition

$$\{\beta_r[(j - k) + (l - m)] + \beta_z(o - p)\} = 2v \tag{3.64}$$

is satisfied, then the denominator term referred to above becomes zero, and the Hamiltonian expression in the nonlinear canonical transformation (the perturbing part, noted above) becomes infinity, resulting in a nonlinear resonance.

The practical consequences of the mathematical condition expressed in Eq. (3.64) may be viewed as follows. If we make the substitutions

$$n_r = (j - k) + (l - m) \tag{3.65}$$

and

$$n_z = o - p \tag{3.66}$$

then Eq. (3.64) becomes

$$n_r \beta_r + n_z \beta_z = 2v. \tag{3.67}$$

Recalling from Eqs. (2.80) and (2.81) that the fundamental components of the secular frequencies are $\omega_r = \beta_r \Omega/2$ and $\omega_z = \beta_z \Omega/2$, Eq. (3.67) becomes

$$n_r \omega_r + n_z \omega_z = v\Omega. \tag{3.68}$$

Thus Eq. (3.68) reveals that a nonlinear resonance may occur when the sum of an integral number of secular frequencies matches a multiple of the fundamental drive frequency Ω; this corresponds to a resonance between overtones of the secular frequencies and the sideband frequencies of the main sinusoidal drive potential. It should be noted that, although Eq. (3.68) [and hence Eq. (3.67)] is a necessary condition for the occurrence of a nonlinear resonance, it is possible, under certain circumstances, for the condition to be satisfied yet no resonance effect to be observed (see later).

The generality of Eq. (3.68) allows us to consider five possible resonance conditions, depending upon the values of v, n_r and n_z; these conditions have been classified by Hagedorn [30] and extended by Franzen et al. [47] as follows.

1. *Pure Coupling Resonance.* For $v = 0$, the term with the RF drive frequency ω vanishes, and n_r and n_z must have different signs. Energy cannot be taken up from the RF drive field. There is only energy exchange between the oscillations in the r and z directions. The amplitudes of the secular oscillations in both directions fluctuate anticyclically between minima and maxima.

For $v = 1$, the following categories exist:

2. *Sum Resonance Condition.* Here n_r and n_z have the same sign, and energy is taken up from the quadrupole RF drive field in both directions at the same time.
3. *Difference Resonance Condition.* Here n_r and n_z have different signs: the ion motion in the r and z directions is coupled and energy is exchanged between both directions. A smaller amount of energy may be taken up from the RF drive field.
4. *The z-Direction Resonance Condition.* When n_r equals zero, energy is taken up from the RF drive field in the z direction only. Under certain circumstances, an ion cloud can leave the ion trap through one or more small holes in the center of one of the end-cap electrodes.
5. *The r-Direction Resonance Condition.* When n_z equals zero, energy is taken up from the RF drive field in the r direction. In an ion trap, the ions may hit the equator of the ring electrode.

If we define a new parameter N, the order of the resonance (which is to be distinguished from the order of the multipole, n), by the relationship

$$N = |n_r| + |n_z| \tag{3.69}$$

then it has been shown [37, 47] that there is a maximum of $2N$ resonance lines for a given value of N and there can exist only resonance lines of order $N, N-2, N-4, \ldots$ for the Nth equation. The strongest resonance lines for a multipole of order n occur for $N = n$, with weaker lines at $N = n - 2, n - 4, \ldots$.

Table 3.1 lists the possible *direction* and *sum* resonance lines for the hexapole and the octopole derived from Eq. (3.67) with $v = 1$. However, as noted above, certain resonance lines may not exist even though the relationship in Eq. (3.68) is satisfied. It may be shown [37, 47] that, because of the rotational symmetry of the trapping potential,

**TABLE 3.1. Possible Nonlinear Resonance Lines for
Hexapole and Octopole Fields Corresponding to
Condition $N = n$**

Multipole	n_r	n_z	Line
Hexapole ($n = 3$)	3	0	$(3\beta_r = 2)$
	0	3	$3\beta_z = 2$
	1	2	$(\beta_r + 2\beta_z = 2)$
	2	1	$2\beta_r + \beta_z = 2$
Octopole ($n = 4$)	4	0	$4\beta_r = 2$
	0	4	$4\beta_z = 2$
	2	2	$2\beta_r + 2\beta_z = 2$
	1	3	$(\beta_r + 3\beta_z = 2)$
	3	1	$(3\beta_r + \beta_z = 2)$

Note: For definitions, see text. Lines in parentheses are not observed
in practice on account of circumstances outlined in text.

resonance lines will not exist when n_r, as defined by Eq. (3.65), is odd on account of
the Hamiltonian becoming zero. Consequently, some of the resonance lines, shown in
parentheses in Table 3.1, should not be observed; one of these $(3\beta_r = 2)$ was erro-
neously postulated in the pioneering work of Dawson and Whetten [34]. Figure 8.6
shows how the remaining "permitted" resonance lines listed in Table 3.1 map onto the
stability diagram. For an account of a highly detailed experimental study showing very
large numbers of nonlinear resonance lines, the reader is directed to the work of Alheit
et al. [50]; interestingly, in this investigation certain nonlinear resonance lines that are
"forbidden" under the Franzen analysis were observed, and it was concluded that this
indicated that the ion trap employed did not have perfect rotational symmetry.

Further discussion of the principal resonance lines observed for the superimposed
hexapole and the octopole fields, as listed in Table 3.1, is given in Chapter 8, illustrating especially how their existence has been tailored to the design of ion trap mass
spectrometers with improved performance. This section concludes with a brief
account of some experimental studies where the existence of nonlinear resonances
gave rise to certain unexpected, and generally undesirable, results that, nevertheless,
led to a greater understanding of this phenomenon.

3.3.2.2. Experimental Observations of Effects Arising from Nonlinear Resonances in the Quadrupole Ion Trap Mass Spectrometer

With the exception of the mass
shift phenomenon (see below), possibly the first reported observation of effects attrib-
utable to the influence of nonlinear resonances on the appearance of the spectrum
from an ion trap mass spectrometer operating in the mass-selective ejection mode is
that by Todd et al. [51] in an investigation into alternative scan modes aimed at
extending the mass/charge ratio range of the commercial instrument available at the
time (650 Th*). These new scan modes were effected by superimposing either a fixed

*The unit Thomson (Th) is defined as a footnote to Table 3.2 and in Chapter 5.

or a variable DC potential upon the RF drive potential so as to afford the means of crossing a stability boundary at a value of q_z significantly below 0.908.

One particular experiment, the reverse scan, involved, first, increasing the RF potential so as to place the q_z value (with $a_z = 0$) of the ion of lowest m/z value of interest just inside the $\beta_z = 1$ boundary (i.e., $q_z < 0.908$); second, superimposing a positive DC potential on the ring electrode so as to move the (a_z, q_z) working point down the stability diagram; and, finally, linearly reducing the DC and RF amplitudes with a constant ratio between them so that the resulting scan line (marked A in Figure 3.8) intersected the $\beta_z = 0$ boundary. As a result, ions were ejected for detection in order of decreasing mass/charge ratio. However, it was noted that sharp peaks, corresponding to the rapid, *spurious*, ejection of ions, occurred prematurely, that is, before the (a_z, q_z) coordinates for ions of known mass encountered the $\beta_z = 0$ boundary.

Further experiments in which a reverse scan was performed in the absence of a superimposed DC potential (i.e., simply scanning the value of q_z from high to low

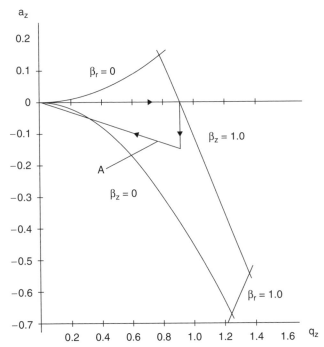

Figure 3.8. Stability diagram for the quadrupole ion trap showing the scan line for the "reverse scan;" the arrows indicate the sequence of changes in the trapping parameters (a_z, q_z) as the scan is effected. (Reprinted from *International Journal of Mass Spectrometry and Ion Processes*, Vol. 106, J. F. J. Todd, A. D. Penman, and R. D. Smith, "Alternative scan modes for mass range extension of the ion trap," pages 117–136, Copyright 1991, with permission from Elsevier.)

References pp. 125–132.

value) followed by a conventional forward analytical scan, showed that known ions that were ejected sharply during the reverse scanning section disappeared from the normal mass spectrum recorded during the analytical scan. It was found that this occurred when the (a_z, q_z) working point crossed the $\beta_z = 0.5$ line, corresponding to ion ejection through interaction with an octopole nonlinear resonance. This effect was only observed during a reverse scan and not during a normal forward scan, that is, as the RF amplitude was increasing. The observation of this spurious ejection of ions was attributed to the fact that, during a reverse scan, the ion energy and oscillation amplitude are decreasing as the RF amplitude is being reduced: this situation places the ions, on average, at greater distances from the center of the trap, where they are more susceptible to the influence of the nonlinear electric fields, which increase rapidly with increasing r and z (see above). It was noted that the occurrence of the spurious peaks during reverse scanning was highly dependent upon the scan rate and upon the helium buffer gas pressure (see also below). Because this spurious effect was not observed during the forward scan, it was not considered to be a problem in operating the commercial ion traps under the normally specified buffer gas conditions, a point to which further reference is made below.

Possibly the most celebrated manifestation of ion losses from the ion trap mass spectrometer caused by nonlinear resonances is the black hole phenomenon first reported by Guidugli and Traldi [52]. In these experiments, it was found that during MS/MS in the ion trap (with zero applied DC potential to the ring electrode), when the product ions had m/z values that fell at $q_z = 0.78$, their intensities dropped practically to zero. In this preliminary report, the results were discussed empirically in terms of investigations relating to the origin of the ions (comparing fragment ions formed directly by EI with the same ions formed by multiple MS/MS experiments), as well as examining the effects of changing the values of different operating parameters: "delta tune" values, tickle time, tickle energy, and buffer gas pressure.

The black hole effect was confirmed by Morand et al. [53], who showed that $[M–Cl]^+$ ions formed by collision-activated dissociation (CAD) of 1,2-dichlorobenzene parent ions showed a sharp loss of intensity near $q_z = 0.78$; they noted that this location in the stability diagram corresponded to $\beta_z = \frac{2}{3}$, that is, where the fundamental component of the secular frequency is one-third that of the main RF drive frequency. They found another black hole at $q_z = 0.64$ (i.e., $\beta_z = \frac{1}{2}$) and suggested that the ion loss was due to ejection by the overtone of the fundamental frequency. It was noted that the phenomenon was absent for trapped ions not produced by collisional experiments; this led to the suggestion that the initial preparation of the ion population is a prerequisite to the observation of the effect, possibly involving the displacement of the ion from the center of the device. This conclusion was supported by observations noted during other experiments on the injection of ions from an external source [54].

A further combined, in-depth, investigation of black holes was carried out collaboratively by the groups of Traldi and Todd, working in Padova and in Canterbury, respectively [55]. Making use of the method of 'dynamically programmed scans', developed in Canterbury [56], in which one or more parameters controlling the operation of the ion-trap can be systematically varied between successive cycles of ionization, ion selection, resonant excitation and spectral acquisition [57], it was possible to show that the black hole observed near to $a_z = 0$, $q_z = 0.78$ was in fact a "black canyon," following the $\beta_z = \frac{2}{3}$ hexapole non-linear resonance line (see Figure 1 in reference [55] and Figure 8.6). In all, three black canyons were observed, coincident with the $\beta_z = \frac{1}{2}$ (one observation) and $\beta_z = \frac{2}{3}$ (two observations) iso-β lines. In agreement with the work of Morand et al. [53] cited above, it was noted that simply tickling a fragment ion formed by EI of the parent sample molecule over a range of values of q_z did not indicate the presence of a black hole, whereas forming the same fragment ion via collisional dissociation of the parent ion at various values of q_z condition did reveal a black hole when the q_z-value of the fragment ion coincided with one of the non-linear resonance lines.

In a further series of experiments, the authors effected collisional dissociation of the selected parent ion by applying *quadrupolar* resonant excitation, rather than the conventional *dipolar* excitation [58]. Experimentally, the difference between these two techniques is that whereas in the latter bipolar auxiliary AC signals, out-of-phase by 180°, are applied, one to each end-cap electrode, in the former a single AC signal is coupled to both end-cap electrodes in parallel (see also below). Simulation studies [59] have shown that under the conditions of quadrupolar excitation the application of a tickle voltage at a frequency ω_T equal to $\beta_z\Omega/2$ causes *radial* excitation of the ion motion, while doubling the frequency ω_T to $\beta_z\Omega$ causes predominantly *axial* excitation. Thus it was found that applying quadrupolar excitation at $\omega_T = \beta_z\Omega/2$ to the parent ion at m/z 148 from 1,2-dichlorobenzene to form m/z 111 showed no evidence of a black hole at $q_z = 0.78$, whereas increasing the quadrupolar tickle frequency to $\beta_z\Omega$ did reveal the existence of a black hole at this value of q_z. These results suggest that for the ejection of the ions via nonlinear resonant excitation, at least on the $\beta_z = \frac{2}{3}$ line, it is necessary that the prior excitation process have been along the z axis of the ion trap. Similar results were obtained by utilizing boundary-activated dissociation [60], in which the parent ions are collisionally activated through excitation of their radial motion by holding the parent ions close to the $\beta_r = 0$ boundary for a period of time (e.g., 100 ms; see also below).

In a very comprehensive theoretical and experimental investigation, Eades et al. [61] examined the effects of storing ions for long periods of time at different nonlinear resonance points. This work arose initially from the observation that peculiarities in CI mass spectra could be observed when the reagent ion was held at a working point that coincided with a nonlinear resonance line. In particular, the mass-selected reagent ion concentration could fall simply because of ejection of the ions before they had reacted with the analyte. Intriguingly, however, another phenomenon was observed, namely the conversion of the reagent ion into another species via collision-induced dissociation brought about by nonlinear resonant excitation. For example, the mass-selected $C_2H_5^+$ (m/z 29) ion formed from methane was found to undergo collisional dissociation to $C_2H_3^+$ (m/z 27), which then reacted with more CH_4 to give $C_3H_5^+$ (m/z 41). This effect was verified both by monitoring the relative amounts of m/z 27, 29, and 41 as a function of the value of q_z at which the $C_2H_5^+$ was held (for 50 ms) (see, e.g., Figure 2 in Ref. 61) and through the differing chemistries

References pp. 125–132.

of the reactions of $C_2H_5^+$ and $C_3H_5^+$ with n-butylbenzene. Thus $C_2H_5^+$ favors proton transfer, whereas $C_3H_5^+$ undergoes hydride abstraction. These results therefore contain a note of caution for those persons utilizing CI, in terms of ensuring that the conditions under which the reagent ions are stored do not coincide with effects arising from nonlinear resonances. Indeed, Eades et al. [61] delineate six sets of conditions under which ion displacement from the center of the ion trap, and hence exposure to nonlinear fields, may occur: high ion population (space charge) causing ion–ion repulsions; absence of buffer gas requiring a longer time for ions to cool to the center of the trap (see below); resonance excitation causing increased axial and/or radial motion (see below); injection of ions near the end-cap or ring electrodes; ion formation near the end-cap or ring electrodes (e.g., laser desorption near the ring electrode); and application of DC voltages which alter axial and/or radial motion.

In a separate study, Mo et al. [49] characterized the appearance of "ghost" peaks observed during the conventional forward analytical scan of an ion trap mass spectrometer. These experiments were carried out with (a) a "normal" ITD with "stretched" geometry (see below) and (b) a trap having further stretched geometry, effected by loosening the assembly bolts. Initially with (a) it was found that additional spurious peaks, of the form seen during the reverse-scanning experiments described above, were observed at m/z values equal to $0.87m_0$, where m_0 is the m/z value of the "genuine" ion in the trap; this observation corresponded to ejection at the q_z value for the hexapole resonance $\beta_z = \frac{2}{3}$. On further stretching to $z_0 = 7.96$ mm (compared to $z_0 = 7.83$ mm for the normal stretched trap), other ghost peaks at $0.67m_0$ (caused by the octopole resonance at $\beta_z = \frac{1}{2}$) and at $0.60m_0$ (caused by the dodecapole resonance at $2\beta_z + \beta_r = 1$) were seen. Verification of these assignments as ghost peaks was carried out by (i) resonantly exciting at the ghost peak m/z value and observing that the ghost peak ion intensity was unaffected and (ii) resonantly exciting at m_0, when it was found that both the intensity of m_0^+ and its ghost peak were diminished together. Further experiments showed that ghost peaks disappeared above a certain helium buffer gas pressure, suggesting greater concentration of the ions at the center of the trap as a result of collisional cooling and, therefore, not subject to nonlinear field effects (compare also the work of Eades et al. [61] described above). It was noted that the octopole ghost peak was quenched at a higher buffer gas pressure than the hexapole ghost peak, reflecting the differing respective dependences of the strengths of these higher order fields upon distance from the center of the trap. It was also found that tickling at the value of m_0 during the normal forward analytical scan enhanced the appearance of the ghost peaks for that ion, again showing the influence of displacing the ions from the center of the trap; as may be anticipated, this resonance enhancement of the ghost peak could be reversed by increasing the buffer gas pressure.

Again, this work highlights a practical issue arising with the use of ion trap mass spectrometers, namely the importance of proper cleaning of the ion trap followed by reassembly to precisely the correct dimensions.

3.3.2.3. Nonlinear Field Effects in Quadrupole Mass Spectrometers and Linear Ion Traps
The foregoing account indicates that, despite the fact that the original discovery of nonlinear resonance effects arose in the study of the behavior of the QMF, the most intensive theoretical and experimental investigations have been associated with

the QIT. For the QMF, the avowed aim of design engineers has been to construct electrode assemblies to exacting mechanical tolerances in order that the contributions from higher order field components are, as far as is possible, minimized [62]. In contrast to this, recent work by Douglas and co-workers [63, 64] has shown how, by analogy with the ion trap, the deliberate inclusion of contributions from nonlinear octopole fields to the quadrupole field within the mass filter and the LIT leads to enhanced performance.

Sudakov and Douglas [63] examined two methods of adding relatively small octopole fields to the main quadrupole field of quadrupoles and LITs with cylindrical rods. The first, "stretching" the quadrupole by moving two rods out from the axis, was found to produce a combination of higher order fields with similar magnitudes in which the octopole field was not necessarily the greatest component. The quadrupole field strength changed significantly, and a large potential appeared on the axis. The second method utilized rod pairs of different diameters and was found to add octopole components of up to several percent while all other higher order fields remained small. An axis potential was also added, but only to the extent of a few percent of the RF voltage and approximately equal to the strength of the octopole field. The axis potential could be removed by moving the larger rod pair out from the axis or by applying an unbalanced RF potential to the two pairs of electrodes.

In the second, more detailed, study Ding et al. [64] examined the performance of QMFs made with round rods with added octopole fields in the range 2.0–4.0%. These added fields were much greater than those normally added to conventional rod sets by mechanical tolerances or construction errors and were generated by making one pair of rods greater in diameter than the other pair. For positive ions, a resolution at half-height of only about 200 was found to be possible if the negative DC output of the quadrupole power supply was connected to the smaller diameter rods. If the positive DC output of the quadrupole power supply was connected to the smaller rods, the resolution improved dramatically; for example, a resolution at half height of 5800 was observed with a rod set with 2.6% added octopole field. For negative ions the best resolution was obtained with the polarity of the DC reversed, that is, with the negative DC applied to the smaller rods.

These findings are totally unexpected when considered against the accepted engineering requirements noted above [62], which are based on the argument that to obtain high mass resolution with a QMF, the contributions from higher order multipole fields must be kept as small as possible. Thus it was found that for instruments that require a rod set that can be used both as a linear trap and as a mass filter (see Chapter 5), employing differently sized pairs of rod electrodes may afford improved trap performance while still being capable of providing conventional mass analysis. Linear ion traps using superimposed higher order multipole fields have also been described by Londry et al. [65] and Franzen and Weiss [66].

3.3.2.4. Mass Shifts

Preamble One of the most important features of any analytical instrument is that it consistently and reproducibly indicates the correct value of the parameter being measured to a precision within the agreed performance specification. In the case of mass spectrometers the two items of data recorded are the mass/charge ratios and the associated abundances of the ions formed from the sample being investigated; of these two measurements, the latter is clearly of little use if the former value is "shifted" from its true value for some reason. This section is concerned with the observation of mass shifts in the commercial version of the QIT, a phenomenon that gave rise to a great saga in the history of ion trapping. This is a tale of a scientific conundrum that led to deep despair among those involved at a critical stage of product development; this despair was alleviated only by an arbitrary modification of the instrument that was kept as a close industrial secret for some years and revealed eventually in an unscheduled announcement at a scientific meeting. Once the secret was announced, the wider mass spectrometry community joined in the search for an explanation of mass shifts.

Shifts in Observed Mass/Charge Ratios   During the summer of 1984, a staff member at Finnigan Corporation realized that something was wrong with the mass spectrum of nitrobenzene produced with the ion trap mass spectrometer [67]: the molecular positive ion mass peak appeared consistently at m/z 122 rather than at m/z 123. Other colleagues in the project team confirmed this observation, and various explanations were considered, for example, that the shift in mass was due to the occurrence of one or more of ion/molecule reactions, that line spreading had occurred due to space charge (see previous discussion of this topic), and that some error existed in the display and acquisition software. However, the project's system chemist, Dennis Taylor, felt that the data could be trusted and, therefore, he looked for another explanation. He acquired several mass spectra of codeine, which has a fragment ion at m/z 124 and, therefore, could be compared directly with the nitrobenzene molecular ion. However, m/z 124 from codeine was found to be assigned correctly. This observation indicated strongly that there were no errors in either the scan generation or spectral acquisition and display software. Further work showed that, while the majority of a multitude of compounds did not exhibit any irregularity in their mass spectra, a small number of compounds did exhibit a mass shift and the mass shifts exhibited could be either positive or negative. The most extreme case to be found was that of the molecular ion of pyrene that sometimes exhibited a mass shift of almost 1 Th, so that this ion was observed at m/z 203, rather than m/z 202. In Table 3.2 is presented a summary of the mass shifts identified by Dennis Taylor while using an ion trap having the theoretical geometry [67].

Inevitably there was much discussion and activity at Finnigan Corporation in an effort to explain and to understand the observed mass shifts in order to produce a marketable instrument. Paul Kelley experimented with the tilting of one of the end-cap electrodes; such modification would introduce odd-order field terms that should not affect the ion ejection process and, indeed, no change in behavior of the ion trap was observed. He then decided to displace outward both end-cap electrodes by the extraordinary amount of 0.75 mm, a little more than 10% of the value of z_0. Upon examination of the mass spectrum of nitrobenzene in the "spaced-out" ion trap, Paul Kelley found that the molecular ion of nitrobenzene was exactly where it ought to be, at 123 Th, and that of pyrene was also at 202 Th! This result produced a profound

TABLE 3.2. Summary of Mass shifts for Ions from a Variety of Compounds as Observed with Ion Trap Having Theoretical Geometry[a]

Parent Compound	Calculated m/z (/Th)[b]	Mass Shift $[\Delta\,(m/z)]$[c]
Nitrobenzene	123	+0.5
Codeine	124	0.0
Pyrene	202	−0.7
Anthracene	178	−0.5
Hexachlorobenzene	282	−0.5
Hexachlorobutadiene	258	0.0
	223	−0.5
Perfluorotributylamine (FC-43)	69	0.0
	100	0.0
	119	+0.2
	131	0.0
	219	0.0
	264	0.0

Source: Table 1 from J. E. P. Syka, in R. E. March and J. F. J. Todd (Eds.), *Practical Aspects of Ion Trap Mass Spectrometry*, Vol. I, CRC Press, Boca Raton, FL, 1995, Chapter 4, p. 190. Copyright 1995 by CRC Press LLC. Reproduced with permission of CRC Press LLC in the format Other Book via Copyright Clearance Center.

[a] Theoretical geometry means hyperbolic electrodes and $r_0^2 = 2z_0^2$ with $r_0 = 1$ cm.
[b] The unit Thomson is defined as 1 Th = 1 u/e_0, where u is the atomic mass unit and e_0 the elementary charge. [From R. G. Cooks, A. L. Rockwood, *Rapid Commun. Mass Spectrom.* **5** (1991) 93.]
[c] The signs of the mass shifts have been reversed from the original publication to comply with the recent definition of mass shift [79] as the true mass of the ion minus the measured mass; hence ions that are ejected early (and have a measured lower mass/charge ratio) are assigned a positive chemical mass shift.

sense of relief in all who were involved with ion trap development. Spaced-out ion traps were assembled and shipments of the Finnigan ITD 700 began in late 1984.

Prior to commencing shipments, which under U.S. and European patent convention (EPC) patent law would constitute public disclosure, the filing of a patent application was considered in order to protect the knowledge relating to the modification of the ion trap geometry. It was argued that knowledge of the modification was of little or no use to users but could be of great value to competitors. On advice of Finnigan's corporate patent counsel, it was decided to retain knowledge of the modification of the electrode geometry as a trade secret.

WORKING IN THE DARK   One of the problems that comes with maintaining secret such a modification is that one must perpetuate a myth that no such modification has been made. In 1988, Diethard Bohme and Ray March purchased an ion trap detector, a Finnigan MAT ITD 800 [68]. They discarded the gas chromatograph and invited a graduate student, Xiaomin Wang, to modify the ITD so as to extend the mass range of the instrument and to carry out resonant excitation of selected ion species. The modification of the instrument progressed well and the mass range was doubled with little problem [69]. However, for the resonant excitation of selected ion species, Wang found that the calculated fundamental axial secular frequencies were 10–12% lower than the experimental values. The problem remained unresolved for about a year and was the source of considerable frustration. Physical measurement of the interelectrode spacing for the end caps was not a trivial task and, indeed, there was no need to check the geometry because the literature abounded with assurances that

$$r_0^2 = 2z_0^2. \quad \text{[Eq. (2.77)]}$$

THE SECRET REVEALED   In the spring of 1992, Randall K. Julian, Jr., of Graham Cooks laboratory at Purdue University, measured the relevant interelectrode spacing and discovered that some modification to the geometry had been made. It was in the summer of that year that the mass spectrometry community learned [70] that early problems of mass assignment with the ion trap had been resolved by "stretching" the separation of the end-cap electrodes. Figure 3.9a shows a cross-sectional view of an ideal QIT where the interelectrode spacing is given by Eq. (2.77); also shown here are the two common asymptotes to the hyperbolas given in Eqs. (2.57) and (2.58) that describe the ring and end-cap electrodes, respectively. The value of r_0 is 1 unit. Figure 3.9b shows the corresponding view of the Finnigan stretched ion trap; the end-cap electrodes have the same shape as in Figure 3.9a [and Eq. (2.58)] and have been moved apart by some 10.8%. Note that the increase in separation in Figure 3.9b is some 30% of z_0 for clarity of presentation. Also shown in Figure 3.9b are the shifted asymptotes for the end-cap electrodes. An alternative arrangement that was available to Finnigan was to reshape the electrodes using the stretched value of z_0, but this was not done. The theoretical shapes of the end-cap electrodes appropriate to the end-cap electrode separation are shown in Figure 3.10; the heavy lines depict the cross section of an ideal ion trap and the light lines show that of an ion trap for which the hyperbolas have been calculated using Eqs. (2.57) and (2.58) with $z_0 = 0.70711 \times 110.6\% = 0.78206$ units and $r_0 = 1$ unit. The width of the ring electrode hyperbola is increased while those of the end-cap electrodes are decreased. To our knowledge such an ion trap has not in fact been constructed, but it is entirely possible that it would have a performance similar to that of the stretched ion trap. It should be possible to calculate ion secular frequencies that would compare well with experimental values.

STRETCHED ION TRAP   Because the value of z_0 had been increased by some 10.6%, the relationship between r_0 and z_0, as given by Eq. (2.77), no longer held true and, in turn, neither did the calculated values of q_z and β_z. A working estimate of the value for q_z in the spaced-out or stretched ion trap, as it came to be known after 1992, is given by

$$q_z = \frac{8eV}{m(r_0^2 + 2z_0^2)\Omega^2}. \quad \text{[Eq. (2.75)]}$$

Both the Finnigan ITD 800 and the Finnigan ITMS mass spectrometers used a stretched ion trap, as did the Varian Saturn GC/MS instrument when it became

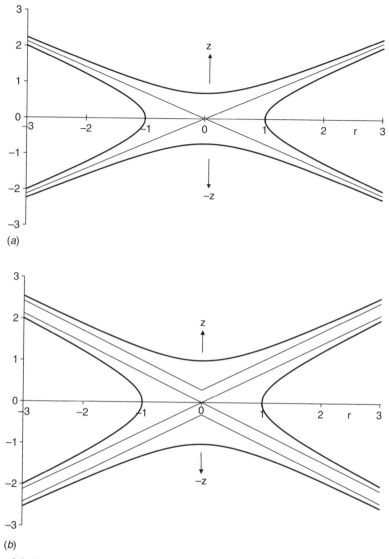

Figure 3.9. Comparison of an ideal quadrupole ion trap and a stretched ion trap: (*a*) A cross-sectional view of an ideal quadrupole ion trap where the inter-electrode spacing is given by $r_0^2 = 2z_0^2$ showing the two common asymptotes to the hyperbolae, the value of r_0 is 1 unit. (*b*) The corresponding cross-sectional view of the Finnigan stretched ion trap where the endcap electrodes have the same shape as in (*a*) and have been moved apart. The increase in separation in (*b*) is some 30% of z_0 for clarity of presentation. Also shown here are the shifted asymptotes for the end-cap electrodes.

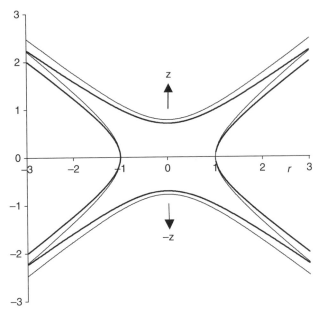

Figure 3.10. Comparison of an ideal quadrupole ion trap and a stretched ion trap where the electrodes have been re-shaped: the heavy lines depict cross-section of an ideal quadrupole ion trap where the inter-electrode spacing is given $r_0^2 = 2z_0^2$; the light lines depict the cross-section of an ion trap for which the hyperbolae have been calculated using Eqs. (2.57) and (2.58) with $z_0 = 0.70711 \times 110.6\% = 0.78206$ units and $r_0 = 1$ unit.

available in 1991. Although the mass shift problem had been solved on a practical basis, mass shifts had not been eliminated entirely.

Physicochemical Properties and Mass Shifts   In 1991, Cooks and co-workers [71] reported that significant mass shifts were found with the ITMS instrument when using axial modulation for mass range extension (see Chapter 2 and below). It was found that the mass shifts were dependent on the frequency and the amplitude of the supplementary voltage applied between the end-cap electrodes; mass peaks were shifted to lower apparent masses, and the mass shifts were observed to be as much as 60 Th for ions of ~22,000 Th.

Traldi and co-workers reported in a series of papers [72–74] that mass shifts could be detected also, if only to a minor extent, in the usual mass range (650 Th) of the ITMS instrument. At low q_z values, strong mass shifts were observed; the magnitude of the mass shift diminished with increasing q_z value and became constant for $q_z \geq 0.4$.

After extensive investigation of the effects of space charge, ion/molecule reactions, and so on, no explanation was forthcoming for an observed difference of 0.1 Th in the mass/charge ratios of the isobaric molecular ions of *o*- and *p*-trifluoromethylbenzonitrile [75]. Perhaps the origin of mass shifts could be related in terms of a physicochemical property of the ions under the influence of the RF field(s) present in the ion trap. Experiments were performed to verify whether such interactions existed between isomeric (and isobaric) molecular ions of o-, m-, and p-chloronitrobenzene and the supplementary AC, or "tickle" voltage. Following isolation of each of the molecular ions in turn and irradiation by a tickle voltage of increasing amplitude, it was found that the molecular ion signal intensity of the para isomer was reduced to 10% of the initial value using a tickle voltage amplitude of 1800 mV. The corresponding amplitudes for reduction of the signal intensities for the meta and ortho isomers were 1900 and 2200 mV, respectively.

In all experiments involving isomeric disubstituted benzene compounds, the magnitudes of the mass shifts were invariably in the order ortho > meta > para. Traldi and co-workers concluded that polarizability was the only property that could account for different mass shifts for isobaric ions having the same composition but differing in structure. Such a conclusion was consistent with an experimentally observed increase in mass shift magnitude with increasing analyte dipole moment. In Figure 3.11 are shown the mass displacements δ_m for a series ($\circ$) of alkylbenzene molecular ions of the form $[C_6H_5-Y]^+$ as a function of the electronic polarizability α_0 [76]; the solid line shows the variation of mass displacement with electronic

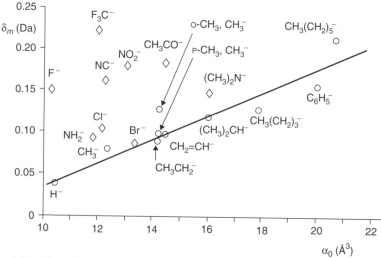

Figure 3.11. Mass displacements, δ_m, as a function of the electronic polarizability, α_0. For ($\diamond$) alkylbenzene compounds of form $[C_6H_5-Y]^+$ ions. (Reprinted from *Organic Mass Spectrometry*, Vol. 28, O. Bortolini, S. Catinella, P. Traldi, "Estimation of the polarizability of organic ions by ion trap measurements," pages 428–432 (1993). © John Wiley & Sons Limited. Reproduced with permission.) For (O) substituted benzene compounds (Reprinted from *Organic Mass Spectrometry*, Vol. 29, O. Bortolini, S. Catinella, P. Traldi, "Evaluation of dipole moments of organic ions in the gas phase," pages 273–276 (1994). © John Wiley & Sons Limited. Reproduced with permission.) The Y-substituents are shown. The solid line shows variation of mass displacement with electronic polarizability for positively-charged molecular ions from alkylbenzene compounds.

polarizability for $[C_6H_5-Y]^+$ ions. The Y substituents are shown. Also shown in this figure are the mass displacements δ_m for the molecular ions of a series ($\Diamond$) of substituted benzene compounds as a function of the electronic polarizability α_0 [77].

Additional support for this hypothesis came from the experimentally observed behavior of isobaric ions (differing in structure) in a QMF [78]. Clear differences were found in the mass shifts for different isomers, proving that the mass displacement effect is a general phenomenon that occurs when ions experience a quadrupole field.

Mass Shifts Explained?   In the preceding paragraphs we have seen how stretching the ion trap geometry away from the ideal relationship

$$r_0^2 = 2z_0^2 \qquad [\text{Eq. (2.77)}]$$

afforded a means of correcting for the observed mass shift phenomenon, yet it appears that compound-dependent, or "chemical," mass shifts may still be observed with certain commercial stretched ion trap mass spectrometers. In a recent extremely elegant and detailed study, Plass et al. [79] have provided a comprehensive description of the chemical mass shift phenomenon and explained its origin. A mass shift is observed when an ion of a particular m/z value is ejected *before* or *after* its expected point as defined by the scanning parameter of the ion trap, for example, amplitude of the RF drive potential. As indicated in a footnote to Table 3.2, the mass shift is defined [p. 239 of Ref. 79] as the true mass of the ion minus the measured mass, so that an ion ejected *early* (and hence having a lower mass/charge ratio assigned to it) is defined as having a *positive* mass shift. Thus it was found that delays in ion ejection times can result from the presence of higher order components caused by field imperfections (see previously) in the vicinity of the holes in the end-cap electrodes; furthermore, the precise ejection conditions may be modified in a compound-dependent fashion by the influence of elastic and inelastic collisions between the ions and the buffer gas atoms. Generally elastic collisions shorten the ejection delay and remove the peak splitting caused by nonlinear field components, and since the differences in collision cross sections for ions of similar masses are typically small, the changes in ejection delay are similar and manifest as very small chemical mass shifts. On the other hand, if inelastic collisions with the buffer gas lead to dissociation of polyatomic ions during the ejection delay caused by the nonlinear fields, then large chemical mass shifts may be observed.

From this work, it is concluded that the addition or subtraction of higher order field components to the fundamental quadrupolar field may deliberately remove or enhance the occurrence of chemical mass shifts, and, in the case of the latter, experiments may be designed that permit the measurement of physicochemical properties of ions together with the capability to distinguish between isomeric species.

3.4. MOTION OF TRAPPED IONS

One aspect that is crucial to the successful operation of the QIT mass spectrometer is an appreciation of the parameters that influence the motion of the trapped ions under certain specified conditions; of special importance is the presence, either intentional or unintentional, of background gas within the trap volume. On the one hand, "elastic" collisions may dampen (i.e., "cool") the kinetic energies of the ensemble of ions and cause migration toward the center of the trap. In addition the ions may experience "reactive" collisions through the occurrence of ion/molecule (or even ion/ion) reactions with suitable reagent species or, when energetically permitted, undergo charge transfer reactions with their colliding partners. On the other hand, *resonant excitation*, generally through the application of an auxiliary AC voltage (usually to the end-cap electrodes), may increase the kinetic energy of mass-selected ions to an extent where, depending upon the amplitude of the auxiliary AC signal and the background gas pressure, sufficient kinetic energy may be imparted to the ions during resonant excitation so as to cause their ejection from the trap. Alternatively, "inelastic" collisions of the ions may occur, resulting in their internal excitation and subsequent CAD; such processes are clearly of potential value in tandem mass spectrometry experiments. These topics have already been mentioned at various stages in the preceding discussion: the aim of this section is to examine briefly the various processes in their own right in order to enhance the reader's understanding of these important effects.

3.4.1. Collision Processes: Collisional Cooling

Historically, the first real physical demonstration of the effect of collisions with background gases in an ion trap is in the pioneering work of Wuerker et al. [1], who studied the containment of charged microparticles. Thus it was observed that with background pressures in the range of 5×10^{-4}–1×10^{-3} Torr the violent motion of the particles was dissipated so that static "crystal arrays" were formed. These arrays could be "melted" and "recrystallized" by changing the frequency of the RF drive potential and the background pressure: the lower the pressure, the longer it took for a static array to reestablish itself. The formation of crystal arrays of electrically charged metallic microparticles suspended in a trap operating at standard atmospheric pressure was also observed by Whetten [80].

The first mathematical description of the collisions of trapped ions was presented by Dehmelt [2] and Major and Dehmelt [81]. For the present discussion, the key results of this work relate to the different collisional effects that depend upon the relative mass of the ion, m_i, and the mass of the background neutral atom or molecule with which it is colliding, m_n. Two extreme sets of conditions were considered: when $m_i/m_n \ll 1$, the neutral species behave as fixed scattering centers, leading to RF heating; when $m_i/m_n \gg 1$, assuming that the background particles are relatively cold, the collisions essentially result in viscous drag which lowers the mean kinetic energy of the ions as a function of time. In this latter case, which clearly applies to the ion trap mass spectrometer operating with helium as a buffer gas, the micromotion is interrupted by the collisions, but only slightly modified in phase and amplitude, while any secular motion is damped out exponentially. The effect of the damping of the secular motion is to cause the ions to migrate toward the center of the trap.

Dawson and Whetten [82] reported upon a series of experiments in which they observed how the addition of neon gas over the pressure range 7×10^{-9}–2×10^{-6} Torr increased the storage capabilities for Hg^+ ions by measuring the signal resulting from the pulse ejection of ions after a preselected trapping time. It should be noted that although there was no mass selection step built into these experiments, it was assumed that the ion signal was in fact due to mercury ions because the charge exchange reaction between Ne and Hg^+ does not occur normally as it is highly endothermic. Similar results were reported by Blatt et al. [83] in a study that showed how the intensity of the fluorescence radiation emitted by trapped Ba^+ ions increased by a factor of 300 over a pressure range of 5×10^{-8}–5×10^{-6} Torr background pressure of helium. Helium was chosen as the buffer gas because, as noted above, the large mass difference between He and Ba leads to viscous drag on the ion motion, resulting in longer storage times. By contrast, Dawson and Lambert [84], using an ion trap in the mass-selective storage mode [85], observed the $N_2^{+\cdot}$ ion peak intensity from nitrogen as a function of added helium, argon, or carbon dioxide, with the nitrogen pressure held at 5×10^{-5} Torr. They found that after an initial rise above a pressure of $\sim 10^{-4}$ Torr of added gas the signal decayed exponentially with a rate constant that increased with the mass of the added gas, presumably due to increased scattering of the ions.

In experiments of a rather different kind, based upon using the ion trap (QUISTOR) as the ion source for a QMF in the manner described by Todd and co-workers [86], March and co-workers [87], using a Monte Carlo simulation model, showed that charge transfer between ions and neutral atoms of the same mass leads to migration to the center of the trap with a reduction in kinetic energy. Specifically, this work suggested that the migration rate is higher at higher values of q_z, so, for example, Ar^{2+} ions migrate faster than $Ar^{+\cdot}$ ions under the same trapping conditions. This migration toward the center of the trap and reduction of the kinetic energy, resulting in a greater extraction efficiency of the ions when using pulse ejection, satisfactorily explained the unexpectedly high intensity ratios of the Ar^{2+} and $Ar^{+\cdot}$ ion signals. Subsequent research by Doran et al. [14] extended this study to include charge exchange reactions of $Ne^{+\cdot}$ and $CO_2^{+\cdot}$ ions in neon and carbon dioxide and confirmed this model for the migration of ions.

The advantages of using helium as a buffer gas at $\sim 10^{-3}$ Torr to improve both the resolution and, *simultaneously*, the sensitivity in the ion trap when used in the mass-selective ejection mode are fully evidenced in the original paper on this instrument by Stafford et al. [88] and are clearly indicated in the corresponding landmark patent for this invention [89]. Normally in mass spectrometry, as in other fields of measurement, resolution and sensitivity work in opposition to one another: an increase in one parameter is invariably accompanied by a decrease in the value of the other. Furthermore, it is generally understood that the background pressure within mass spectrometer analyzers should be kept as low as possible in order to minimize the probability of collisions: here, the pressure of helium should ideally be maintained at a value some 1000 times greater than the "norm"!

The explanation given for this improved performance of the ion trap mass spectrometer [88] focused on three consequences arising from the viscous damping effect of the collisions between the ions and the helium buffer gas atoms. (1) The ion trajectories are caused to collapse to the center of the trap in both the radial and axial directions, so that ions spend most of their time near the center and are not therefore subject to any nonlinear fields (see earlier) arising from field imperfections due to mechanical errors, holes in the electrodes, and so on. (2) Normally, as the amplitude of the RF drive potential is increased, the maximum displacement will increase, but the buffer gas damping effect opposes this, except when the ion trajectories actually become unstable in the axial direction; in the radial direction, the ions are still tightly bunched and can pass through the exit holes in the end-cap electrodes. (3) In a collision with a low-mass damping atom gas there is a small change in momentum of the ion, and therefore no significant scattering, and hence there are no abrupt changes in direction, as might be the case if the ion collided with a high-mass species, such as a sample or other background gas molecules.

Not surprisingly, much work has been directed toward improving our understanding of the role of buffer gas in optimizing the performance of ion trap mass spectrometers through the phenomenon of *collisional cooling*. For a concise account of this topic the reader is referred to the survey by Brodbelt [90]. For example, experiments have shown how changing the helium buffer gas pressure affects the detection efficiency of benzene molecular ions by cooling the ions down so that they are located near the center of the trap, as noted above; collisional cooling rate constants have been determined, showing that there is an essentially linear collisional cooling rate up to a pressure of ~0.7 mTorr of helium. The linearity of collisional cooling rate with helium pressure indicates that the absolute number of collisions required to maximize the signal remains relatively constant at different buffer gas pressures. Further studies compared the ejected ion current at different q_z values without cooling and with cooling for a period of 100 ms and showed that in the latter series of experiments the benzene ion signal was some three to four times greater than that observed without cooling, with greater signal enhancement being observed with increasing values of q_z up to ~0.7. The application of supplemental resonant excitation (see below) also shows the effects of cooling at different helium pressures, and, in particular, the effects of collisional cooling on collision-activated dissociation processes have been explored (see also below). The reader is also referred to the discussion of the effect of buffer gas pressure on the appearance of ghost peaks, discussed earlier in this chapter.

3.4.2. Collision Processes: Trapping Injected Ions

The first commercial instruments based on the principle of mass-selective ejection employed "internal" ionization, that is, the analyte ions were created *inside* the trapping volume by means of electron or chemical ionization. However, it soon became apparent that for increased analytical versatility, using both these two methods and other ionization techniques such as electrospray ionization and matrix-assisted laser desorption ionization (MALDI), the capability to trap externally generated ions was essential. In his pioneering publication, Fischer [5] pointed out that, on the basis of the underlying theory of ion containment, it was self-evident that for successful trapping the ion must be formed from a molecule in the space between the electrodes and it should be impossible to trap ions that enter the field from outside. Kishore and Ghosh

[91] proposed a method whereby ions are injected into an ion trap while the RF drive potential is set at zero followed by rapidly increasing the amplitude to an appropriate value after a suitable delay. Todd et al. [92] employed a method based upon phase space analysis to support this approach and considered alternative means for directing the ions into the trapping volume. This method of gating the RF potential during ion injection was also considered by O and Schuessler [93–95]. However, the key to successfully trapping ions injected from an external source lies in the use of a buffer gas to remove their excess energy via collisional cooling, a method first described by March and co-workers [96, 97], who employed a system in which an ion trap was mounted in the detector housing of a Kratos MS-30 double-focusing mass spectrometer.

The first systematic study of the injection of ions into a QIT mass spectrometer operating on the principle of mass-selective ejection is that by Louris et al. [98] in which the effects of several different damping gases were compared. Two separate sets of experiments on the axial injection of ions were performed, one in which various organic compounds were ionized by means of an external EI source and the other in which metallic surfaces were mounted in a laser desorption ionization source. An elaborate series of tests showed that the ions under study were indeed being formed externally and successfully contained within the trapping volume. One important observation arose from a series of experiments in which ions of different mass were injected into the ion trap at different amplitudes of the RF drive potential, namely that the higher the mass/charge ratio value, the lower the threshold value of q_z at which ions may be trapped.

In explaining these observations, the authors drew on the model of pseudopotential wells developed by Dehmelt [2] and Major and Dehmelt [81] (originally developed for conditions in the absence of collisions, see above). The effect is probably best viewed in terms of Eq. (3.15):

$$\bar{D}_z = \frac{4eV^2z_0^2}{m(r_0^2 + 2z_0^2)^2\Omega^2} = \frac{mz_0^2\Omega^2}{16e}q_z^2. \qquad \text{[Eq. (3.15)]}$$

Rearranging this expression, we have

$$\frac{mz_0^2\Omega^2}{16}q_z^2 = e\bar{D}_z = \bar{W}_z \qquad (3.70)$$

where $\bar{W}_z$ represents the kinetic energy of the ion oscillating within a pseudopotential well of depth $\bar{D}_z$. If we assume that all ions of the same charge are injected with the same kinetic energy, independent of their mass, and that they must be collisionally cooled so that their kinetic energy in the z direction is less than $\bar{W}_z$, then for a fixed value of the drive frequency Ω

$$q_z \propto \sqrt{\frac{1}{m}}. \qquad (3.71)$$

This inverse square-root dependence [Eq. (3.71)] appeared to be compatible with the experimental data. Further studies suggested that neon is about as good as helium for promoting trapping, but with argon and xenon the signals were only 10–20% as intense as with helium.

In a more recent very detailed investigation, Quarmby and Yost [99] carried out a combined modeling and experimental study of the injection of ions formed by the electrospray technique into the QIT. Using the SIMION 6.0 program [100] to model the details of the end-cap electrodes, they showed that the holes weaken the RF trapping field near the holes, an important effect when the ions have large axial trajectories, as during ion injection. The holes were found also to cause RF field penetration out toward the incoming ions, which may thereby be accelerated or decelerated as they approach the trap depending upon the phase of the RF drive potential. In the main, this field penetration was found to increase the percentage of ions trapped, especially with slower moving high-mass ions. Generally, good agreement was obtained between experimental and modeled results for the percentage of trapped ions as a function of q_z at the time of injection for high-mass ions.

A key aspect of this work was that it showed that helium buffer gas is essential to stop externally injected ions over a narrow range of phase angles; however, at most of the phase angles of the RF drive potential, it is impossible to trap the injected ions at all. Making allowance for end-cap holes in the model suggested that ions may be trapped over two RF phase angles; however, in practice, only one favorable phase angle was observed. These apparently favorable phases correspond to conditions under which, in the absence of buffer gas, the ions may undergo many oscillations before being ejected. These ions were termed *pseudostable* ions and were first noted in modeling studies performed by O and Scheussler [101]. Computations by Quarmby and Yost [99] showed such ions may travel as far as 1500 mm before ejection and that, in the presence of buffer gas, there are sufficient collisions for the ion to be cooled and trapped.

3.4.3. Resonant Excitation

So far in this section we have considered how collisional cooling of the ions can reduce their kinetic energy and cause the ions to migrate toward the center of the ion trap. Fortunately it is found that, although the amplitudes of the secular motions of the ions in the axial and radial directions are reduced, the frequencies of oscillation remain essentially the same. This means that the equations

$$\omega_{u,\,n} = (n + \tfrac{1}{2}\beta_u)\Omega \qquad 0 \le n < \infty \qquad \text{[Eq. (2.80)]}$$
$$\omega_{u,\,n} = -(n + \tfrac{1}{2}\beta_u)\Omega \qquad -\infty < n < 0 \qquad \text{[Eq. (2.81)]}$$

remain valid, and taking the fundamental frequency ($n = 0$) for the axial component of motion ($u \equiv z$), Eq. (2.80) becomes the well known expression (3.11):

$$\omega_z = \tfrac{1}{2}\beta_z\Omega. \qquad \text{[Eq. (3.11)]}$$

As a consequence, if we can apply an additional, "supplementary," oscillating electric field along the axis of the ions trap with a frequency equal to ω_z, it should be possible to couple resonantly with the fundamental secular frequency of those ions that have the correct value of β_z. In an ideal quadrupole field and in the absence of any collisional effects, these ions will therefore receive additional energy, their trajectories will expand, and ultimately they will collide with or pass through holes in one of the end-cap electrodes. Since β_z is directly related to the trapping parameters a_z and q_z [by Eq. (2.82)], under specified conditions of U, V, and Ω this resonant excitation process will only affect those ions having a particular mass/charge ratio. This process is the same as that employed by von Busch and Paul [27] to mass selectively eject $^{24}Mg^+$ and $^{26}Mg^+$ from a QMF in order to enrich ^{25}Mg, as described earlier in this chapter. Sample calculations of ω_z are given in Chapter 2, from which it is evident that typically they fall into the tens to hundreds of kilohertz range; the additional oscillating signal with which resonant excitation is effected is frequently termed a *supplementary AC potential*, or more colloquially a *tickle potential* (voltage).

This supplementary AC potential may be connected to the ion trap in a variety of ways. The most common method is *dipolar* coupling, in which the two outputs from a bipolar AC supply are applied, one to each end-cap electrode. *Monopolar* coupling occurs when a single AC signal is connected to one end-cap electrode, while the other end-cap electrode is held at a constant potential, usually zero. While the additional electric fields to which the trapped ions are subjected are different for these two methods of applying the supplementary oscillating electric field, it is possible to excite selectively the axial secular motion of the ions, according to foregoing equations. The third method is either to connect a single AC potential to both end caps in parallel or, alternatively, to superimpose it upon the RF drive potential applied to the ring electrode. In either case, the additional field has a quadrupolar symmetry and, as noted previously in this chapter, simulation studies [59] have shown that it is possible to excite either the radial or the axial secular motion of the ions, depending upon the magnitude of the tickle frequency, ω_T, applied. It should be noted that, although the impression given by the above account is that the resonant excitation process should occur sharply, at a specified frequency, in fact, the observed resonances can be broad, resulting from, among other things, local space charge effects within the trapped ion cloud and the influence of nonlinear field components which cause the frequency of the secular motion to change as the ion moves further away from the center of the trap (see above).

Earlier in this section, when introducing the topic of resonant excitation, it was implied that the ion trajectories would grow only to a point where the ions either are lost on the electrodes or are ejected in the absence of collisions with background gas. In practice, provided the amplitude of the supplementary AC signal is sufficient, typically around 3–6 V, ions will be successfully ejected from the trap. At lower AC amplitudes, depending upon the pressure, collisions with the buffer gas atoms will cause the ions to be retained within the trap but may possibly lead to CAD [alternatively called collision-induced dissociation (CID)]. The discussion which follows will concentrate upon the application of a supplementary AC potential to effect *resonant ejection*; CAD resulting from resonant excitation is considered later.

3.4.3.1. Resonant Excitation: Ion Ejection

We have seen from the foregoing account that ions of a given charge number may be mass-selectively ejected from the ion trap by the superposition of a supplementary AC potential of the appropriate frequency on the main RF trapping field. This form of resonant excitation is generally achieved in a monopolar or dipolar fashion by means of suitable connections to the end-cap electrodes, so as to effect excitation of the ion trajectories along the axis of the trap. All commercial QIT mass spectrometers currently employ resonant ion ejection at one or more points within their duty cycle of operation.

Historically, the first application of resonant excitation to the modern QIT mass spectrometer was the use of *axial modulation* to improve the spectral quality of the RF-only mass-selective ejection of ions [102]. In the basic mode of operation, as the amplitude of the RF drive potential applied to the ring electrode is increased linearly, the values of q_z of the trapped ions increase and move sequentially across the $\beta_z = 1$ stability boundary, whereupon the ions are ejected for detection. However, it is found that the presence of ions of higher mass/charge ratio in the ion trap can cause peak broadening of the ejected ions, especially if the masses are closely separated. This loss of signal quality can be regained through axial modulation, in which a supplementary AC signal of approximately $6\,\mathrm{V_{p\text{-}p}}$ is applied at a frequency slightly less than half of the frequency of the RF drive potential. The choice of this frequency follows from Eq. (3.11), in which the value of β_z selected is slightly less than unity. As a consequence, just as the ions are being ejected, their secular motion enters into resonance with the supplementary field so that the ions are energized as they suddenly "come into step" and are therefore much more tightly bunched as they are ejected, thereby substantially improving the spectral resolution.

From applying the axial modulation signal at the $\beta_z = 1$ stability boundary, it is a small step to consider reducing the frequency of the supplementary AC signal to a value at which ions having a lower value of β_z will enter into resonance: provided that the amplitude of the AC signal is sufficiently high, resonant energy absorption will cause unstable trajectories to develop, resulting in the ions being ejected from the trap and detected. In this way, the mass/charge ratio range of the ion trap mass spectrometer may be increased. For example, if 650 Th is the maximum mass/charge ratio that can be ejected at $\beta_z = 1$, then operating at, say, $\beta_z = 0.1$ will increase the mass/charge ratio range to 6500 Th; to generate a mass spectrum under these conditions, an axial modulation frequency of one-twentieth of the RF drive frequency would be applied while ramping linearly the amplitude of the RF drive potential so as to successively bring the ions of increasing mass/charge ratio into resonance. Ion ejection is brought about at fixed values of β_z and q_z [see Eq. (2.82)] and we note from Eq. (3.17) that at constant q_z the depth of the pseudopotential well, $\overline{D}_z$, trapping the ions increases linearly with the amplitude V of the RF drive potential. Thus, during a scan, ions about to be ejected are trapped in increasingly deeper potential wells. As a result, it is found that increasing the amplitude of the supplementary AC signal during scanning improves the efficiency of ion ejection. A further discussion on axial modulation and mass range extension is included in Chapter 8, which also contains details of the use of combined dipolar and nonlinear fields to improve mass spectrometer performance.

References pp. 125–132.

The discussion immediately above has been concerned with the use of resonant excitation to improved mass spectral quality through the use of axial modulation. A further use of a supplementary AC field is to eject preferentially unwanted ions as a prior step to some other experiment within the ion trap, for example, tandem mass spectrometry. Thus a suitable combination of supplementary AC frequency and limited scanning of the amplitude of the RF drive potential offers a means of removing ions having mass/charge ratios below or above that of the selected precursor ion. Perhaps a more sophisticated means of achieving precursor ion isolation is to employ the method of filtered noise fields (FNFs), developed by Kelley [103]. Here a broadband supplementary AC power supply is programmed to provide a signal comprising many frequency components over a specified range but containing one or more "notches" where the amplitude at that frequency is zero, corresponding to the mass/charge ratio(s) of ions that it is desired to retain in the trap. Application of a notch filter can be especially valuable during ion accumulation when, for example, there are interfering background ions resulting from solvent, matrix, or column bleed or from the air, as might be the case when ions are being admitted from an atmospheric glow discharge source. Further details of the use of FNFs and other techniques for ion selection may be found in the publication by Goeringer et al. [104] and the account by Yates et al. [105] (see also SWIFT, Sections 5.3.4 and 6.3.3.4).

3.4.3.2. Resonant Excitation: Collision-Activated Decomposition

From the preceding sections we have seen how both the cooling of the kinetic energy of the trapped ions by collisions with low-mass buffer gas atoms and the resonant excitation of the motion of mass-selected ions can significantly improve the sensitivity and resolution of the QIT mass spectrometer. In this section we consider how these collision and excitation processes may be combined to cause dissociation of specified precursor ions as part of a tandem mass spectrometry experiment.

We noted earlier that in order to eject ions at the typical helium buffer gas pressures employed in the ion trap mass spectrometer ($\sim 10^{-3}$ Torr) supplementary AC amplitudes of approximately 6 V_{p-p} are employed. However, if low-amplitude tickle voltages of approximately a few hundred millivolts are applied for, say, 10 ms, then it may be found that the selected ions have dissociated to form product ions of lower mass/charge ratio which have been trapped and may be analyzed by the normal mass-selective ejection scanning method. This procedure therefore forms the basis of a tandem-in-time mass spectrometry/mass spectrometry (MS/MS) analytical method [106]. Ions are first created within or injected into the ion trap, a precursor ion is selected by one of the methods described earlier, the isolated ion is collisionally dissociated by resonant excitation, and the resulting product ions are then ejected mass selectively for detection. Indeed, further product ion isolation and dissociation steps may be incorporated into the duty cycle to effect sequential MS^n capability.

The internal excitation energy required for CAD is typically of the order of 5–6 eV, so the question arises as to how applying a supplementary AC amplitude of, say, 200 mV during resonant excitation can impart so much internal energy to the precursor ions. Two possible answers are that either very many low-energy collisions with the buffer gas cause the ions to become excited progressively through a series of intermediate states until the decomposition threshold energy is reached or there are fewer higher energy collisions of ions that have become excited translationally through interaction with the RF drive potential.

If the latter explanation is accepted, then one can envisage a complex interplay between low-energy elastic collisions and high-energy inelastic collisions of ions with buffer gas atoms. Thus, the application of the tickle voltage forces the mass-selected ions away from the center of the trap to displacements near the boundary of the trapping volume. If the buffer gas pressure is insufficiently large or the tickle amplitude is too great, these ions will, of course, be resonantly ejected from the trap and lost (see above) unless elastic "stopping" collisions analogous to those that permit the trapping of externally injected ions ensure that they remain confined. However, these resonantly "displaced" ions are now at values of r and z where they are subjected to the intense oscillating electric fields associated with the RF drive potential and, at the correct phase angle, will be accelerated to kinetic energies of up to several tens of electron-volts (see specimen calculations earlier in this chapter). Thus high-energy inelastic collisions with the buffer gas atoms can then lead to significant internal excitation. Obviously, for any given mass-selected ion this cycle of events may occur many times, so that the final excited state before decomposition may still be reached through a succession of intermediate excited states. At the same time, elastic collisions with buffer gas atoms may cool the accelerated ion before it has acquired sufficient kinetic energy to undergo a successful inelastic collision, and indeed a long-lived internally excited ion may possibly undergo collisional deexcitation with a buffer gas atom.

This model for the collisional activation of ions by resonant excitation appears to be consistent with a number of general observations. (a) The efficiency of conversion of a precursor ion into its product ion(s) is generally much higher (up to 100% in some cases) compared to that observed with the same species in a conventional tandem-in-space instrument, for example, a triple-quadrupole mass spectrometer. This observation is consistent to there being a multicollision/ion accumulation environment within the trap, in contrast to the single-collision regime within the collision cell of a triple-stage quadrupole mass spectrometer. (b) To a first approximation, in tandem mass spectrometry with the ion trap the efficiency of conversion is often determined by the product of the tickle amplitude and tickle duration, sometimes called the *fluence* [58], rather than by the amplitude and duration separately. (c) Higher energy dissociation processes are favored at lower buffer gas pressures, as for example illustrated by observing the ratio of the yields of [m/z 91]/[m/z 92] from the collisionally activated *n*-butylbenzene precursor ion [90] (although longer cooling times may be necessary to compensate for the loss of sensitivity at lower pressures, see above). This last observation is consistent with the model that the precursor ions may acquire greater kinetic energy for conversion in an inelastic collision when the probability of "interfering" elastic buffer gas collisions occurring is lower.

The mass of the collision gas may have a significant impact on the internal energy available for dissociation and the corresponding efficiency of conversion from precursor into product ions. The effect of the mass of the background neutral species relative to that of the ion has already been considered in the discussion on collisional cooling, earlier in this chapter. Here, it was noted, from the work of Major and

References pp. 125–132.

Dehmelt [81], that collisions with high-mass neutral atoms or molecules could lead to ion scattering. On the other hand, the relative masses of the colliding partners may have a significant effect on the energy available for conversion to internal excitation as a result of an inelastic collision. This can be seen from the following simple treatment of collision energies.

3.4.3.3. Collision Energies   Consider a bimolecular "hard-sphere" collision (i.e., ignoring intermolecular forces) between an ion of mass m_i having velocity v_i with a neutral atom of mass m_n and velocity v_n. In the so-called laboratory (i.e., experimental) frame of reference, the total kinetic energy of the colliding species, E_L, is given by

$$E_L = \tfrac{1}{2}m_i v_i^2 + \tfrac{1}{2}m_n v_n^2.$$ (3.72)

Now the total kinetic energy of the system can be considered to comprise two components: (a) the overall kinetic energy of motion of the center of mass of the colliding particles and (b) the relative kinetic energies of the two particles along the line of centers joining the two particles (that also passes through the center of mass). From the principle of conservation of momentum, the momentum of the center of mass equals the sum of the momenta of the two colliding species, so we can write

$$(m_i + m_n)v_C = m_i v_i + m_n v_n$$ (3.73)

where v_C is the velocity of the center of mass. Also we can write

$$v_R = v_i - v_n$$ (3.74)

in which v_R is the relative velocity of the ion and the neutral gas atom. Hence from Eqs. (3.73) and (3.74) we have

$$m_n(v_i - v_R) + m_i v_i = (m_i + m_n)v_C.$$ (3.75)

An expression for E_L [defined in Eq. (3.72)] may now be obtained in terms of v_R and v_C by using the substitutions

$$v_i = v_C + \frac{m_n}{m_n + m_i}v_R$$ (3.76)

from Eq. (3.75) and

$$v_n = v_C + \frac{m_i}{m_n + m_i}v_R$$ (3.77)

from Eqs. (3.74) and (3.76), such that

$$E_L = \frac{1}{2}m_i\left(v_C + \frac{m_n}{m_n + m_i}v_R\right)^2 + \frac{1}{2}m_n\left(v_C - \frac{m_i}{m_n + m_i}v_R\right)^2.$$ (3.78)

Upon squaring each of the velocity terms in Eq. (3.78), the cross terms in $v_C v_R$ cancel and, collecting terms in v_C^2 and v_R^2, one obtains

$$E_L = \frac{1}{2}(m_i + m_n)v_C^2 + \frac{1}{2}\frac{m_i m_n}{m_i + m_n}v_R^2. \tag{3.79}$$

In Eq. (3.79), the first term on the right-hand side represents the kinetic energy of the center of mass, which in the absence of external forces is constant, whereas the second term corresponds to the relative kinetic energy of the colliding particles, E_C, within the moving center-of-mass frame of reference. In terms of any chemical consequences of the collision between the two species, it is the quantity E_C that is of interest. Hence we can write

$$E_C = \frac{1}{2}\frac{m_i m_n}{m_i + m_n}v_R^2. \tag{3.80}$$

Let us consider the special case where the initial kinetic energy of the ion is considerably in excess of that of the thermal kinetic energy of the neutral atom, so that, as an approximation, we can assume that the latter species is at rest, that is, $v_n = 0$. From Eq. (3.72), recalling that now $v_R = v_i$, we therefore have

$$E_L = \tfrac{1}{2}m_i v_R^2 \tag{3.81}$$

which, upon substitution into Eq. (3.80), gives

$$E_C = E_L\frac{m_n}{m_i + m_n}. \tag{3.82}$$

Consequently, under these conditions, the fraction of the initial kinetic energy of the energetic ion that is partitioned into the center-of-mass energy is essentially determined by the mass of the neutral gas atom. If it is assumed that, as a result of the collision, a fraction of the center-of-mass energy E_C is converted into internal excitation energy of the ion, then Eq. (3.82) suggests that, for a given value of E_L, the degree of internal excitation will increase as the mass of the neutral gas species increases. Thus, if all the available center-of-mass kinetic energy is converted to internal energy, when $m_i = 100$ and $m_n = 1$, this represents only $\sim 1\%$ of the original laboratory kinetic energy, whereas if the neutral species has $m_n = 40$, then $\sim 29\%$ of the value of E_L may be available to internally excite an ion of the same mass.

Several groups have examined the effects of increasing the mass of the buffer gas; these studies have generally taken the form of either adding a small percentage of the higher mass species to the normal buffer gas pressure of helium or pulsing a burst of high-mass gas to a system operating with helium at the normal helium pressure during only the resonant excitation stage of the duty cycle. Recently, Glish and coworkers [107] have reported upon an extensive study in which they compared the performance of three different ion trap systems working with helium, air, and argon as "pure" buffer gases; the instruments were the Finnigan ITMS and LCQ and the

Bruker Esquire ion traps. Although there were some initial difficulties in directly obtaining comparable measurements of the buffer gas pressures in the different experimental arrangements, they found that the use of argon increased the sensitivity and the CID efficiency but significantly reduced the resolution. Attempts were made to compare the relative activation and cooling rates with the three gases; it was found that the Esquire trap operated more effectively with argon than did the ITMS, whereas the LCQ model approached Esquire performance when the argon pressure was reduced and the value of β_z at which resonant ejection was applied was lowered to $\beta_z = \frac{2}{3}$. It was found that, to achieve the same efficiency of CAD, it was necessary to use a higher tickle voltage with argon than with helium, which suggests that argon is better than helium at cooling via elastic collisions or more efficient at collisionally "quenching" already excited ions inelastically. The authors [107] attributed the improved performance of the Esquire ion trap with heavier buffer gas to the more efficient and faster ion ejection process that is thought to be achieved by combining the use of dipolar resonant excitation with the presence of the hexapole nonlinear resonance "line" that exists at $\beta_z = \frac{2}{3}$. Overall, it is not immediately clear why one would want to replace helium by a heavier gas, given the distinct loss of resolution noted above, unless there were some other significant advantage. This could, for example, be the need to try to access higher energy states of the precursor ion through improved collisional energy transfer (see above) or the use the ion trap in a field application where it may be desirable to rely on air as the buffer gas.

Tandem mass spectrometry with the ion trap is considered further in Chapter 7, and the reader is also referred to other detailed accounts that abound in the literature [108].

3.4.4. Boundary-Activated Dissociation (BAD)

So far we have considered the application of a supplementary AC potential to one or both of the end-cap electrodes to effect resonant excitation as being the only means of causing CAD. However, in 1991 Paradisi et al. [109] noted that if an ion was isolated selectively in the ion trap and then the RF and DC amplitudes of the drive potentials adjusted so that the (a_z, q_z) working point of the ion lay just inside either the $\beta_r = 0$ or the $\beta_z = 0$ boundaries, fragment ions appeared which could be trapped (provided that their mass/charge ratios were above the corresponding value of the LMCO) and then analyzed by the normal mass-selective instability scan. It was concluded that, as the working point approached a region of instability, the kinetic energy of the ion increased such that CAD occurred in a manner akin to the conventional resonant excitation method described above. Numerous quantitative tests were carried out using both helium and argon as the buffer gas. As noted above, argon was found to give more extensive fragmentation than was helium; indeed, it was noted that at certain (a_z, q_z) coordinates, away from the boundaries, fragmentation was found when using argon whereas none was observed with helium, again suggesting that the cooling collisions with the heavier gas may be able to deposit sufficient internal energy to exceed the threshold for dissociation. In practical terms, BAD does not appear to have become widely used as an analytical technique, possibly because of the difficulty of establishing the correct operating conditions on commercial instruments; however, the effect has been well characterized and the reader is directed to Ref. 110 for further details.

3.4.5. Surface-Induced Dissociation

Another technique, also dating from 1991, is that of surface-induced dissociation (SID), developed by Lammert and Cooks [111]. The method, which follows from previous work by the Cooks group on SID using other types of mass spectrometers, employs a short (<5-μs), fast-rising (<20-ns-rise-time), high-voltage DC pulse (e.g., 325 V) which is applied to the end caps of a standard QIT. This method is in contrast to the application of the supplementary AC potential normally used to excite resonantly and to dissociate ions in the trap (see above). The effect of the DC pulse is to cause previously mass-selected precursor ions to become rapidly unstable in the radial direction and subsequently to collide with the ring electrode. Sufficient internal energy is acquired in this collision to cause high-energy fragmentations of relatively intractable molecular ions such as pyrene and benzene. The product ions are then captured within the trapping field and analyzed by the conventional mass-selective ejection mode. Simulation studies showed that the surface collisions occur at kinetic energies in the range of tens to hundreds of electronvolts. Results obtained with the pyrene molecular ion (m/z 202) showed low yields (ranging from 0.2 to 3% of the precursor ion intensity) of fragment ions at m/z 122, 150, 151, 174, and 175. These results contrast with normal CAD via resonant excitation, in which the only processes observed corresponded to losses of H$^{\cdot}$ and H$_2$. Other studies showed that the appearance energies for m/z 122 and 150 are 17 and 11 eV, respectively, indicating the high levels of internal excitation that are possible with the SID technique.

3.5. CONCLUSION

In the study of the QIT, it is of importance to have a degree of understanding of the theory of ion trap operation and an awareness of the applications of the ion trap as discussed in later chapters. It is even of value to understand the possibilities for simulation of ion behavior as an adjunct to experimentation. However, it is vital that one has some understanding of what is going on in the ion trap, that is, the dynamics of ion trapping as have been discussed in this chapter.

REFERENCES

1. R. F. Wuerker, H. Shelton, R. V. Langmuir, Electrodynamic containment of charged particles, *J. Appl. Phys.* **30** (1959) 342–349.
2. H. G. Dehmelt, Radiofrequency spectroscopy of stored ions I: Storage, in D. R. Bates (Ed.), *Advances in Atomic and Molecular Physics*, Vol. 3, Academic, New York, 1967, pp. 53–72.

3. J. F. J. Todd, R. M. Waldren, R. E. Mather, The quadrupole ion store (QUISTOR). Part IX. Space-charge and ion stability. A. Theoretical background and experimental results, *Int. J. Mass Spectrom. Ion Phys.* **34** (1980) 325–349.

4. M. F. Finlan, R. F. Sunderland, J. F. J. Todd, The quadrupole mass filter as a commercial isotope separator, *Nucl. Instrum. Methods* **195** (1982) 447–456.

5. E. Fischer, Die dreidimensionale Stabilisirung von Ladungsträgern in einem Vierpolfeld, *Z. Phys.* **156** (1959) 1–26.

6. C. Schwebel, P. A. Möller, P. T. Manh, Formation et confinement d'ions multicharges dans un champ quadrupolaire a haute fréquence, *Rev. Phys. Appl.* **10** (1975) 227–239.

7. E. P. Sheretov, V. A. Zenkin, V. F. Samodurov, Three-dimensional accumulation quadrupole mass spectrometer, *Zh. Tekh. Fiz.* **43** (1973) 441.

8. H. A. Schuessler, Physics of atoms and molecules, in W. Hanle and H. Kleinpoppen (Eds.), *Progress in Atomic Spectroscopy, Part B*, Plenum, New York, 1979, p. 999.

9. J. E. Fulford, D.-N. Hoa, R. J. Hughes, R. E. March, R. E. Bonner, R. F. Wong, Radiofrequency mass selective excitation and resonant ejection of ions in a three-dimensional quadrupole ion trap, *J. Vac. Sci. Technol.* **17** (1980) 829–835.

10. P. K. Ghosh, A. P. Chattopadhyay, Ion stability and many particle interactions in a QUISTOR, *Int. J. Mass Spectrom. Ion Phys.* **46** (1983) 75–78.

11. A. P. Chattopadhyay, P. K. Ghosh, QUISTOR: A study of many particle systems, *Int. J. Mass Spectrom. Ion Phys.* **49** (1983) 253–263.

12. D. C. Burnham, D. Kleppner, Practical limitations of the electrodynamic ion trap, *Bull. Am. Phys. Soc. Ser. II* **11** (1968) 70.

13. R. F. Bonner, R. E. March, The effect of charge exchange collisions on the motion of ions in three-dimensional quadrupole electric fields. Part II. Program improvements and fundamental results, *Int. J. Mass Spectrom. Ion Phys.* **25** (1977) 411–431.

14. M. C. Doran, J. E. Fulford, R. J. Hughes, Y. Morita, R. E. March, R. F. Bonner, Effects of charge-exchange reactions on the motion of ions in three-dimensional quadrupole electric fields. Part III. A two-ion model, *Int. J. Mass Spectrom. Ion Phys.* **33** (1980) 139–158.

15. H. G. Dehmelt, Radiofrequency spectroscopy of stored ions I: Spectroscopy, in D. R. Bates (Ed.), *Advances in Atomic and Molecular Physics*, Vol. 5, Academic, New York, 1969, pp. 109–154.

16. F. Vedel, J André, Influence of space charge on the computed statistical properties of stored ions cooled by a buffer gas in a quadrupole rf trap, *Phys. Rev. A* **29** (1984) 2098–2101.

17. S. Guan, A. G. Marshall, Equilibrium space charge distribution in a quadrupole ion trap, *J. Am. Soc. Mass Spectrom.* **5** (1994) 64–71.

18. G.-Z. Li, S. Guan, A. G. Marshall, Comparison of equilibrium ion density distribution and trapping force in Penning, Paul, and combined ion traps, *J. Am. Soc. Mass Spectrom.* **9** (1998) 473–481.

19. R. E. March, R. J. Hughes, J. F. J. Todd, *Quadrupole Storage Mass Spectrometry*, Wiley Interscience, New York, 1989.

20. G. C. Stafford, D. M. Taylor, S. C. Bradshaw, J. E. P. Syka, M. Uhrich, Enhanced sensitivity and dynamic range on an ion trap mass spectrometer with automatic gain control (AGC). Proc. 35th Ann. ASMS Conf. on Mass Spectrometry and Allied Topics, Denver, CO, May 24–29, 1987, pp. 775–776.

21. G. C. Stafford, Jr., D. M. Taylor, S. C. Bradshaw, Method of mass analysing a sample, European Patent 0,237,268 (1991).

22. G. C. Stafford, Jr., D. M. Taylor, S. C. Bradshaw, Method of increasing the dynamic range and sensitivity of a quadrupole ion trap mass spectrometer, U.S. Patent 5,107,109 (1992).

23. J. W. Hager, Method of reducing space charge in a linear ion trap mass spectrometer, U.S. Patent 6,627,876 (2003).

24. J. F. J. Todd, R. M. Waldren, R. F. Bonner, The quadrupole ion store (QUISTOR). Part VIII. The theoretical estimation of ion kinetic energies: A comparative survey of the field, *Int. J. Mass Spectrom. Ion Phys.* **34** (1980) 17–36.

25. R. M. Waldren, J. F. J. Todd, The use of matrix methods and phase-space dynamics of the modelling of RF quadrupole-type device performance, in D. Price and J. F. J. Todd (Eds.), *Dynamic Mass Spectrometry*, Vol. 5, Heyden, London, 1978, pp. 14–40.

26. J. F. J. Todd, D. A. Freer, R. M. Waldren, The quadrupole ion store (QUISTOR). Part XI. The model ion motion in a pseudo-potential well: An appraisal in terms of phase-space dynamics, *Int. J. Mass Spectrom. Ion Phys.* **36** (1980) 185–203.

27. F. von Busch, W. Paul, Isotope separation by an electrical mass filter, *Z. Phys.* **164** (1961) 581–587.

28. G. Lüders, Über den einfluss von fehlern de magnetischen feldes auf die betatron-schwingungen in synchrotron mit starker stabilisierung, *Suppl. Nuovo Cimento (Series X)* **2** (1955) 1075–1146.

29. F. von Busch, W. Paul, Non-linear resonances in the electric mass filter as a consequence of field errors, *Z. Phys.* **164** (1961) 588–594.

30. R. Hagedorn, *Stability and Amplitude Ranges of Two Dimensional Non-Linear Oscillations with Periodical Hamiltonian Applied to Betatron Oscillations in Circular Particle Accelerators, Part I: General Theory and Part II: Applications*, CERN Reports 57-1 (1957).

31. R. Hagedorn, A. Schoch, *Stability and Amplitude Ranges of Two Dimensional Non-Linear Oscillations with Periodical Hamiltonian Applied to Betatron Oscillations in Circular Particle Accelerators, Parts III: Non-Linear Resonance Curves and Maximum Amplitudes for 3rd Order Subresonance*, CERN Reports 57-14 (1957).

32. A. Schoch, *Theory of Linear and Non-Linear Perturbations of Betatron Oscillations in Alternating Gradient Synchrotrons, Appendix I: Equations of Motion in Alternating Gradient Synchrotrons*, CERN Reports 57-021 (1957).

33. M. Barbier, A. Schoch, *Study of Two-Dimensional Non-Linear Oscillations by Means of an Electromechanical Analogue Model, Applied to Particle Motion in Circular Accelerators*, CERN Reports 58-5 (1958).

34. P. H. Dawson, N. R. Whetten, Non-linear resonances in quadrupole mass spectrometers due to imperfect fields I. The quadupole ion trap, *Int. J. Mass Spectrom. Ion Phys.* **2** (1969) 45–59.

35. P. H. Dawson, N. R. Whetten, Ion storage in three-dimensional, rotationally symmetric, quadrupole fields. II. A sensitive mass spectrometer, *J. Vac. Sci. Technol.* **5** (1968) 11–18.

36. Y. Wang, J. Franzen, The non-linear resonance QUISTOR. Part 1. Potential distribution in hyperboloidal QUISTORs, *Int. J. Mass Spectrom. Ion Processes* **112** (1992) 167–178.

37. Y. Wang, J. Franzen, K. P. Wanczek, The non-linear resonance ion trap. Part 2. A general theoretical analysis, *Int. J. Mass Spectrom. Ion Processes* **124** (1993) 125–144.

38. P. H. Dawson, N. R. Whetten, Non-linear resonances in quadrupole mass spectrometers due to imperfect fields. II. The quadrupole mass filter and the monopole mass spectrometer, *Int. J. Mass Spectrom. Ion Phys.* **3** (1969) 1–12.

39. M. H. Friedman, A. L. Yergey, J. E. Campana, Fundamentals of ion motion in electric radio-frequency multipole field, *J. Phys. E Sci. Instrum.* **15** (1982) 53–61.

40. I. Szabo, New ion-optical devices utilizing oscillatory electric fields. I. Principle of operation and analytical theory of multipole devices with two-dimensional electric fields, *Int. J. Mass Spectrom. Ion Processes* **73** (1986) 197–235.

41. C. Hägg, I. Szabo, New ion-optical devices utilizing oscillatory electric fields. II. Stability of ion motion in a two-dimensional hexapole field, *Int. J. Mass Spectrom. Ion Processes* **73** (1986) 237–275.

42. C. Hägg, I. Szabo, New ion-optical devices utilizing oscillatory electric fields. III. Stability of ion motion in a two-dimensional octopole field, *Int. J. Mass Spectrom. Ion Processes* **73** (1986) 277–294.

43. C. Hägg, I. Szabo, New ion-optical devices utilizing oscillatory electric fields. IV. Computer simulations of the transport of an ion beam through an ideal quadrupole, hexapole, and octopole operating in the RF-only mode, *Int. J. Mass Spectrom. Ion Processes* **73** (1986) 295–312.

44. S. C. Davis, B. Wright, Computer modelling of fragmentation processes in radio frequency multipole collision cells, *Rapid Commun. Mass Spectrom.* **4** (1990) 186–197.

45. J. Franzen, Simulation study of an ion cage with superimposed multipole fields, *Int. J. Mass Spectrom. Ion Processes* **106** (1991) 63–78.

46. Y. Wang, J. Franzen, The non-linear ion trap. Part 3. multipole components in three types of practical ion trap, *Int. J. Mass Spectrom. Ion Processes* **132** (1994) 155–172.

47. J. Franzen, R.-H. Gabling, M. Schubert, Y. Wang, Nonlinear ion traps, in R. E. March and J. F. J. Todd (Eds.), *Practical Aspects of Ion Trap Mass Spectrometry*, Vol. 1, CRC Press, Boca Raton, FL, 1995, Chapter 3.

48. J. D. Jackson, *Classical Electrodynamics,* 3rd ed., Wiley, New York, 1999, Chapter 3.

49. W. Mo, M. L. Langford, J. F. J. Todd, Investigation of "ghost" peaks caused by non-linear fields in the ion trap mass spectrometer, *Rapid Commun. Mass Spectrom.* **9** (1995) 107–113.

50. R. Alheit, D. Kleineidam, F. Vedel, M. Vedel, G. Werth, Higher order non-linear resonances in a Paul trap, *Int. J. Mass Spectrom. Ion Processes* **154** (1996) 155–169.

51. J. F. J. Todd, A. D. Penman, R. D. Smith, Alternative scan modes for mass range extension of the ion trap, *Int. J. Mass Spectrom. Ion Processes* **106** (1991) 117–136.

52. F. Guidugli, P. Traldi, A phenomenological description of a black hole for collisionally induced decomposition products in ion-trap mass spectrometry, *Rapid Commun. Mass Spectrom.* **5** (1991) 343–348.

53. K. L. Morand, S. A. Lammert, R. G. Cooks, Concerning "black holes" in ion-trap mass spectrometry, *Rapid Commun. Mass Spectrom.* **5** (1991) 491.

54. J. D. Williams, H.-P. Reiser, R. E. Kaiser, Jr., R. G. Cooks, Resonance effects during ion injection into an ion trap mass spectrometer, *Int. J. Mass Spectrom. Ion Processes* **108** (1991) 199–219.

55. F. Guidugli, P. Traldi, A. M. Franklin, M. L. Langford, J. Murrell, J. F. J. Todd, Further thoughts on the occurrence of "black holes" in ion-trap mass spectrometry, *Rapid Commun. Mass Spectrom.* **6** (1992) 229–231.

56. A. D. Penman, J. F. J. Todd, D. A. Thorner, R. D. Smith, The use of dynamically programmed scans to generate parent-ion tandem mass spectra with the ion-trap mass spectrometer, *Rapid Commun. Mass Spectrom.* **4** (1990) 108–113.

57. A. M. Franklin, J. F. J. Todd, Dynamically programmed scans, in R. E. March and J. F. J. Todd (Eds.), *Practical Aspects of Ion Trap Mass Spectrometry*, Vol. 3, CRC Press, Boca Raton, FL, 1995, Chapter 11.

58. R. E. March, A. W. McMahon, F. A. Londry, R. J. Alfred, J. F. J. Todd, F. Vedel, Resonance excitation of ions stored in a quadrupole ion trap. Part 1. A simulation study, *Int. J. Mass Spectrom. Ion Processes* **95** (1989) 119–156.

59. R. E. March, F. A. Londry, R. L. Alfred, A. D. Penman, J. F. J. Todd, F. Vedel, M. Vedel, Resonance excitation of ions stored in a quadrupole ion trap. Part 2. Further simulation studies, *Int. J. Mass Spectrom. Ion Processes* **110** (1991) 159–178.

60. C. Paradisi, J. F. J. Todd, P. Traldi, U. Vettori, Boundary effects and collisional activation in a quadrupole ion trap, *Org. Mass Spectrom.* **27** (1992) 251–254.

61. D. M. Eades, J. V. Johnson, R. A. Yost, Nonlinear resonance effects during ion storage in a quadrupole ion trap, *J. Am. Soc. Mass Spectrom.* **4** (1993) 917–929.

62. W. E. Austin, A. E. Holme, J. H. Leck, The mass filter: design and performance, in P. H. Dawson (Ed.), *Quadrupole Mass Spectrometry and Its Applications*, Elsevier: Amsterdam, 1976. Chapter VI. Reprinted as an "American Vacuum Society Classic" by the American Institute of Physics. (ISBN 1563964554)

63. M. Sudakov, D. J. Douglas, Linear quadrupoles with added octopole fields, *Rapid Commun. Mass Spectrom.* **17** (2003) 2290–2294.

64. C. Ding, N. V. Kononkov, D. J. Douglas, Quadrupole mass filters with octopole fields, *Rapid Commun. Mass Spectrom.* **17** (2003) 2495–2502.

65. F. Londry, B. A. Collings, W. R. Stott, Fragmentation of ions by resonant excitation in a high order multipole field, low pressure ion trap, U.S. Patent Application No. 2003/0189171 (2003).

66. J. Franzen, G. Weiss, Nonlinear resonance ejection from linear ion traps, U.S. Patent Application No. 2004/0051036 (2004).

67. J. E. P. Syka, Commercialization of the quadrupole ion trap, in R. E. March and J. F. J. Todd (Eds.), *Practical Aspects of Ion Trap Mass Spectrometry*, Vol. 1, CRC Press, Boca Raton, FL, 1995, Chapter 4; J. E. P. Syka, The geometry of the finnigan ion trap: History and theory, paper presented at the Ninth Asilomar Conference on Mass Spectrometry, September 1992.

68. R. E. March, Quadrupole ion trap mass spectrometry: A view at the turn of the century, *Int. J. Mass Spectrom.* **200** (2000) 285–312.

69. X. Wang, D. K. Bohme, R. E. March, Extension of the mass range of a commercial ion trap using monopolar resonance ejection, *Can. J. Appl. Spectrosc.* **38**(2) (1993) 55–60.

70. J. Louris, unscheduled announcement on ion trap geometry, at the 40th Ann. ASMS Conf. on Mass Spectrometry and Allied Topics, Washington, DC, May 31–June 5, 1992.

71. R. E. Kaiser, R. G. Cooks, G. C. Stafford, J. E. P. Syka, P. H. Hemberger, Operation of a quadrupole ion trap mass spectrometer to achieve high mass/charge ratios, *Int. J. Mass Spectrom. Ion Processes* **106** (1991) 79–115.

72. P. Traldi, O. Curcuruto, O. Bortolini, Mass displacement in ion trap mass spectrometry: A unique and valuable tool in ion structural studies, *Rapid Commun. Mass Spectrom.* **6** (1992) 410–412.

73. O. Bortolini, S. Catinella, P. Traldi, Mass displacements in ion trap mass spectrometry: Can they be related to electronic properties of the substitutent groups of the ions under investigation? *Org. Mass Spectrom.* **27** (1992) 927–928.

74. P. Traldi, D. Favretto, S. Catinella, O. Bortolini, Mass displacements in quadrupole field analysers, *Org. Mass Spectrom.* **28** (1993) 745–751.

75. O. Bortolini, P. Traldi, Evaluation of the polarizability of gaseous ions, in R. E. March and J. F. J. Todd (Eds.), *Practical Aspects of Ion Trap Mass Spectrometry*, Vol. II, CRC Press, Boca Raton, FL, 1995, Chapter 4.

76. O. Bortolini, S. Catinella, P. Traldi, Estimation of the polarizability of organic ions by ion trap measurements, *Org. Mass Spectrom.* **28** (1993) 428–432.

77. O. Bortolini, S. Catinella, P. Traldi, Evalutation of dipole moments of organic ions in the gas phase, *Org. Mass Spectrom.* **29** (1994) 273–276.

78. P. Traldi, D. Favretto, Mass displacements in a quadrupole mass filter: A tool for evaluation of intrinsic dipole moments of ionic species, *Rapid Commun. Mass Spectrom.* **6** (1992) 543–544.

79. W. R. Plass, H. Li, R. G. Cooks, Theory, simulation and measurement of chemical mass shifts in RF quadrupole ion traps, *Int. J. Mass Spectrom.* **228** (2003) 237–267.

80. N. R. Whetten, Macroscopic particle motion in quadrupole fields, *J. Vac. Sci. Technol.* **11** (1974) 551–518.

81. F. G. Major, H. G. Dehmelt, Exchange-collision technique for the rf spectroscopy of stored ions, *Phys. Rev.* **170** (1968) 91–107.

82. P. H. Dawson, N. R. Whetten, The three-dimensional quadrupole ion trap, *Naturwissenschaften* **56** (1969) 109–112.

83. R. Blatt, U. Schmeling, G. Werth, On the sensitivity of ion traps for spectroscopic applications, *Appl. Phys.* **20** (1979) 295–298.

84. P. H. Dawson, C. Lambert, High-pressure characteristics of the quadrupole ion trap, *J. Vac. Sci. Techonol.* **12** (1975) 941–942.

85. P. H. Dawson, J. W. Hedman, N. R. Whetten, A simple mass spectrometer, *Rev. Sci. Instrum.* **40** (1969) 1444–1450.

86. R. F. Bonner, G. Lawson, J. F. J. Todd, Ion-molecule reactions studies with a quadrupole ion storage trap, *Int. J. Mass Spectrom. Ion Phys.* **10** (1972) 197–203.

87. R. F. Bonner, R. E. March, J. Durup, Effect of charge exchange reactions on the motion of ions in three-dimensional quadrupole electric fields, *Int. J. Mass Spectrom. Ion Phys.* **22** (1976) 17–34.

88. G. C. Stafford, Jr., P. E. Kelley, J. E. P. Syka, W. E. Reynolds, J. F. J. Todd, Recent improvements in an analytical application of advanced ion trap technology, *Int. J. Mass Spectrom. Ion Processes* **60** (1984) 85–98.

89. G. C. Stafford, P. E. Kelley, D. R. Stephens, Method of mass analyzing a sample by use of a quadrupole ion trap, U.S. Patent 4,540,884, Claim 17 (1985).

90. J. S. Brodbelt, Effects of collisional cooling on detection, in R. E. March and J. F. J. Todd (Eds.), *Practical Aspects of Ion Trap Mass Spectrometry*, Vol. 1, CRC Press, Boca Raton, FL, 1995, Chapter 5.

91. M. N. Kishore, P. K. Ghosh, Trapping of ions injected from an external source into a three-dimensional RF quadrupole field, *Int. J. Mass Spectrom. Ion Phys.* **29** (1979) 345–350.

92. J. F. J. Todd, D. A. Freer, R. M. Waldren, The quadrupole ion store (QUISTOR). Part XII. The trapping of ions injected from an external source: A description in terms of phase-space dynamics, *Int. J. Mass Spectrom. Ion Phys.* **36** (1980) 371–386.

93. C.-S. O, H. A. Schuessler, Confinement of pulse-injected external ions in a radiofrequency quadrupole ion trap, *Int. J. Mass Spectrom. Ion Phys.* **40** (1981) 53–66.

94. C.-S. O, H. A. Schuessler, Confinement of ions injected into a radiofrequency quadrupole ion trap: pulsed ion beams of different energies, *Int. J. Mass Spectrom. Ion Phys.* **40** (1981) 67–75.

95. C.-S. O, H. A. Schuessler, Confinement of ions injected into a radiofrequency quadrupole ion trap: Energy-selective storage of pulse-injected ions, *Int. J. Mass Spectrom. Ion Phys.* **40** (1981) 77–86.

96. M. Ho, R. J. Hughes, E. Kazdan, P. J. Mathews, A. B. Young, R. E. March, Isotropic collision induced dissociation studies with a novel hybrid instrument. Proc. 32nd Ann. ASMS Conf. on Mass Spectrometry and Allied Topics, San Antonio, TX, May 27–June 1, 1984, p. 513–514.

97. J. E. Curtis, A. Kamar, R. E. March, U. P. Schlunegger, An improved hybrid mass spectrometer for collisionally activated dissociation studies. Proc. 35th Ann. ASMS Conf. on Mass Spectrometry and Allied Topics, Denver, CO, May 24–29, 1987, pp. 237–238.

98. J. N. Louris, J. W. Amy, T. Y. Ridley, R. G. Cooks, Injection of ions into a quadrupole ion trap mass spectrometer, *Int. J. Mass Spectrom. Ion Processes* **88** (1989) 97–111.

99. S. T. Quarmby, R. A. Yost, Fundamental studies of ion injection and trapping of electrosprayed ions on a quadrupole ion trap, *Int. J. Mass Spectrom.* **190/191** (1999) 81–102.

100. D. J. Dahl, *SIMION 3D Version 6.0*, Idaho National Engineering Laboratory Chemical Materials and Processes Department, Lockheed Idaho Technologies Company, Idaho Falls, ID, 1995.

101. C.-S. O, H. A. Scheussler, Confinement of ions created externally in a radio-frequency ion trap, *J. Appl. Phys.* **52** (1981) 1157–1166.

102. D. B. Tucker, C. H. Hameister, S. C. Bradshaw, D. J. Hoekman, M. Weber-Grabau, The application of novel ion trap scan modes for high sensitivity GC/MS. Proc. 36th Ann. ASMS Conf. on Mass Spectrometry and Allied Topics, San Francisco, CA, June 5–10, 1988, pp. 628–629.

103. P. E. Kelley, Mass spectrometry method using notch filter, U.S. Patent 5,134,286 (1992).

104. D. E. Goeringer, K. G. Asano, S. A. McLuckey, D. Hoekman, S. W. Stiller, Filtered noise field signals for mass-selective accumulation of externally formed ions in a quadrupole ion trap, *Anal. Chem.* **66** (1994) 313–318.

105. N. A. Yates, M. M. Booth, J. L. Stephenson, Jr., R. A. Yost, Practical ion trap technology: GC/MS and GC/MS/MS, in R. E. March and J. F. J. Todd (Eds.), *Practical Aspects of Ion Trap Mass Spectrometry*, Vol. III, CRC Press, Boca Raton, FL, 1995, Chapter 4.

106. J. N. Louris, R. G. Cooks, J. E. P. Syka, P. E. Kelley, G. C. Stafford, Jr., J. F. J. Todd, Instrumentation, applications and energy deposition in quadrupole ion trap MS/MS spectrometry, *Anal. Chem.* **59** (1987) 1677–1685.

107. R. M. Danell, A. S. Danell, G. L. Glish, R. W. Vachet, The use of static pressures of heavy gases within a quadrupole ion trap, *J. Am. Soc. Mass Spectrom.* **14** (2003) 1099–1109.

108. R. E. March and J. F. J. Todd (Eds.), *Practical Aspects of Ion Trap Mass Spectrometry*, Vol. III, CRC Press, Boca Raton, FL, 1995.

109. C. Paradisi, J. F. J. Todd, P. Traldi, U. Vettori, Boundary effects and collisional activation in a quadrupole ion trap, *Org. Mass Spectrom.* **27** (1992) 251–254.

110. P. Traldi, S. Catinella, R. E. March, C. S. Creaser, Boundary excitation, in R. E. March and J. F. J. Todd (Eds.), *Practical Aspects of Ion Trap Mass Spectrometry*, Vol. 1, CRC Press, Boca Raton, FL, 1995, Chapter 7.

111. S. A. Lammert, R. G. Cooks, Surface-induced dissociation of molecular ions in a quadrupole ion trap mass spectrometer, *J. Am. Soc. Mass Spectrom.* **2** (1991) 487–491.

4

SIMULATION OF ION TRAJECTORIES IN THE QUADRUPOLE ION TRAP

4.1. INTRODUCTION

An early stimulus to calculate the trajectory of an ion in a QIT was the attractive photomicrograph of Wuerker et al. [1] showing a charged macroparticle of aluminum dust that had executed a Lissajou-like figure-of-eight (or infinity symbol) trajectory in the r–z plane. A 2:1 Lissajous trajectory was observed upon which was superimposed the drive frequency Ω. Application of a small auxiliary potential across the end-cap electrodes permitted visual measurement of the resultant frequency of ion motion in the z direction in resonance with the applied potential. Initially, it was reported [1] that ion motion was violent and mixed up; however, when the background pressure was increased to several micrometers, thus dissipating the initial particle kinetic energy, the particles were seen to take up stable arrays. Here was proof that the trajectory of an ion could sample the major part of the storage volume of an ion trap as it moved in three dimensions in response to the applied RF field, that increased ion trajectory excursions in the z direction occurred as a result of ion axial resonance, and that ion trajectories were modified by ion/neutral collisions. An objective was then to calculate the trajectory so that the simulated trajectory looked like (or at least resembled) the behavior of the charged macroparticle trajectory.

There are several aspects to the simulation of ion trajectories. The simplest simulation is to calculate the trajectory of a single ion in an ideal quadrupole field in the absence of buffer gas, as is shown in Figure 4.1. The three-dimensional trajectory indicated with an arrow is, indeed, a rippled Lissajou figure of which the outline resembles a boomerang. The figures-of-eight are seen more clearly in the trajectory projections onto the x–z and y–z planes. The trajectory projection onto the x–y plane is an ellipse. Figure 4.1 demonstrates the problem of the presentation of the great deal of data obtained in the simulation of the trajectory; while all of the information is contained in the trajectory indicated with an arrow in Figure 4.1, comprehension of the data requires planar projections, as shown together with, for example, temporal variations of ion excursions and kinetic energies in each of the x, y, and z directions.

The single ion trajectory simulation can be repeated for any a_z, q_z coordinate or working point within (or beyond) the stability diagram. Resonance excitation (or "tickling") of ions can be explored by the addition of a supplementary AC potential to the RF drive potential. Application of, say, a negative-going supplementary potential to the RF drive potential applied to the ring electrode is equivalent electrically to the application of a positive-going supplementary potential to each of the end-cap electrodes. Resonant excitation carried out in this manner is described as quadrupolar excitation and can be simulated readily. Such is not the case for dipolar excitation, where equal though out-of-phase supplementary potentials are applied to each of the end-cap electrodes (see below). The next major problem is to incorporate the effects of ion/neutral collisions so as to simulate the collisional process of momentum dissipation and concomitant ion focusing to the center of the ion trap. Thereafter, resonant excitation and resonant ejection may be simulated. All of the above simulation stages may be repeated within a quadrupole field that is no longer ideal in that higher-order terms are present due to electrode imperfections and truncation, and to the presence of orifices for injection and ejection of charged species.

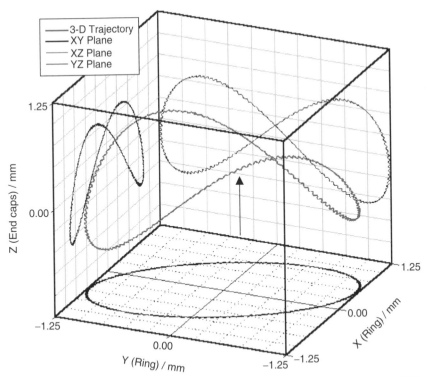

Figure 4.1. Two- and three-dimensional representations of trajectory of an ion (m/z 100) at $q_z = 0.2$ for a period of 125 μs generated by SIMION 3D. The three-dimensional trajectory (indicated by arrow) is projected on each of the x–y, x–z, and y–z planes. *Macro*motion (Lissajous figure) and *micro*motion (minute oscillation) components of ion's trajectory are clearly visible. (Reprinted from the *Journal of Mass Spectrometry*, vol. 34, M. W. Forbes, M. Sharifi, T. R. Croley, Z. Lausevic, R. E. March, "Simulation of ion trajectories in a quadrupole ion trap: A comparison of three simulation programs," Plate 1, 1219–1239 (1999). © John Wiley & Sons Limited. Reproduced with permission.)

The next logical step is to repeat the above simulations in traps of different geometry such as cylindrical ion traps.

While much useful information can be gleaned from the study of single ion trajectories, the investigation of spatial and energy distributions requires consideration of the advent of enhanced computational facilities that permit the simultaneous calculation of the trajectories of more than 600,000 ions. In such calculations, a wide range of mass/charge ratios can be accommodated so as to permit the simulation of mass spectra and ion kinetic energy distributions. Simulation of the trajectories of a large number of ions permits the evaluation of space-charge effects. Ion trajectories through complete instruments, from ion source through the ion trap to the detector, can now be simulated together with the in-trap processes of charge exchange, chemical ionization, cluster ion formation (as in proton-bound dimers), and ion fragmentation.

Simulation studies prior to 1983 have been discussed in Chapter 3, Section 3.2 of the first edition of this work [2]. Here, we present an introduction to the theory underlying the various types of ion trajectory simulations that may be carried out, from ion trajectories in a pure quadrupole field in the absence of collisions with neutral particles to relatively complex simulations of resonantly-excited ion trajectories in nonideal ion traps in the presence of buffer gas.

4.2. RECENT APPLICATIONS OF SIMULATIONS

There has been little activity in the development of new simulation programs because the use of commercial packages for the calculation of ion trajectories has largely precluded research in this area. An excellent review of the application of simulations to the understanding of the QIT has been given in Ref. 3. The Cooks group has carried out multiparticle simulations [4], a simulation study of nondestructive detection of ions in a QIT using a DC pulse to force coherent ion motion [5], and a multiparticle simulation of ion injection under the influence of helium buffer gas using short injection times and DC pulse potentials [6]. Yoshinari [7] has carried out a numerical analysis of the behavior of ions injected into an ion trap, and the March group has reported a comparison of three simulation programs, ITSIM, SIMION, and ISIS (see below) for the calculation of ion trajectories in an ion trap [8]. André and co-workers have carried out simulation studies of a new operating mode for a QIT [9]. A simulation study of the effect of geometric aberration, space charge, dipolar excitation, and damping on ion axial secular frequencies in a nonlinear Paul trap has been reported by Sevugarajan and Menon [10]. A user program for SIMION has been developed for the simulation of ion trajectories in a digital ion trap (see Chapter 8). Three-dimensional random ion/neutral collisions were modeled with a Maxwellian velocity distribution for the buffer gas. A mass-selective ion ejection scan method is described and simulated, where both the trapping quadrupole field and the dipolar excitation field are driven digitally and their frequencies are scanned proportionally. In addition, simulation studies were carried out of the influence of space charge, duty cycle modulation as an alternative ejection scan method, and ion injection into the ion trap [11].

4.3. THEORETICAL BACKGROUND

The calculation of ion trajectories within an ideal quadrupole field and quadrupolar excitation of ion trajectories can be carried out by direct integration of the Mathieu equation. To proceed beyond this stage to the calculation of ion trajectories in an ion trap wherein the field is nonideal or to simulate the resonant excitation of ion trajectories by dipolar excitation, it is necessary to calculate the field throughout the ion trap. Once the means to calculate the field have been achieved, ion trajectories can be calculated for a wide variety of ion traps and under a range of conditions. Approaches must be developed for the simulation of collisions of an ion with neutral species and later for the simulation of ion–ion interactions. Once these developments have been made, simulations can be extended to ensembles of many ions for the simulation of, for example, mass spectra.

4.3.1. Numerical Integration of the Mathieu Equation

It is possible to calculate the trajectory of a charged particle in an ideal quadrupole field by numerical integration of the Mathieu equation. This method offers the advantage that the details of the trajectory can be calculated directly at any point in time with great accuracy. Such simulation studies are carried out generally for the condition where the RF drive potential is applied to the ring electrode and the end-cap electrodes are grounded. Because the simulation of resonance excitation in quadrupolar mode involves the simple addition of the auxiliary potential to the RF drive potential, resonantly-excited ion trajectories can be calculated directly by numerical integration of the Mathieu equation also.

In the calculation of an ion trajectory in an ideal QIT of known dimension r_0, one must define the initial positions of ion position and velocity $(x, y, z, \dot{x}, \dot{y}, \dot{z})$, the RF drive amplitude (zero-to-peak) V of phase ξ_0 and radial frequency Ω, and the mass m of the singly-charged ion. Once the initial conditions have been determined, the ionic parameters a_z, q_z, β_z, C_{2n}, A, and B may be calculated, as described in Chapter 2. A simulation program for quadrupolar resonance, such as SPQR [12, 13], assumes that the potential applied to the ring electrode, Φ_0^R, is of the form

$$\Phi_0^R = U + V\cos(\Omega t + \xi_0) \tag{4.1}$$

and the potential applied in phase (quadrupolar mode) to the end-cap electrodes, Φ_0^E, is of the form

$$\Phi_0^E = U' + V'(\Omega' t + \xi_0') \tag{4.2}$$

where U' is zero and V' is the amplitude of the tickle voltage of frequency Ω' and of phase ξ_0'. The potential obtained is a form of Eq. (2.79):

$$\Phi(r, \phi, z) = A_0^0 + A_1^0 z + A_2^0(\tfrac{1}{2}r^2 - z^2) + A_3^0 z(\tfrac{3}{2}r^2 - z^2) + A_4^0(\tfrac{3}{8}r^4 - 3r^2 z^2 + z^4) + \cdots \tag{4.3a}$$

where the values $n = 0$, 1, 2, 3, 4 correspond to the monopole, dipole, quadrupole, hexapole, and octopole components, respectively, of the potential field Φ. When Eq. (4.3a) is expressed in Cartesian coordinates, one obtains

$$\Phi(x, y, z) = A_0^0 + A_1^0 z + A_2^0[\tfrac{1}{2}(x^2 + y^2) - z^2] + A_3^0 z[\tfrac{3}{2}(x^2 + y^2) - z^2]$$
$$+ A_4^0[\tfrac{3}{8}(x^4 + 2x^2 y^2 + y^4) - 3z^2(x^2 + y^2) + z^4] + \cdots. \tag{4.3b}$$

By substitution of the potential from Eq. (4.3b) into Eq. (4.4) and analogous equations in y and z [14],

$$\frac{d^2x}{dt^2} = -\frac{e}{m}\frac{\partial\phi}{\partial x} \tag{4.4}$$

the equations of motion for a singly-charged positive ion, expressed in Cartesian coordinates, are obtained as

$$\frac{d^2x}{dt^2} + \frac{e}{m}[A_2^0 + 3A_3^0 z + A_4^0(\tfrac{3}{2}x^2 + \tfrac{3}{2}y^2 - 6z^2)]x = 0 \tag{4.5}$$

$$\frac{d^2y}{dt^2} + \frac{e}{m}[A_2^0 + 3A_3^0 z + A_4^0(\tfrac{3}{2}x^2 + \tfrac{3}{2}y^2 - 6z^2)]y = 0 \tag{4.6}$$

and

$$\frac{d^2z}{dt^2} + \frac{e}{m}[A_1^0 - 2A_2^0 z + \tfrac{3}{2}A_3^0(x^2 + y^2 - 2z^2) - 6A_4^0(x^2 + y^2 - \tfrac{3}{2}z^2)]z = 0. \tag{4.7}$$

Ion trajectories are calculated by numerically integrating Eqs. (4.5)–(4.7) for which the three second-order differential equations are cast in the form of six first-order differential equations; for example, Eq. (4.5) can be expressed as

$$\frac{dx}{dt} = \dot{x} \quad \text{and} \quad \frac{d\dot{x}}{dt} = -\frac{e}{m}[A_2^0 + 3A_3^0 z + A_4^0(\tfrac{3}{2}x^2 + \tfrac{3}{2}y^2 - 6z^2)]x. \tag{4.8}$$

The resulting set of six first-order differential equations are solved using the Bulirsch–Stoer method for numerical integration with adaptive step size and energy control [15]. With this technique, it is possible to take large time steps of the order of one cycle of the RF drive potential. In general, large time steps can be used during relatively uninteresting segments of a simulation while smaller time steps, ranging from one-twentieth to one-fiftieth of an RF cycle, are used for segments where details of the ion's motion are sought.

In this manner, many informative simulations can be carried out that will show the wide variety of ion trajectories, resonance excitation, relative efficiencies of resonance excitation at different frequencies, ion isolation at the upper apex of the stability diagram, and differentiation between ion trajectories that are unstable and those for which the maximum excursions exceed the dimensions of the device. From the theoretical treatment above, it can be seen that regardless of the percentage of hexapole and octupole field components that are present in the field within the ion trap, an analytic, differentiable expression can be expressed that describes exactly the potential for quadrupolar auxiliary-potential application. This condition does not hold for the monopolar and dipolar resonance excitation modes; that is, a closed expression cannot be obtained for the force experienced by an ion within the ion trap when subjected to either monopolar or dipolar resonance excitation.

4.3.2. Calculation of Electrostatic Fields

For the simulation of ion trajectories in a nonideal quadrupole ion trap of specified geometry and deviations of the electrodes from ideality (by virtue of electrode truncation, electrode perforations, etc.), the electrostatic potential at each point in the ion trap must be determined. The two methods by which such determinations can be made are the direct method of electrostatic field calculation and the interpolation of the potential using overrelaxation techniques.

4.3.2.1. Direct Method   The direct method is by far the faster of the two methods in that direct calculation of electrostatic fields acting on an ion at every time step can be made by modern PCs equipped with high-speed processors while sacrificing little computational time. The direct method uses a multipole expansion to calculate the electric field. Potentials applied to the ring electrode (Φ_0) and any auxiliary potential (Φ_{aux}) have the general form given by Eqs. (4.1) and (4.2), respectively. For a symmetrical electrode configuration such as the QIT, an expression for the potential $\Phi(\rho, \theta, \phi)$ in spherical polar coordinates [Eq. (2.78)] becomes

$$\Phi(r, \phi, z) = A_0^0 + A_1^0 z + A_2^0 [\tfrac{1}{2}r^2 - z^2] + A_3^0 z [\tfrac{3}{2}r^2 - z^2]$$
$$+ A_4^0 [\tfrac{3}{8}r^4 - 3r^2 z^2 + z^4] + \cdots \qquad \text{[Eq. (4.3a)]}$$

upon expansion of Eq. (2.78) and expression in cylindrical polar coordinates. As previously described for the A_n^0 constants, the values $n = 0, 1, 2, 3, 4$ correspond to the monopole, dipole, quadrupole, hexapole, and octopole components, respectively, of the potential field Φ. For a pure quadrupole field, the constants A_3^0 and A_4^0 are zero; A_1^0 is also zero when the applied auxiliary potential is in quadrupolar mode (as opposed to the dipolar or monopolar mode). ITSIM makes all calculations of the potential using the direct method such that all of the higher-order fields associated with the nonlinear stretched ion trap (see Chapter 3) and dipolar auxiliary fields can be conquered.

The significance of Eq. (4.3a) is that the electric potential (E) at any point in the ion trap can be determined from the negative gradient of the electric potential Φ:

$$E = -\nabla\Phi \tag{4.9}$$

$$E_\rho = -[A_2^0 \rho + 3A_3^0 \rho z + A_4^0(\tfrac{3}{2}\rho^3 - 6\rho z^2)] \tag{4.10}$$

$$E_z = -[A_1^0 - 2A_2^0 z - 3A_3^0 z^2 + A_3^0 - 3A_3^0 z^2 - 2A_4^0(3\rho^2 z - 2z^3)]. \tag{4.11}$$

4.3.2.2. Matrix Field Interpolation   A second method of field calculation is known as matrix field interpolation wherein the electrostatic forces in an ion trap are calculated by an indirect method. The basic process involves the software creating, in memory, a three-dimensional matrix. Each member of the array corresponds to a potential (P) in the ion trap. Once the matrix has been established, the potential array must be refined in that the potential on each nonelectrode pixel in the array is calculated. The method by which the potential is refined is an iterative process known as the overrelaxation method. The potential on each nonelectrode pixel must satisfy the Laplace condition [16] that the electrostatic gradient must be zero for all pixels in the array. The electrostatic potential at each pixel is interpolated from its nearest neighbors in the array, as is illustrated in Figure 4.2.

Typically, for the QIT having cylindrical symmetry, there are five unique types of pixels, arising from their location and the geometry of the ion trap:

1. points on the electrodes;
2. interior points that are equally spaced from the neighboring points, as in Figure 4.2;
3. points adjacent to an electrode;
4. points on the symmetry axis that are equally spaced from the neighboring points; and
5. points on the symmetry axis that are also adjacent to an electrode.

In contrast, SIMION [17] identifies six types of pixels:

1. interior points,
2. left/right-edge points,
3. interior axis points,

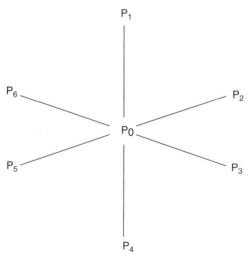

Figure 4.2. Electrostatic potential at pixel P_0 is calculated from potentials on each neighboring pixel $(P_1 - P_6)$. In each iteration of the overrelaxation method, every pixel in array is calculated as for P_0 and iterations continue until the voltage calculated for every P_0 changes by less than the specified value. (Reprinted from the *Journal of Mass Spectrometry*, vol. 34, M. W. Forbes, M. Sharifi, T. R. Croley, Z. Lausevic, R. E. March, "Simulation of ion trajectories in a quadrupole ion trap: A comparison of three simulation programs," Fig. 5, 1219–1239 (1999). © John Wiley & Sons Limited. Reproduced with permission.)

4. corner axis points,

5. upper corner points, and

6. top-edge points.

The gradient on each pixel in the array is calculated in one step of the overrelaxation process and the changes made in each step are propagated to the next iteration. This process continues until all of the pixels come into equilibrium and do not change by more than a given potential gradient, the "convergence objective". At this stage, the electrostatic field at each point in the ion trap is known and can be used to calculate the trajectory of an ion.

4.3.3. Computer Simulation Programs

Three computer programs have been used extensively for the calculation of ion trajectories; these programs are ION and electron optics SIMulation package (SIMION) [17], Ion Trajectory SIMulation (ITSIM) [18], and Integrated System for Ion Simulation (ISIS) [14]. ITSIM and ISIS are self-contained packages that were designed for and are restricted to simulations of ion trajectories within quadrupolar devices, particularly the QIT. SIMION 3D is a versatile software package that allows the user to simulate ion trajectories in virtually any electrostatic or magnetic field. It was designed initially for the study of ion optics and was developed later for application to the QIT.

4.3.3.1. ITSIM   ITSIM was developed in the laboratory of R. G. Cooks at Purdue University and has been used with great success in developing three-dimensional visual representations of ion trajectories [19] in addition to phase space and Poincaré plots. ITSIM appeared originally for a DOS-based PC environment but has since been made available in a Windows platform with an enhanced user interface and extended simulation capability. ITSIM has been used also for the study of the effects of helium buffer gas, RF phase angle and DC pulse potentials on ion-trapping efficiency during ion injection [20], as well as the trajectories of ions undergoing DC and resonant excitation [4].

4.3.3.2. ISIS   ISIS was developed in the laboratory of R. E. March at Trent University from a series of modules for calculating ion trajectories; these modules included a program for the direct integration of the Mathieu equation (MA), the field interpolation method (FIM), and a simulation program for quadrupolar resonance (SPQR) [14, 21]. ISIS has been used extensively in the study of kinetic energy effects [22–24] and DC and RF fields [25] during axial modulation as well as the method of mass-selective isolation [26] and frequency absorption analyses of resonantly excited ions [27].

4.3.3.3. SIMION   In that SIMION was based on the study of ion optics, it has been applied widely for the examination of ion trajectories in ion sources and beam instruments. SIMION has the means for the creation of custom electrode geometries controlled by user programs, that is, algorithms developed by the user to control the potentials applied to the electrodes. Charged particle trajectories through a concatenation of custom-designed electrodes can be modeled directly by SIMION which can take into account electrode truncation, holes in electrodes, and field penetration through such holes. SIMION has been used for the simulation of ion behavior in a number of devices, including QMFs [28], time-of-flight mass spectrometers [29], ion cyclotron resonance mass spectrometers, and QITs [29]. Doroshenko and Cotter have used SIMION extensively in their studies of ion injection [30] into a QIT, as have Arkin and Laude in comparisons of the axial electric fields generated by the hyperbolic, hybrid, and cylindrical ion traps [31]. In addition, SIMION has been shown to be powerful for modeling ion optics or lenses whereby externally-created ions are guided from an ion source such as an electrospray ionization apparatus (see Chapter 8) into an ion trap. Both simple Einzel lenses with static DC voltages [32] and complex ion funnels with RF voltages have been studied [33].

4.3.3.4. Dialogue and Operating Platform   Although all three programs run on conventional PCs and all are computationally rigorous, the most noticeable differences between SIMION 6.0, ITSIM 4.1 and ISIS are their characteristics in user platform. Each of the programs performs at a different rate and has different display functions. The time required to complete a given simulation varied significantly, with the dominating factor being the number of ions in the simulation. ITSIM and SIMION permit simultaneous calculations of the trajectories of ensembles of ions while ISIS performs successive single-ion trajectories. While clouds of thousands of ions can be simulated, in practice, SIMION will run at an acceptable rate of only some 100–200 ions.

ITSIM was written in C++ and makes use of Windows 32-bit memory-sharing capabilities [18]. ITSIM 4.1 combines input files for ion definition, voltage programming, simulation run time, and data output into a user-friendly visual interface that allows simulations to be run easily through single input and executable files. ITSIM allows the user both to define all of the experimental conditions prior to running the simulation and to change parameters during a simulation. ITSIM is a true multiparticle simulator and can be used to characterize both single-ion trajectories and those of large ensembles of ions in timely fashion.

ISIS runs in a DOS-based environment and the simulation parameters are compiled from a series of input files for scan functions, ion definition, and data output options. There is no option to view ion trajectory data online or in real time. The various modules of ISIS were written in either BASIC or FORTRAN [14] and are compiled separately.

SIMION 6.0 differs from ITSIM and ISIS in both design and user format. It offers an interactive dialogue in a DOS-based environment, but it is driven by a graphical user interface (GUI) of layered menus with mouse-activated buttons to guide the user from one option to another. While SIMION requires that the user must develop his or her own electrode designs and must write algorithms for the control of voltages and simulating collisions, it has the versatility to model virtually any type of mass spectrometer.

4.3.3.5. Electrode Design SIMION requires the user to create custom electrode geometries whereas ITSIM and ISIS are self-contained packages capable of calculating ion trajectories in the ideal, stretched (see Chapter 3), and cylindrical ion trap geometries. ITSIM 5.0 allows externally-generated ions to fly into the ion trap volume through holes in end-cap electrodes and permits the simulation of ion trajectories when ions are ejected through such holes and strike a plane of detection [34–36].

The first component of SIMION's electrode design involves producing a structure that approximates the shape and dimensions of a real QIT. Tools are available to create various geometric shapes, so once the dimensions of the ion trap are known, accurate hyperbolic electrodes can be created as given in Eqs. (2.45) and (2.46) for the ring and end-cap electrodes, respectively.

The second component of SIMION's electrode design is the continuity of the surface; while rigorous hyperbolas can be drawn, it is the resolution of the pixel array that determines the precision of the electrode surface. Each pixel consumes 10 bytes of conventional random access memory (RAM) such that a three-dimensional cubic array $100 \times 100 \times 100$ (10^6) pixels will be used and 10 Mbytes of RAM allotted in memory. Roughness in electrode design translates into roughness in the calculated field and may introduce a significant source of error into the calculated ion trajectory.

For the comparison [8] of the performances of ITSIM, ISIS, and SIMION (see below), the desired QIT geometry was created in SIMION, as illustrated in Figure 4.3.

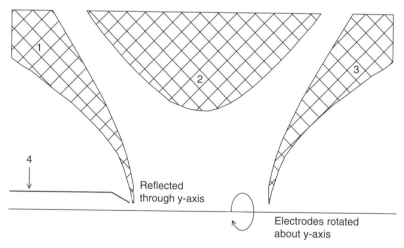

Figure 4.3. Electrode array created in SIMION to construct QIT. Electrodes 1 and 3 are entrance and exit end-cap electrodes, respectively; electrode 2 is ring electrode and electrode 4 is the ion gate used to inject ions into ion trap. The two-dimensional array was reflected through the *y* axis to generate a three-dimensional structure. (Reprinted from the *Journal of Mass Spectrometry*, vol. 34, M. W. Forbes, M. Sharifi, T. R. Croley, Z. Lausevic, R. E. March, "Simulation of ion trajectories in a quadrupole ion trap: A comparison of three simulation programs," Fig. 3, 1219–1239 (1999). © John Wiley & Sons Limited. Reproduced with permission.)

Four two-dimensional electrodes were created: two hyperbolic end-cap electrodes with holes at the center (electrodes 1 and 3), one hyperbolic ring electrode (electrode 2), and one conical ion gate at the entrance of the left end-cap electrode (electrode 4). The potential array was 900 pixels wide by 450 pixels high, reflected through the y axis, then rotated about the y axis to yield a three-dimensional array $900 \times 900 \times 900$ pixels. The design in Figure 4.3 was scaled (a SIMION option) by a factor of 4 mm/90 pixels in arriving at the ideal ion trap geometry specified as $r_0^2 = 2z_0^2$ with $r_0 = 10$ mm. Figure 4.4 is a schematic of the complete SIMION electrode array and shows a magnified view of the individual pixels comprising the electrodes; the electrodes of the ion trap have a resolution of 0.0444 mm per grid unit, as determined by the scaling factor. The significance of the grid resolution becomes apparent when the electrostatic fields in the ion trap are calculated.

Whereas ions can exist only within the bounds of the ion trap with ITSIM (Version 4.1) and ISIS, ions can be created outside of the ion trap with SIMION. Figure 4.4 illustrates how ions can be created at the mouth of the ion optic or gate and guided through the entrance end-cap electrode for subsequent confinement, that is, a simulation of external ion creation followed by ion injection and ion confinement.

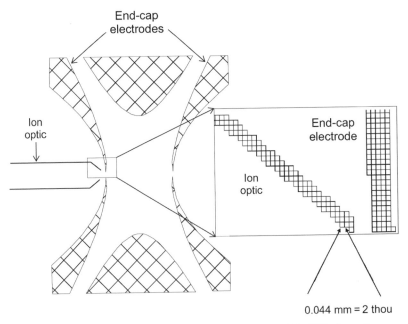

Figure 4.4. Analysis of the electrode resolution obtained in a SIMION electrode array following application of a scaling factor of 0.0444 mm/pixel. Pixel size corresponds to precision of 0.002 in. Note that the ratio of r_0 to z_0 has been distorted to accommodate enlargement of the electrode structure. (Reprinted from the *Journal of Mass Spectrometry*, vol. 34, M. W. Forbes, M. Sharifi, T. R. Croley, Z. Lausevic, R. E. March, "Simulation of ion trajectories in a quadrupole ion trap: A comparison of three simulation programs," Fig. 4, 1219–1239 (1999). © John Wiley & Sons Limited. Reproduced with permission.)

4.3.3.6. Scan Functions and User Programs Once the electrostatic potential matrix has been completed, the simulators apply the voltages to the electrodes in order to realize an operating mode of an instrument. While all three programs can be made to perform the same tasks, ITSIM and ISIS are designed with built-in, time-dependent modules tailored specifically to allow the user to perform complex manipulations, whereas SIMION requires the user to write *user programs* to control the voltages.

ITSIM is the most user friendly of the three programs and offers three options for voltage programming in the main menu. The RF drive voltage may be programmed with variable frequencies and amplitude ramping so as to simulate the basic mass-selective operation of the ion trap. The AC potentials may be applied with variable amplitude and frequency for performing axial modulation (see Chapter 3) and monopolar, dipolar, and quadrupolar excitation. In addition, DC potentials are available for the simulation of DC pulsing and DC trapping.

The voltage programming entails the construction of a scan table in which the user specifies the amplitude, frequency, and duration of application; a schematic of such a scan table wherein the temporal variation of all of the necessary potentials is portrayed is described as a scan function (see Chapter 6, Figure 6.4). The preferred units of time, voltage, and frequency can be adjusted such that the amplitude of the RF drive voltage can be programmed as a real voltage, for example, $V_{RF} = 789.6 \text{ V}_{0\text{-p}}$; as a function of the q_z value for a given ion species, for example, $q_z = 0.3$ for m/z 100; or as a function of the LMCO (see Chapter 2) that gives the lower limit for mass/charge ratio of ions that may be confined.

ISIS follows a voltage programming method that is similar to that of ITSIM. The user defines separate modules in the overall *scan function*—the *segment zero* and the *current segments*. When there are no changes over the course of a simulation, the current segment zero will be rerun successively for the duration of the simulation. A different *current segment* module must be defined for each change in the trapping parameters. The segment files are ASCII text code and they can be adjusted either through the ISIS interface or directly. ISIS uses the basic International System of units (SI) for all input parameters and the voltages are assigned in terms of q_z values for the given ion under the specific trapping conditions.

SIMION requires that all dynamic voltages applied to the ion trap electrodes be controlled from *user programs*, that is, algorithms similar to BASIC code that calculate and apply the voltages. The user programs are called at the beginning of a run when the user decides to "Fly'm," SIMION's descriptor for initiating a simulation. SIMION alone permits the user to adjust any or all of the variables during the flight. For SIMION to perform scan functions in a manner comparable to those of ITSIM and ISIS, a number of modifications were implemented; a complete record of the user program is given in the Supplementary Material to Ref. 8. SIMION accepts and returns values in the user program in units of atomic mass units, volts, electron-volts, millimeters, and microseconds; any conversion to other conventions must be incorporated in the user program.

References pp. 159–160.

4.3.3.7. Ion Definition   The identity of an ion is defined by the parameters mass (m), charge (e), initial position (x, y, z), initial velocity in Cartesian coordinates ($\dot{x}$, $\dot{y}$, $\dot{z}$) or polar three-dimensional coordinates (kinetic energy E_k, azimuth θ, elevation ϕ), time of creation (t), and cross-sectional area (σ). Of these parameters, SIMION and ISIS consider all but σ. Because ITSIM can simulate ion/neutral collisions with a hard-sphere collision model, σ must be defined for each ion.

The three methods for defining ion properties of exact ion definition, uniform ion distribution, and randomized distributions are all possible with the three simulation packages. For each package, the exact ion definition method requires the user to input individually the coordinates, velocities, masses, and so on, for each ion. In ITSIM and SIMION, groups of ions can be saved for reuse later. Figure 4.5a, in which any ion property (P) is specified precisely for each ion, can characterize the exact definition of ions.

In SIMION's *Define* dialogue, exact ions can be defined using the *Define Ions Individually* option. A large number of ions can be created and the ion trajectories are simulated in the order in which they were defined, allowing the user to observe individual trajectories and the effects of altering initial properties. Exact distributions

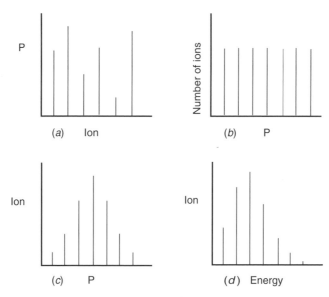

Figure 4.5. Histograms illustrating exact ion definition: (a) user inputs value of a given ion property P for each ion separately; (b) method of uniform distribution, whereby an equal number of ions will have a particular value for given property and number of ions is distributed evenly across range of values; (c) Gaussian distribution of ion properties; (d) Boltzmann distribution of ion properties. Any property P can be generated using a Gaussian distribution in ITSIM and kinetic energies of ions can be created from a Maxwell–Boltzmann distribution in ISIS. (Reprinted from the *Journal of Mass Spectrometry*, vol. 34, M. W. Forbes, M. Sharifi, T. R. Croley, Z. Lausevic, R. E. March, "Simulation of ion trajectories in a quadrupole ion trap: A comparison of three simulation programs," Fig. 6, 1219–1239 (1999) © John Wiley & Sons Limited. Reproduced with permission.)

of ions can also be created in ITSIM where the ions are defined in families and the user may specify the number of ions in each family. Exact ions can also be specified in ISIS and a file is created for each ion. Exact ion definition is a tedious process. It is often more efficient to create distributions of ions by another method.

A second method for creating ensembles of ions is the uniform definition of ions and is illustrated in Figure 4.5*b*. SIMION provides an option to *Define Ions by Groups* in which any or all of an ion's properties can be incremented to create a uniform group. For example, defining the first ion and specifying a mass increment of 1 unit can create an ensemble of 11 ions of mass/charge ratios 95–105 with identical positions, velocities, and so on. SIMION allows the user to randomize ion properties such as ions' positions, kinetic energies, initial cone angle (SIMION's term for direction), and time of birth.

In ISIS, groups of ions can be defined in the *Edit an Ion* dialogue by allowing the user to maintain a certain default set of parameters (mass, charge, velocity) while the variable parameter is adjusted incrementally.

ITSIM has adopted a somewhat different method for generating a uniform distribution of ions. In the *Ion Generation Parameters* dialogue, the user is prompted to define families of ions with a specified number of members. A high and low value for each of x, y, z, $\dot{x}$, $\dot{y}$, $\dot{z}$ and time of birth can be specified and the ions will be distributed uniformly throughout the specified ranges. For example, 10 ions can be created over one RF base cycle by specifying a start time (low value) of $0\,\mathrm{s}$ and an end time (high value) of $9.0909 \times 10^{-7}\,\mathrm{s}$. The ion generator uses a random-number seed to determine the distribution of the variable property for the ion ensemble. For a small number of ions, the ensemble will not necessarily be evenly populated but, as larger ensembles are created, the distribution tends to approximate the form of Figure 4.5*b* to a greater extent.

A third method of ion creation generates ions with properties varied according to a Gaussian distribution, as illustrated in Figure 4.5*c*, or a Maxwell–Boltzmann-like distribution, as shown in Figure 4.5*d*. ISIS includes a function by which the initial kinetic energies of an ensemble of ions can be generated from a Boltzmann distribution of energies. Real populations of ions under set conditions will have energies that can be described by the function

$$n_i = \frac{Ne^{-\varepsilon_i/kT}}{\sum_j g_j e^{-\varepsilon_j/kT}}. \tag{4.12}$$

The number of ions n_i with energy ε_i is related to the total number of ions in the ensemble (N), the temperature (T), the energy of the allowed levels (ε_j), and the degeneracy of the population of level j (g_j). Such a treatment is perhaps the most accurate thermodynamic means for generating ensembles of ions with realistic energies. ITSIM is capable of generating ions with any or all of their initial positions (x, y, z) and/or velocities $(\dot{x}, \dot{y}, \dot{z})$ fitted to a Gaussian distribution. A Gaussian distribution is helpful when more realistic groups of ions are desired: for example, a beam of ions focused by an ion lens prior to injection into an ion trap and an ion ensemble after an

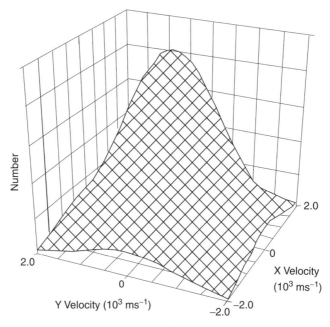

Figure 4.6. Representation of the population of ions with radial velocities varied according to Gaussian distribution. The range of velocities is limited by a minimum ($-2000\,\mathrm{ms}^{-1}$) and a maximum ($2000\,\mathrm{ms}^{-1}$) and is centered about the mean velocity ($0\,\mathrm{ms}^{-1}$). In this example, the standard deviation was assigned a value of $1000\,\mathrm{ms}^{-1}$. (Reprinted from the *Journal of Mass Spectrometry*, vol. 34, M. W. Forbes, M. Sharifi, T. R. Croley, Z. Lausevic, R. E. March, "Simulation of ion trajectories in a quadrupole ion trap: A comparison of three simulation programs," Fig. 7, 1219–1239 (1999). © John Wiley & Sons Limited. Reproduced with permission.)

initial cooling period [37–40]. This latter example obviates simulation of ionization and cooling processes. It would be unrealistic, for example, to specify an ensemble of ions starting in the same radial plane with velocities in the axial direction only because, in practice, nearly all of the ions will have some component of radial velocity as they pass the end-cap electrode. Thus the Gaussian distribution can be used to generate a group of ions with x and y velocities fitted to a Gaussian curve by specifying high and low limits, the mean velocity, and the standard deviation; then, by overlapping the two distributions, a hypothetical population of ions with x and y velocities could be generated. The majority of the ions will not have substantial components of kinetic energy in the radial plane (as would be expected when an effective ion lens is used), but the deviations from ideality associated with a real chemical system have not been ignored. Figure 4.6 illustrates a hypothetical population of ions that would be generated with a symmetrical variation of radial velocities.

4.3.3.8. Calculation of an Ion Trajectory   The trajectory of an ion is calculated by numerical integration of Newton's equations of motion at specified time intervals for the duration of the simulation. This process is carried out in two stages. First, the ion's initial position is used to determine the electrostatic forces (E) acting upon the ion at the time t at which the calculation is made. The forces are then used to determine the ion's new acceleration (a_x, a_y, a_z). Second, a numerical integration is performed to determine the ion's position at the next calculation time, $t + \Delta t$, such that the ion's current velocity ($v_{x,t}$, $v_{y,t}$, $v_{z,t}$) is adjusted by the acceleration terms to yield three new velocity terms (e.g., $v_{x,(t+\Delta t)}$) and the ion flies in a straight line. A standard fourth-order Runge–Kutta algorithm is used for all three software packages, although ITSIM allows the user to select fifth- and eighth-order Runge–Kutta algorithms, both of which support automatic error control and step size adjustment. Two types of error, round-off error and truncation error, are present in a Runge–Kutta algorithm, but steps can be taken to minimize the contribution of either one [14].

While completion of the two stages constitutes one integration step, it is the magnitude of the time interval Δt, or integration step size, that determines the quality of the ion's trajectory. When Δt is too large, the calculated path of the ion can be extremely rough and the trajectory will not show minute oscillations or components of ion motion that may occur between t_n and t_{n+1}. Conversely, when Δt is made excessively small, beyond the point at which significant data are captured, then the simulation will run more sluggishly and round-off errors begin to mount. Hence there is an optimum integration time for which simulations run at an acceptable rate while accurate trajectories are calculated. Both ISIS and ITSIM have adjustable static integration times; an integration time of 10 ns has been found to return acceptable results.

In contrast, SIMION has an built-in algorithm that adjusts automatically the integration time at each step. One disadvantage of static integration times is illustrated by the following analogy. When driving on a rough road with steep hills, the control of vehicle speed should be based on driver reaction time to unforeseen obstacles. In this manner, when a steep downward slope is approached, the vehicle's momentum will be sufficiently low that the vehicle remains on the road and is not propelled over the edge of the road. In the same way, an ion trajectory simulator can blindly propel an ion over a large potential barrier and miss an important component of an ion's motion. To obviate this problem, SIMION calculates one integration step ahead and, if the ion's kinetic energy is found to change by more than an allowed amount, the integration step size is halved.

4.3.3.9. Data Collection and Display

When direct comparisons of computational abilities are to be made, it is convenient to extract and to manipulate the experimental data obtained. In addition, real-time visual representations of ion trajectories can prove to be useful in helping to characterize the motions of ions in the ion trap. ITSIM and SIMION have several options for viewing ion trajectory data during a simulation, whereas the ISIS package supports only a graphical user interface. ITSIM has the more advanced graphical user interface for analyzing ion properties in novel forms. Display options include spatial and kinetic energy distributions with time, voltage scan functions, two-dimensional animations, phase space (variation of the ion's velocity in a given direction with its position), and Poincaré plots that strobe the ion's velocity and position at a predefined frequency.

A second means for analyzing simulation data is to capture various properties of the ion (time, position, velocity, etc.) at specified intervals. All three software packages are capable of capturing simulation data for each ion at every integration step or at larger time intervals. The data are written to a file in comma- or tab-delimited ASCII format such that they can be imported into spreadsheet or scientific graphing software packages. SIMION allows the capture of numerical data for each ion flown in a multiparticle simulation, whereas ITSIM will write the data for one ion at a time. Thus, single-ion trajectories must be run if the user wishes to export data for an ion ensemble.

A major concern when analyzing extracted data is that the files can become extremely large for a short experimental run. For example, a typical quadrupole ion trap will perform an analytical scan in a time frame of some hundreds of milliseconds. For an ion trajectory calculation using an integration time of 10 ns, a simulation using only one ion would generate 100 000 records ms^{-1} in a data file. Hence an effort must be made to decrease the quantity of data collected without diminishing significantly the trajectory quality. It has been determined that the error introduced by presenting data at 100-ns intervals was considered insignificant, particularly when compared with the advantage of decreasing the volume of data.

4.3.3.10. Collision Models
Operation of a QIT requires the presence of a buffer gas, normally helium. When ions are permitted to attain large radial orbits, few ions will be transmitted successfully through the holes in an end-cap electrode and strike a detector during an analytical scan. Kinetic cooling of ions can be induced with the introduction of a relatively high pressure (10^{-3} Torr) of helium; ions suffer collisions and lose kinetic energy so that the ion cloud is focused near the center of the ion trap. As noted previously, one of the first illustrations of kinetic cooling was published by Wuerker et al. [1] and shows particles of aluminum dust trapped under the influence of a quadrupolar electrostatic field. At high background pressures, the motions of the particles were seen to crystallize and to migrate slowly toward the center of the ion trap while maintaining stable trajectories.

A number of collisional models have been proposed, but ion velocities in a QIT are extremely variable such that no dominant collision model has emerged thus far. The models that have been studied include a biased random-walk statistical model [41], a three-dimensional, ion/neutral, hard-sphere collisional model [29], an inelastic molecular dynamics model [42], and a Monte Carlo, high-energy model [43]. The two essential criteria for a collision model are that the properties of a buffer gas (nature, velocity, pressure) must be approximated [44] and the model should not be excessively taxing on simulation time.

ISIS utilizes a collision model based on the Langevin collision theory while ITSIM is equipped with both hard-sphere and Langevin collision models. In contrast, one of SIMION's demonstration user programs includes a somewhat less refined velocity-damping model. Londry et al. have discussed in detail the theory [14] and effects [26] of collisional cooling (see Chapter 3). Julian et al. [4] have pointed out that, at low ion velocities, the Langevin collision model will best approximate collisions whereas at higher velocities a hard-sphere model affords a more accurate approach.

Langevin Collision Model   For the Langevin collision model, the probability (P) of an ion suffering a collision can be calculated as a function of its charge (e), the permittivity of a vacuum (ε_0), the polarizability of the buffer gas (α), the reduced mass of the ion and target (μ), the pressure of the buffer gas (p), the Boltzmann constant (k), the temperature (T), and the step size (dt):

$$P = \frac{e}{2\varepsilon_0} \sqrt{\frac{\alpha}{\mu} \frac{p}{kT}} \, dt. \qquad (4.13)$$

The energy transfer parameter that governs the ion energy gain or loss in a collision is a function of the velocities of the colliding partners and a randomly-selected scattering angle (p. 271 in Ref. 3); however, the probability that a collision will occur is independent of velocity, as Eq. (4.13) indicates.

Hard-Sphere Collision Model   The hard-sphere collision model in ITSIM is a velocity-dependent treatment of the probability of a collision occurring in unit time wherein the velocity is scaled by a time-dependent collisional damping term (A); the probability of a collision is calculated as follows [45]:

$$P = \frac{\sigma v_r p}{kT} \, dt \qquad (4.14)$$

where σ is the collision cross section, v_r is the relative velocity between an ion and a buffer gas atom, p is the buffer gas pressure, k is the Boltzmann constant, T is the buffer gas temperature, and dt is the step size.

Collision Factor   SIMION uses a different method to calculate collision probability. At each integration, the distance that an ion has traveled is calculated from the ion's velocity (v) and the time step (t). A factor called the *mean-free path* (MFP) is then used to calculate a collision factor [(CF) as in Eq. (4.14)]:

$$CF = 1 - e^{-vt/MFP}. \qquad (4.15)$$

The calculated velocity-dependent CF is compared with a random number between zero and one; when the random number is >CF, a collision occurs and the ion's velocity is adjusted in accordance with Eq. (4.15) to obtain the new velocity, v_{new}:

$$v_{new} = v \left[\frac{m_{ion} - m_{buffer}}{m_{ion} + m_{buffer}} \right] \qquad (4.16)$$

where m_{ion} and m_{buffer} are the masses of the ion and buffer gas, respectively.

4.3.4. Comparison of Simulators

Comparison of the three simulators was carried out using identical ions in a collision-free system followed by examination of the ions' spatial (radial and axial) variations with time, kinetic energy (E_k) variations with time, and secular frequencies. In a further comparison, identical ions were subjected to collisional cooling followed by examination of ion kinetic energy variation with time. Finally, a comparison was made of 16 ions injected into an ion trap from an external ion source.

References pp. 159–160.

4.3.4.1. Single-Ion Trajectories in a Collision-Free System   The trajectory of a single defined ion in an ion trap was calculated by each simulator for a period of 25 µs. The data were collected at an output increment of 10 ns because the duration of the simulation was short and an exact comparison was sought. In Table 4.1 is presented a summary of the trapping parameters together with the initial ion properties for the simulations. In Table 4.2 are presented some numerical results. That these

TABLE 4.1. Summary of Trapping Parameters and Initial Ion Properties Used for Simulation of Single-Ion Trajectory

Parameter	Value
r_0, z_0 (mm)	10.00, 7.071
f (MHz)	1.1
γ (rad)	0
q_z, a_z	0.400, 0.000
P_{He} (Torr)	0
m/z (Th)	100
x, y, z (mm)	0.500, 0.500, 0.500
$\dot{x}$, $\dot{y}$, $\dot{z}$ (ms^{-1})	0.000, 200.000, −100.000
Output increment (µs)	0.01

Source: Reprinted from the *Journal of Mass Spectrometry*, vol. 34, M. W. Forbes, M. Sharifi, T. R. Croley, Z. Lausevic, R. E. March, "Simulation of ion trajectories in a quadrupole ion trap: A comparison of three simulation programs," Table 1, 1219–1239 (1999). © John Wiley & Sons Limited. Reproduced with permission.

TABLE 4.2. Numerical Simulation Results for Single-Ion Comparison

Parameter[a]	SIMION	ITSIM	ISIS
Integration time (µs)	0.007	0.01	0.01
z_{min} (mm)	−1.0631	−1.0206	−1.0100
z_{max} (mm)	1.0421	1.0264	1.0000
ω_z (kHz)	161.1	161.1	161.0
r_{min} (mm)	0.1791	0.1786	0.1740
r_{max} (mm)	1.1477	1.1475	1.1390
ω_r (kHz)	78.7	78.4	78.4
$E_{k,min}$ (eV)	0.0050	0.0054	0.0050
$E_{k,max}$ (eV)	1.6089	1.5251	1.5298

Source: Reprinted from the *Journal of Mass Spectrometry*, vol. 34, M. W. Forbes, M. Sharifi, T. R. Croley, Z. Lausevic, R. E. March, "Simulation of ion trajectories in a quadrupole ion trap: A comparison of three simulation programs," Table 2, 1219–1239 (1999). © John Wiley & Sons Limited. Reproduced with permission.

[a]The subscripts max and min refer to the maximum and minimum values of a particular parameter over the course of the simulation.

basic calculations enjoy such a high degree of agreement is a tribute to those who spent so many hours writing, testing, and honing the three simulator packages.

Axial and Radial Variations   The temporal variations of the selected ion's axial and radial excursions from the center of the ion trap are shown in Figures 4.7a and b, respectively. There is remarkably little difference between the three ion trajectories. The most notable differences are observed in the regions of the curves where the ion's motion underwent a change of direction. The ion trajectory calculated by SIMION showed deviations from those of ITSIM and ISIS; these deviations may be indicative of the differences in computational methods. The differences between the trajectories at the maxima and minima in both the r and z directions were <0.1 mm in all cases.

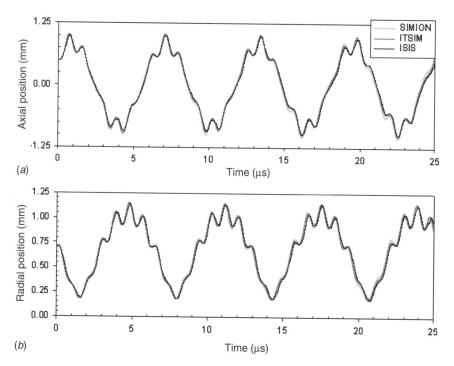

Figure 4.7.   Simulation of a single-ion trajectory (m/z 100 at $q_z = 0.4$ and collision free) by each of SIMION, ITSIM, and ISIS. Simulation was run for 25 μs. (a) Overlaid plots of axial position with time. (b) Overlaid plots of radial position with time. Details of the trapping parameters and initial ion properties are given in Table 4.1 and the results of the simulations are given in Table 4.2. (Reprinted from the *Journal of Mass Spectrometry*, vol. 34, M. W. Forbes, M. Sharifi, T. R. Croley, Z. Lausevic, R. E. March, "Simulation of ion trajectories in a quadrupole ion trap: A comparison of three simulation programs," Fig. 11, 1219–1239 (1999). © John Wiley & Sons Limited. Reproduced with permission.)

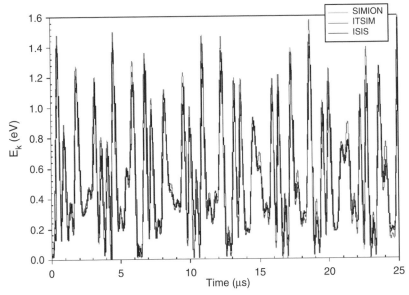

Figure 4.8. Plot of the kinetic energy variation with time of a single ion calculated by each of SIMION, ITSIM, and ISIS for the conditions in Figure 4.7. Maximum kinetic energy attained by the ion was ~1.5 eV. (Reprinted from the *Journal of Mass Spectrometry*, vol. 34, M. W. Forbes, M. Sharifi, T. R. Croley, Z. Lausevic, R. E. March, "Simulation of ion trajectories in a quadrupole ion trap: A comparison of three simulation programs," Fig. 12, 1219–1239 (1999). © John Wiley & Sons Limited. Reproduced with permission.)

Ion Kinetic Energy Variation   In Figure 4.8 are shown three overlapping plots of the temporal variations of the ions' kinetic energies. It is quite extraordinary that virtually no differences among the kinetic energies calculated by the three simulators were observed.

The minor differences between the trajectory calculated by SIMION and those calculated by ITSIM and ISIS are attributed to the matrix calculation method employed by SIMION. There is no conclusive evidence to indicate that the matrix method was responsible for these minor differences; rather it is by a process of elimination that the interpolation method is identified as the source of the minor trajectory differences. Great effort had been expended in the construction of the electrode geometry in SIMION in order to minimize sources of error associated with the field interpolation method (e.g., resolution of the array, refining process).

Conversely, a drawback of the direct-calculation method is that it does not consider discontinuities in the electrode surfaces. A grid resolution of 0.002 in. in the SIMION electrode structure is comparable to the precision with which electrodes can be machined for real instruments [3]. However, the electrodes in a real instrument are relatively small, the surfaces are truncated, and holes are drilled through the end-cap electrodes, so it is suggested that the matrix method may give more realistic results if the effects of such discontinuities are significant. Quarmby and Yost [46] have reported that end-cap perforations can weaken the local RF-trapping field and ions having large axial excursions are susceptible to this distortion of the field. Although the trajectories of SIMION appear to stand apart from those of ITSIM and ISIS, it should not be concluded that SIMION calculations are any more or less accurate than those of ITSIM and ISIS.

Frequency Analysis   Frequency analysis of the ion's trajectory calculated by each program was carried out. For ITSIM and SIMION simulations from which numerical data were obtained, a power spectral Fourier transform frequency analysis was performed on both the axial and radial data to elucidate ω_z and ω_r. For ISIS, the ion's frequencies were calculated manually; the times at which an ion passed a particular position were recorded and the average of the differences between each pair of points was used to determine the secular frequency. Plots of the power spectra of the radial and axial frequencies of the ion trajectory simulated by ITSIM are presented in Figure 2.18. The results in Table 4.2 indicate that, again, there is a remarkable degree of agreement in that the calculated secular frequencies ω_z and ω_r were shifted by <0.01% from the theoretical values.

4.3.4.2. Collisional Cooling   As discussed in Chapter 3, collisional cooling of ions confined in a QIT is essential in order to obtain high mass resolution and sensitivity. Because SIMION treats ion/neutral collisions in a unique manner, a comparison is made first of ITSIM with ISIS and then collisional cooling with SIMION and with ITSIM is compared.

Comparison of ITSIM with ISIS   The trajectory of a single ion of *m/z* 100 was simulated in both ITSIM and ISIS for 1 ms so as to compare the effects of collisional cooling using the Langevin collision model. The ion's initial conditions are given in Table 4.3. The higher than normal bath gas pressure used here to reduce computational time did not impair the simulation [14]. The results generated by ISIS and by ITSIM [8] showed the trends in condensation of ion motion and reduction of ion kinetic energy to be in good agreement. From these two calculations, it can be extrapolated that an ensemble of ions filling much of the ion trap volume would be condensed to a small ion cloud near the center of the ion trap and having low kinetic energy. The similarity of the models is illustrated by the observation that the ITSIM simulation involved 151 collisions while that of ISIS involved 149 collisions.

Comparison of SIMION with ITSIM   A comparison, similar to that above, has been made between the velocity-damping model of SIMION and the hard-sphere model of ITSIM; the simulation parameters are given in Table 4.3. The results generated by SIMION and by ITSIM [8] showed the expected trends in condensation of ion motion and reduction of ion kinetic energy. The MFP setting for SIMION (40) was found to approximate the effects of a pressure of helium buffer gas of 15 mTorr in

TABLE 4.3. Summary of Initial Properties and Trapping Parameters Used for Comparisons of Collisional Cooling

Parameter[a]	Value
m/z (Th)	100
x, y, z (mm)	0, 0, 0
$(\dot{x}, \dot{y}, \dot{z})$ (ms^{-1})	982.227, $-$694.57, 694.57
P_{He} (mTorr)	15
α (Å^3)	0.20495
T (K)	493.0
MFP (SIMION)	40
q_z, a_z	0.400, 0.000
f (MHz)	1.100

Source: Reprinted from the *Journal of Mass Spectrometry*, vol. 34, M. W. Forbes, M. Sharifi, T. R. Croley, Z. Lausevic, R. E. March, "Simulation of ion trajectories in a quadrupole ion trap: A comparison of three simulation programs," Table 3, 1219–1239 (1999). © John Wiley & Sons Limited. Reproduced with permission.

[a]Pressure, polarizability and temperature apply to the collision models used in ITSIM and ISIS, whereas the MFP factor applies to SIMION.

ITSIM using a collisional cross section $\sigma = 50$ Å^2. The major difference arises from the limitation of SIMION's model in that ion energy increase from a collisional encounter is not allowed. Although the hard-sphere model induces a more rapid decline in ion kinetic energy as a result of the first 15 or so collisions with ITSIM, ion kinetic energy after 100 or so collisions with SIMION was excessively low.

4.3.4.3. Ion Injection There is an enormous experimental advantage to be gained by trapping ions generated externally to the ion trap and injected subsequently into the ion trap [47] (see Chapter 8). However, the relatively low efficiency of ion ejection and subsequent ion confinement have stimulated a number of simulations of this process. Doroshenko and Cotter [30] have used SIMION 6.0 to study in great detail the effects of ion kinetic energy and RF phase angle (γ) on the trapping efficiency, as have He and Lubman [29]. Weil et al. [6] used ITSIM to show the influence of various pressures of helium buffer gas and pulsed DC potentials on the trapping efficiency of injected ions.

DC Pulse on the Exit End-Cap Electrode Enhanced trapping efficiency of externally-generated ions can be achieved by the application of a DC pulse to the exit end-cap electrode [8]. In a simulation carried out with ITSIM, 20 ions of m/z 100 were created in the vicinity of the entrance end-cap electrode of the ion trap immediately prior to the application of a +250 DC pulse for 1 μs. No collisions were simulated. During the pulse, the motion of the ion ensemble acquired a degree of coherence and ion kinetic energy maxima ranged from 35 to 130 eV. Once the pulse was removed, ion kinetic energy maxima were generally less than 35 eV; all of the ions were confined within the ion trap during the simulation period of 10 µs and the motion coherence was retained.

Trapping Efficiency as a Function of RF Phase Angle γ   A series of simulations using ITSIM were run for ensembles of 10^5 ions of m/z 100 in an investigation of the effects of RF phase angle γ at creation on the percentage trapping efficiency ε in a collisional system. Ensembles of ions were run at kinetic energies of 10, 80, and 300 eV, described as low, intermediate, and high, respectively. In each simulation, the ions were created uniformly over one RF cycle of duration 0.9091 µs, commencing at $t = 0$. Ions were created on the z axis at $z = -7.070$ mm and were directed along the z axis with uniform distributions (± 0.6 mm) in the x and y directions. The z velocities were exactly 4000, 12,000, and 30,000 ms^{-1} with uniform distributions of x and y velocities (± 3000 ms^{-1}). Ion/neutral collisions were simulated using the Langevin collision model at a pressure of 15 mTorr of helium. A DC pulse of +250 V was applied to the exit end-cap electrode for 0.5 µs from $t = 1.5$ µs to $t = 2.0$ µs. The duration of the simulation was 10 µs. On average, some 278 ions were injected over each degree of the RF cycle.

For ions of low and intermediate kinetic energy, some 14–35% of ions can be trapped; however, no ions having kinetic energies of 300 eV were trapped. For ions of low kinetic energy, the RF phase angle window during which ions could be trapped was $5\pi/4$–$3\pi/2$ rad, that is, only one-eighth of an RF cycle. For ions of intermediate kinetic energy, ions could be trapped during two windows of equal width from 0 to $2\pi/5$ rad and $8\pi/5$ to 2π rad, that is, about one-third of an RF cycle.

Trajectory Calculations for Injected Ions   A simulation of the injection of ions located initially well upstream of the entrance end-cap electrode was carried out based on the electrode arrangement shown in Figure 4.4. An array of 16 ions was created at the mouth of the ion optic in Figure 4.4 and was directed with a kinetic energy of 25 eV parallel to but not coincident with the z axis. SIMION can calculate ion trajectories both outside and inside the ion trap, while ITSIM (Version 4.1 but not Version 5.0) and ISIS are restricted to trajectory simulations inside the ion trap. Thus, ion trajectories were calculated by SIMION initially until each ion passed through the entrance end-cap electrode whereupon the ion properties were captured; subsequent trajectories were calculated by SIMION (with MFP = 40), ITSIM, and ISIS using the captured ion properties as initial conditions. Ion trajectories within the ion trap were calculated for 50 µs.

Because the trajectory of each ion was calculated by each simulator, a comparison can be made of the fate (ejected, ✗; trapped, ✓) of each ion as determined by each simulator. In Table 4.4 is given a summary of the fates of each ion together with the percentage trapping efficiency, ε(%), for each simulator. The three simulators indicated that the same eight were trapped and the same two ions were ejected; for the remaining six ions there was disagreement among the simulators. The differences in the observed trapping efficiencies are hardly significant given that a number of factors could have contributed to the outcome of the simulation.

TABLE 4.4. Summary of Fates of Each Ion for 50-μs Simulations of Ion Injection

Ion	SIMION Fate[a]	ITSIM Fate[b]	ISIS Fate[c]
1	✓	✓	✓
2	✗	✓	✓
3	✗	✗	✗
4	✓	✓	✓
5	✓	✓	✓
6	✓	✓	✗
7	✓	✓	✓
8	✗	✗	✗
9	✓	✓	✓
10	✓	✓	✓
11	✓	✓	✓
12	✓	✗	✗
13	✓	✓	✓
14	✗	✓	✓
15	✗	✓	✓
16	✗	✓	✓

Source: Reprinted from the *Journal of Mass Spectrometry*, vol. 34, M. W. Forbes, M. Sharifi, T. R. Croley, Z. Lausevic, R. E. March, "Simulation of ion trajectories in a quadrupole ion trap: A comparison of three simulation programs," Table 5, 1219–1239 (1999). © John Wiley & Sons Limited. Reproduced with permission.

[a]Percentage trapping efficiency, $\varepsilon(\%)$, is 62%.
[b]Percentage trapping efficiency, $\varepsilon(\%)$, is 81%.
[c]Percentage trapping efficiency, $\varepsilon(\%)$, is 75%.

4.4. CONCLUSIONS

The field of ion trajectory simulation is now well advanced and the SIMION Version 6.0 and ITSIM Version 5.0 simulation packages are used with confidence. From a comparison of the performance, design, and operation of the ion trajectory simulators SIMION-3D, ITSIM, and ISIS, it was found that there are many similarities and some differences. The greatest similarity was observed for the simulation of a single ion in a collision-free system; the calculated spatial trajectory components, kinetic energies, and secular frequencies were virtually identical. The results of the ion injection simulations under collisional conditions are indicative of the complexity that can be introduced readily into simulations. Random effects such as collisions of ions with buffer gas atoms (or molecules) and accumulated calculation errors together with the different approaches to field calculation are believed to have contributed to the differences in ion-trapping efficiency.

SIMION is the simulator of choice for the simulation of ion trajectories in hybrid instruments and in custom-designed assemblies of electrodes. For the QIT, ITSIM is the best choice on the basis of computational speed for running multiparticle simulations and user friendliness.

REFERENCES

1. R. F. Wuerker, H. Shelton, R. V. Langmuir, *J. Appl. Phys.* **30**(3) (1959) 342–349.

2. R. E. March, R. J. Hughes, J. F. J. Todd, *Quadrupole Storage Mass Spectrometry*, Chemical Analysis Series, Vol. 102, Wiley, New York, 1989.

3. R. K. Julian, Jr., R. G. Cooks, R. E. March, F. A. Londry, Ion trajectory simulations, in R. E. March and J. F. J. Todd (Eds.), *Practical Aspects of Ion Trap Mass Spectrometry*, Vol. I, CRC Press, Boca Raton, FL, 1995, Chapter 6.

4. R. K. Julian, M. Nappi, C. Weil, R. G. Cooks, *J. Am. Soc. Mass Spectrom.* **6** (1995) 57.

5. R. G. Cooks, C. D. Cleven, L. A. Horn, M. Nappi, C. Weil, M. H. Soni, R. K. Julian, Jr., *Int. J. Mass Spectrom.* **146/147** (1995) 147.

6. C. Weil, M. Nappi, C. D. Cleven, H. Wollnik, R. G. Cooks, *Rapid Commun. Mass Spectrom.* **10** (1996) 742.

7. K. Yoshinari, *Rapid Commun. Mass Spectrom.* **14** (2000) 215.

8. M. W. Forbes, M. Sharifi, T. R. Croley, Z. Lausevic, R. E. March, *J. Mass Spectrom.* **34** (1999) 1219.

9. P. Perrier, T. Nguema, M. Carette, J. André, Y. Zerega, G. Brincourt, R. Catella, *Int. J. Mass Spectrom.* **171** (1997) 19.

10. S. Sevugarajan, A. G. Menon, *Int. J. Mass Spectrom.* **197** (2000) 263.

11. L. Ding, M. Sudakov, S. Kumashiro, *Int. J. Mass Spectrom.* **221** (2002) 117.

12. R. E. March, A. W. McMahon, F. A. Londry, R. L. Alfred, J. F. J. Todd, F. Vedel, *Int. J. Mass Spectrom. Ion Processes* **95** (1989) 119.

13. R. E. March, A. W. McMahon, E. T. Allinson, F. A. Londry, R. L. Alfred, J. F. J. Todd, F. Vedel, *Int. J. Mass Spectrom. Ion Processes* **99** (1990) 109.

14. F. A. Londry, R. L. Alfred, R. E. March, *J. Am. Soc. Mass Spectrom.* **4** (1993) 687.

15. W. H. Press, B. P. Flannery, S. A. Teukolsky, W. T. Vetterting, *Numerical Recipes*, Cambridge University Press, New York, 1986, p. 563.

16. D. Corson, P. Lorrain, *Introduction to Electromagnetic Fields and Waves*, W. H. Freeman, San Francisco, 1962, p. 36.

17. D. A. Dahl, *SIMION 3D Version 6.0 User's Manual*, Idaho National Engineering Laboratory Chemical Materials and Processes Department, Lockheed Idaho Technologies Company, Idaho Falls, ID, 1995.

18. H. A. Bui, R. G. Cooks, *J. Mass Spectrom.* **33** (1998) 297.

19. M. Nappi, C. Weil, C. D. Claven, L. A. Horn, H. Wollnik, R. G. Cooks, *Int. J. Mass Spectrom. Ion Processes* **161** (1995) 77.

20. C. Weil, M. Nappi, C. D. Claven, H. Wollnik, R. G. Cooks, *Rapid Commun. Mass Spectrom.* **10** (1996) 742.

21. R. E. March, F. A. Londry, R. L. Alfred, J. F. J. Todd, A. D. Penman, F. Vedel, M. Vedel, *Int. J. Mass Spectrom. Ion Processes* **110** (1991) 159.

22. R. E. March, M. R. Weir, M. Tkaczyk, F. A. Londry, R. L. Alfred, A. M. Franklin, J. F. J. Todd, *Org. Mass Spectrom.* **28** (1993) 499.

23. M. Splendore, F. A. Londry, R. E. March, R. J. S. Morrison, P. Perrier, J. André, *Int. J. Mass Spectrom. Ion Processes* **156** (1996) 11.

24. R. L. Alfred, F. A. Londry, R. E. March, *Int. J. Mass Spectrom. Ion Processes* **125** (1993) 171.

25. M. Splendore, M. Lausevic, Z. Lausevic, R. E. March, *Rapid Commun. Mass Spectrom.* **11** (1997) 228.

26. R. E. March, M. Tkacyzk, F. A. Londry, R. L. Alfred, *Int. J. Mass Spectrom. Ion Processes* **125** (1993) 9.

27. R. E. March, F. A. Londry, R. L. Alfred, A. M. Franklin, J. F. J. Todd, *Int. J. Mass Spectrom. Ion Processes* **112** (1992) 247.

28. K. Blaum, Ch. Geppert, P. Muller, W. Nörtershäuser, E. W. Otten, A. Schmitt, N. Trautmann, K. Wendt, B. A. Bushaw, *Int. J. Mass Spectrom.* **181** (1998) 67.

29. L. He, D. M. Lubman, *Rapid Commun. Mass Spectrom.* **11** (1997) 1467.

30. V. M. Doroshenko, R. J. Cotter, *J. Mass Spectrom.* **31** (1997) 602.

31. C. R. Arkin, D. A. Laude, A hybrid quadrupole ion trap for Fourier transform analysis. Proc. 46th Ann. ASMS Conf. on Mass Spectrometry and Allied Topics, Orlando, FL, May 31–June 4, 1998, p. 506.

32. E. E. Gard, M. K. Green, H. Warren, E. J. O. Camara, F. He, S. G. Penn, C. B. Lebrilla, *Int. J. Mass Spectrom. Ion Processes* **157/158** (1996) 115.

33. S. A. Shaffer, K. Tang, G. A. Anderson, D. C. Prior, H. R. Udseth, R. D. Smith, *Rapid Commun. Mass Spectrom.* **11** (1997) 1813.

34. E. R. Badman, R. C. Johnson, W. R. Plass, R. G. Cooks, *Anal. Chem.* **10** (1998) 4896.

35. J. M. Wells, W. R. Plass, G. E. Patterson, Z. Ouyang, E. R. Badman, R. G. Cooks, *Anal. Chem.* **11** (1999) 3405.

36. J. M. Wells, W. R. Plass, G. E. Patterson, Z. Ouyang, E. R. Badman, R. G. Cooks, Chemical mass shifts in ion trap mass spectrometry. Proc. 47th Ann. ASMS Conf. on Mass Spectrometry and Allied Topics, Dallas, TX, June 13–18, 1999, p. 568.

37. P. H. Hemberger, N. S. Nogar, J. D. Williams, R. G. Cooks, J. E. P. Syka, *Chem. Phys. Lett.* **191** (1992) 405.

38. C. D. Cleven, R. G. Cooks, A. W. Garrett, N. S. Nogar, P. H. Hemberger, *J. Phys. Chem.* **100** (1996) 40.

39. I. Siemers, R. Blatt, T. Sauter, W. Neuhauser, *Phys. Rev. A* **38** (1988) 5121.

40. S. Guan, A. G. Marshall, *J. Am. Soc. Mass Spectrom.* **15** (1994) 64.

41. R. G. Gilbert, *J. Chem. Phys.* **80** (1984) 11.

42. B. H. Mahan, *J. Chem. Ed.* **C51** (1974) 308.

43. M. Ho, Honours Thesis, Trent University, Peterborough, ON, 1983.

44. V. Moriwaki, M. Tachikawa, Y. Maeno, T. Shimizu, *Jpn. J. Appl. Phys.* **31** (1992) L1640.

45. R. G. Cooks, personal communication, 1998.

46. S. T. Quarmby, R. A. Yost, *Int. J. Mass Spectrom. Ion Processes* **190/191** (1999) 81.

47. P. Kofel, Injection of mass-selected ions into the radiofrequency ion trap, in R. E. March and J. F. J. Todd (Eds.), *Practical Aspects of Ion Trap Mass Spectrometry*, Vol. II, CRC Press: Boca Raton, FL, 1995, Chapter 2.

5

LINEAR QUADRUPOLE ION TRAP MASS SPECTROMETER

Quadrupole Ion Trap Mass Spectrometry, Second Edition, By Raymond E. March and John F. J. Todd
Copyright © 2005 John Wiley & Sons, Inc.

5.1. INTRODUCTION

The conventional views concerning operation of two- and three-dimensional quadru-pole instruments were that the former are used for transmission of ions with or with-out mass selection while the latter are used for ion trapping and subsequent mass selection. Since 1983, the two-dimensional QMF has been eclipsed to some extent by the high versatility and performances of successive generations of three-dimensional QITs. Yet the development of two-dimensional quadrupole instruments has continued apace such that present-day QMFs have a mass range from 2–4000 Th* with unit mass resolution. Furthermore, in triple-stage quadrupole mass spectrometers, the transmis-sion of ions from one rod array to the next can be optimized by suitable adjustment of the amplitude and phase of the RF drive voltage applied to each rod array. With the advent of atmospheric pressure ionization sources came the challenge of improving ion transmission from the ion source, through a series of chambers of successively lower pressure, and into the first analyzing QMF. Two-dimensional RF-only quadrupole rod arrays are used extensively as ion pipes for the transmission of ions through each pres-sure region. Thus, in a present-day triple-stage quadrupole mass spectrometer there may be as many as three RF-only ion pipes, the first QMF with Brubaker [1, 2] lenses before and after, the RF-only quadrupole collision cell, and the second QMF with Brubaker lenses before and after also, for a total of 10 rod arrays. Originally, a Brubaker lens consisted of four short rods located immediately upstream of a QMF and to which only RF voltages were applied. In a later version [3], appropriate DC voltages were applied to the four short rods ahead of the mass filter in order to nullify partially the DC fringing fields; such lenses are used widely for enhanced ion transmission.

*In this work, we have drawn upon many publications wherein the authors have ascribed the unit of thomson (Th) to the mass/charge ratio of an ion. In deference to this common usage that has become acceptable to edi-tors of mass spectrometry journals, we have used the unit of thomson here. However, we are mindful that the unit of thomson has yet to be recognized by the International Union of Pure and Applied Chemistry (IUPAC).

Two areas of development of two-dimensional quadrupole instruments are discussed here: the first area is the development of LITs wherein ions are confined within a two-dimensional quadrupole field and ejected mass selectively subsequently; the second area concerns the development of ion tunnels for the efficient transmission of ions in a pressure region.

5.1.1. History

The early years of QMF development [4] and the basic patents belong to the Paul group at the University of Bonn [5]. One of the accomplishments of von Zahn, a member of the Paul group, was the construction of a mass filter 5.82 m in length in an attempt to achieve a mass resolution of the order of 16,000 [6]; another of von Zahn's accomplishments was the first publication on the monopole [7]. The development of quadrupole devices was pursued feverishly during the 1960s [8] because of the demand for compact, simple, partial-pressure analyzers for upper atmosphere and space research. Brubaker [1–3] dominated the mass filter field during this period by virtue of his studies and design innovations. However, in 1969, it was suggested [9] that a continuing but slower development of quadrupole devices might be expected.

In 1959, Langmuir and co-workers reported on their development of an ion trap based on a six-electrode structure with cubic geometry [10, 11]; the trap consisted of six planar sheets of metal. Later versions of this type of ion trap consisted of sets of six annuluses [12, 13]. This type of structure is introduced here because a variation of this structure has been proposed as a rectilinear ion trap that is discussed later.

5.1.1.1. Mass Discrimination It had been recognized by Brubaker [1] and Dawson [14] that *mass discrimination* in the mass filter, wherein the transmission efficiency of high-mass ions is less than that of low-mass ions, is due to the effect of the transverse components of the fringing fields at the entrance and exit to the mass filter. The Brubaker lens [1–3] was intended to reduce such mass discrimination. Of course, ion acceptance into an RF-only quadrupole rod array is greater than that into a QMF to which RF and DC voltages are applied.

5.1.1.2. Variable Retarding Field In 1972, Brinkmann reported [15, 16] on the application of a variable retarding field located behind (or downstream) of the exit aperture of a mass filter operated in the RF-only mode. The retarding field was induced by the application of a variable DC voltage to a hemispherical mesh electrode; behind the mesh was a funnel-shaped detector. Because the mass filter was RF only such that it worked as a high-pass filter, there was virtually zero loss of ions of high mass number. Mass separation of the ions contained in the RF-only rod array was achieved by analysis of ion energies that are influenced strongly by the longitudinal components of the exit fringing field.

Brinkmann considered that ions at the exit to the quadrupole field have different energies according to their working point in the stability diagram. As the amplitude of the RF drive potential is increased, the LMCO (see Chapter 2) of the mass filter and the q_x value for a given ion species increase also. As the q_x value for a given ion species approaches $q_{x, \text{max}}$ (and the LMCO approaches the mass/charge ratio of the given ion species), ion trajectories grow rapidly so that ions reach the fringing field region away from the mass filter axis and gain kinetic energy in the fringing field where the field strength has its maximum value. The variable retarding field can be set to transmit only those ions of relatively high kinetic energy. Because of the linear relation between the RF voltage amplitude and the LMCO, the kinetic energy gained increases linearly with ion mass. At 500 Th, a resolving power greater than 1000 was obtained.

5.1.1.3. Spectroscopic Studies   Two-dimensional RF multipole ion traps have been used for many years for spectroscopic studies [17–19], but such studies are beyond the scope of this work.

5.1.1.4. Resonant Ejection   Although resonant ejection of ions confined in a QIT [20] is carried out routinely, relatively few reports of resonant excitation and/or ejection of ions flowing through a quadrupole rod set have been made. Paul et al. [21] applied an auxiliary field to a QMF operated close to the RF-only mode in order to achieve high resolution in an isotope separator. Watson et al. [22] selectively ejected ions in the range m/z 54 to m/z 196 from an RF-only quadrupole collision cell of a tandem quadrupole mass spectrometer. The ions, which were confined for ~500 ms, were ejected using an auxiliary potential amplitude of ≤ 500 mV. Cousins and Thomson [23] have demonstrated fragmentation of m/z 609 in the quadrupole collision cell corresponding to MS^3 and MS^4 using resonance excitation square-wave modulated to 4 Hz combined with background subtraction. March et al. [24] have investigated the resonance excitation leading to fragmentation of NaI cluster ions in the range m/z 172 to m/z 2872 in a modified quadrupole collision cell of a Quattro LC triple-stage quadrupole (TSQ) mass spectrometer (Micromass, Manchester, UK). Parenthetically, most commercial TSQ instruments are fitted with hexapole collision cells for enhanced transmission. However, the application of a supplementary potential to one or more rods of a hexapole rod array does not induce excitation of ion motion. When a quadrupole rod array prepared by Micromass was substituted as the collision cell, in place of the original hexapole collision cell, no change in transmission for MS/MS operation with Glu-fibrinopeptide B was observed [24].

5.2. LINEAR ION TRAP

Two new ion trap mass spectrometers were reported recently that make use of the basic structure of a QMF, that is, of an assembly of four rods in a parallel array, for ion trapping, ion trajectory manipulation, and mass-selective ion ejection. These new instruments make use of the ion-trapping properties of two-dimensional quadrupole fields and are described as LITs.

5.2.1. Thermo Finnigan Linear Ion Trap

The Thermo Finnigan LIT was first disclosed publicly [25] in 1995 in U.S. Patent 5,420,425, filed on May 27, 1994. In an article published online on April 26, 2002

[26], this new LIT instrument was described by J. C. Schwartz, M. W. Senko, and J. E. P. Syka of Thermo Finnigan as a two-dimensional QIT mass spectrometer. With the Thermo Finnigan LIT, mass-selective ion ejection from the LIT occurs radially.

5.2.1.1. Advantages of a Linear Ion Trap

The principal advantage of a LIT is that a greater number of ions can be confined in the physically larger device than can be confined in a three-dimensional QIT. Thus the onset of space charge repulsion and accompanying loss of mass resolution and change in mass assignment is experienced only at greater ion loading of the trapping device. Virtually all of the useful operating characteristics of the three-dimensional QIT are retained in the LIT, that is, collisional focusing, resonant excitation and ejection, multiple stages of mass selectivity for tandem mass spectrometry (MS^n), axial modulation, and variation of mass resolution as a function of mass-selective ion ejection scanning rate.

5.2.1.2. Description of the Thermo Finnigan Linear Ion Trap

The Thermo Finnigan LIT [26], shown schematically in Figure 5.1, is based on an LCQ QIT mass spectrometer platform in which the ion beam travels from left to right. The QIT of the LCQ has been replaced by three quadrupole rod arrays wherein the rods have hyperbolic geometry. An ion beam from an electrospray source is directed through a heated capillary and two successive rod arrays (or ion pipes), through a front lens, and into a LIT composed of three sections of quadrupole hyperbolic rod arrays with $r_0 = 4$ mm. Radial confinement of ions is effected by the RF trapping potential well in the center section and axial confinement by DC potentials applied to the front and rear lenses. The basic design of the LIT is shown in Figure 5.2, where the front and rear sections are of length 12 mm and the center section is of length 37 mm. An RF potential at 1 MHz was applied to the LIT, and an auxiliary AC potential was applied in dipolar mode across the rods in the x direction of the center section.

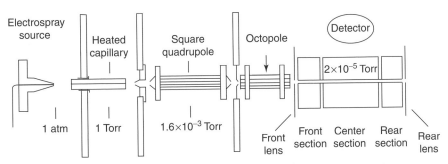

Figure 5.1. Schematic diagram of the Thermo Finnigan LIT mass spectrometer. The ion beam travels from left to right. Typical operating pressures are given. (Reprinted by permission of Elsevier from "A two-dimensional quardupole ion trap mass spectrometer," by J. C. Schwartz, M. W. Senko, J. E. P. Syka, Fig. 5, *Journal of the American Society for Mass Spectrometry,* **13** (2002) 659–669, by the American Society for Mass Spectrometry.)

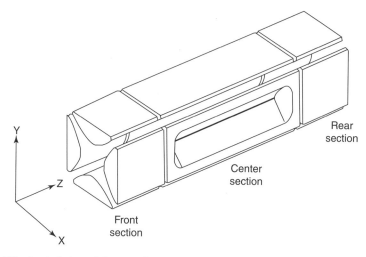

Figure 5.2. Angled view of three sections of the two-dimensional LIT. The detector faces the center section. (Reprinted by permission of Elsevier from "A two-dimensional quardupole ion trap mass spectrometer," by J. C. Schwartz, M. W. Senko, J. E. P. Syka, Fig. 1, *Journal of the American Society for Mass Spectrometry,* **13** (2002) 659–669, by the American Society for Mass Spectrometry.)

In this two-dimensional version of the three-dimensional LCQ instrument, mass-selective ion ejection is accomplished by radial resonant excitation of the stored ions from the center section. For the ions to impinge upon an external detector, a 0.25-mm-high slot was cut along the middle 30 mm of the center section of one rod, as shown in Figure 5.2. A standard detector system of conversion dynode and electron multiplier was mounted in front of the slot shown in Figure 5.2. Incompatibility between the form of present detectors and the axially extended beam of rectangular cross section must be addressed. (See also the Appendix to this chapter.)

5.2.2. MDS SCIEX Linear Ion Trap

The MDS SCIEX LIT mass spectrometer was first disclosed publicly [27] in 2001 in U.S. Patent 6,177,668, filed on June 1, 1998. In an article published online on February 4, 2002 [28], James W. Hager described the LIT mass spectrometer as being a totally new approach to ion confinement and to mass-selective ion ejection. In the MDS SCIEX instrument, mass-selective ion ejection occurs axially through coupling of radial and axial motion in the exit fringing field. In addition, ion trapping in the MDS SCIEX instrument can occur either in the pressurized collision cell region or in a low-pressure quadrupole rod array downstream of the collision cell.

5.2.2.1. Description of the MDS SCIEX Linear Ion Trap The MDS SCIEX instrument shown schematically in Figure 5.3 is based on a triple-stage quadrupole mass spectrometer wherein, normally, the ion beam travels from left to right. Ions are focused collisionally close to the axis in Q_0, mass selected in Q_1, and collisionally dissociated

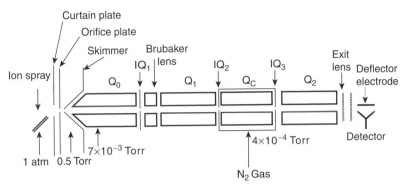

Figure 5.3. Schematic diagram of the SCIEX LIT mass spectrometer. Ion trapping can be wrought in either Q_c or Q_2. Apparatus is based on the ion path in the triple-stage quadrupole mass spectrometer and the ion beam travels from left to right. Typical operating pressures are given. (Reprinted from *Rapid Communications in Mass Spectrometry,* Vol. 16, "A new linear ion trap mass spectrometer," by J. W. Hager, Fig. 1, 512–526 (2000). © John Wiley & Sons Limited. Reproduced with permission.)

to fragment ions in Q_c. A product ion mass spectrum of the precursor ion is obtained from a mass scan of Q_2. Ions generated by an Ionspray source travel through a countercurrent curtain gas, an orifice, and into a differentially-pumped interface region maintained at a pressure of 2 Torr. Ions pass through a skimmer and into an RF-only quadrupole ion guide (Q_0) wherein the pressure is maintained at 7×10^{-3} Torr. The rods of Q_0 are 120 mm in length and they are coupled capacitively to the RF drive voltage of Q_1 so that the transmission through Q_0 can be optimized for the ion species mass selected in the mass analyzer Q_1. IQ_1 is the first interquadrupole aperture that separates the Q_0 chamber from the analyzer chamber. Immediately downstream of IQ_1 is an RF-only Brubaker lens (quadrupole) of length 24 mm coupled capacitively to the RF drive voltage of Q_1 and followed by the RF/DC quadrupole mass filter Q_1 of length 127 mm.

The collision cell, Q_c, is a quadrupolar array of round rods of length 127 mm enclosed in a pressurized container to which the collision gas, normally nitrogen, is introduced. Collision gas escapes through the orifices of diameter 3 mm in the aperture lenses IQ_2 and IQ_3; the ambient collision gas pressure in Q_c is determined by the admission of gas, the conductances of IQ_2 and IQ_3, and the combined pumping speed of the turbomolecular pumps. The collision cell Q_c is coupled capacitively to the second QMF, Q_2, wherein the rods are of length 127 mm also. All quadrupole rod arrays were composed of round rods fabricated with a field radius r of 4.17 mm. Downstream of Q_2 are located two lenses; the first lens has a mesh-covered aperture of diameter 8 mm and the second lens has a clear aperture of diameter 8 mm. The mesh-covered lens is referred to as the exit lens. The detector is shown in Figure 5.3.

5.2.2.2. Ion Trapping in Collision Cell Q_c
When Q_c is operated as a LIT, an RF drive voltage of 1 MHz is applied directly to Q_c and Q_2 was coupled capacitively to

it; when Q_2 is operated as a LIT, the arrangement is reversed. An auxiliary AC potential is added to the RF drive voltage and can be applied in quadrupolar fashion to the rod arrays of either Q_c or Q_2 so as to excite ions at secular frequencies. Because the same polarity of the auxiliary AC potential is applied to a pair of opposite rods, the excitation is termed quadrupolar [29].

Ions are trapped radially in the LIT Q_c by the trapping potential well created from the RF potential applied to the rod array similar to the trapping of ions in a three-dimensional ion trap. The trapping potential well is active whenever an RF potential is applied. Ions are prevented from passing axially through the rod array by DC bias potentials applied to the aperture plates, IQ_2 and IQ_3. Thus, a linear ion-trapping potential resembles a bathtub having a parabolic cross section and near-vertical ends. Ions are admitted over a period of time while IQ_2 is "open" and IQ_3 is "closed"; IQ_2 is then closed to complete the trap. The efficiency of trapping approaches 100% because of the high acceptance of ions into a linear trap that does not have a quadrupole field along the z axis. Ions enter the LIT close to the zero-field centerline of the device and encounter a series of momentum-dissipating collisions with nitrogen collision gas at a pressure of 4×10^{-4} Torr. Ion radial kinetic energy is reduced to a level corresponding to less than the radial potential well depth and ion axial kinetic energy is reduced prior to encountering the end electrodes.

5.2.2.3. Mass-Selective Axial Ion Ejection

The position of the deflection or conversion electrode above the Y-shaped detector at the right-hand side (RHS) of the instrument shown in Figure 5.3 indicates that the emerging mass-selected ion beam has been ejected axially from Q_c and transmitted through Q_2. It was found that the presence of an auxiliary quadrupole or AC field applied to bring about radial resonant excitation effected axial ion ejection when the ion radial secular frequency matched that of the auxiliary AC field. Near the exit aperture of the LIT, ion radial and axial motions become coupled as ions traverse the fringing field, resulting in enhanced axial kinetic energy and axial ejection. Ions having large radial amplitudes (and higher kinetic energies) near the exit aperture are affected to a greater degree than those confined near the axis of the trap. The dynamics of axial ion ejection have been examined in greater detail recently using a combination of analytic theory and computer modeling [30].

5.2.2.4. Ion Accumulation in Q_0

Let us examine once more the apparatus shown in Figure 5.3, where Q_0 is an RF-only quadrupole in which the ion beam is focused collisionally close to the axis so as to enhance the transmission of ions from Q_0 through IQ_1 and into Q_1, which is an RF/DC QMF. With the application of a positive DC potential to IQ_1, ions can be accumulated in Q_0 so as to enhance the subsequent flow rate of ions into Q_1. Here, Q_1 can act mass selectively in order to determine the precursor ion species for transmission to Q_c, where ions can be confined and subjected to resonant dissociation and the product ions can be ejected axially and mass selectively as described above. In this MS/MS experiment where collision-induced dissociation (CID) is brought about by resonant excitation in Q_c, Q_2 serves merely to transmit the product ions at low pressure to the detector.

5.2.2.5. CID by Variation in Precursor Ion Axial Kinetic Energy   An alternative arrangement is for the ions to be passed from Q_1 to Q_c in such a fashion that ions fall through a potential difference so that CID occurs in Q_c by virtue of the preselected ion kinetic energy. The nascent product ions together with undissociated precursor ions are accumulated in the LIT Q_c, then scanned out mass-selectively to the detector. This MS/MS experiment utilizes variation in precursor ion axial kinetic energy in the execution of CID.

5.2.2.6. Ion Trapping in RF-Only Quadrupole Mass Filter Q_2   In yet a further arrangement made possible by the serendipitous selection of the original experimental breadboard platform, ions are again passed from Q_1 to Q_c in such a fashion that ions fall through a potential difference so that CID occurs in Q_c by virtue of the preselected ion kinetic energy. Here, the nascent product ions together with undissociated precursor ions pass through Q_c, which acts as a standard collision cell; ions pass into the LIT Q_2, which is operated at RF only and with suitable DC potentials applied appropriately to IQ_3 and the exit lens. Ions are accumulated in the LIT Q_2. Upon the application in quadrupolar mode of an auxiliary AC potential to the pair of rods in the *x* direction, the ions confined in Q_2 are excited radially and, in the fringing field at the exit of Q_2, are ejected axially and mass-selectively. The LIT Q_2 is not pressurized directly and the ambient pressure in Q_2 is established by the leakage of gas from Q_c and the conductance of the pumping system. This MS/MS experiment utilizes the first two stages of a standard triple-stage quadrupole mass spectrometer employed in the normal fashion together with a quadrupolar trapping device with all of its inherent performance capabilities, such as MS^n impeded only, perhaps, by the relatively low ambient pressure. This hybrid instrument is identified as the low-pressure LIT.

5.2.3. Ion Confinement Theory

To appreciate the performance capabilities of a quadrupolar LIT, let us recall from Chapter 2 some aspects of the theory of the confinement of charged particles in a quadrupole trapping field. Ions stored in a LIT are subject to a two-dimensional quadrupole field where the potential ϕ at any point (x, y) is expressed as

$$\phi_{x, y} = \frac{\phi_0}{2r_0^2}(x^2 - y^2) \qquad \text{[Eq. (2.25)]}$$

where ϕ_0 is given as $2(U + V \cos \Omega t)$ in Eq. (2.30), where U is the DC component (and is zero in the LIT), V is the zero-to-peak amplitude of the RF potential, r_0 is the radius of the inscribed circle to the rod array, Ω is the radial frequency of the RF potential, and *t* is time. The q_x trapping parameter expressed in terms of experimental properties is given by

$$q_x = -\frac{4eV}{mr_0^2\Omega^2}. \qquad \text{[Eq. (2.40)]}$$

The two-dimensional stability diagram that is shown in Figure 2.7 is obtained from the solutions to the Mathieu equation and is similar in shape to the upper part of the stability diagram for the QIT. The stability diagram is composed of a continuum of iso-β_u lines, where β_u is a complex function of a_u and q_u. From β_u is derived the spectrum of resonant frequencies, $\omega_{u, n}$, that characterize radial ion motion,

$$\omega_{u, n} = (n + \tfrac{1}{2}\beta_u)\Omega \qquad \text{[Eq. (2.80)]}$$

which, for the x direction and $n = 0$, can be written as

$$\omega_x = \tfrac{1}{2}\beta_x\Omega . \tag{5.1}$$

Thus ions trapped in a LIT can be excited resonantly by the application of an auxiliary AC potential to some of the rods of the rod array. Ions excited thus can be caused to dissociate following collisions with nitrogen gas molecules and, in the limit, can be driven radially from the LIT. This situation is similar to that in the QIT in that ions can be ejected mass-selectively to as to isolate a given species in the LIT. Resonant excitation of isolated ions can lead to ion dissociation and subsequent storage of product ions. The cycle of ion isolation and CID can be repeated several times.

5.2.4. Ion Trap Capacities

A simple model proposed by Campbell et al. [31] for the comparison of ion trap capacities was based on the ratio of trapping volumes such that

$$N_{2D, A} : N_{2D, B} : N_{3D} = (\pi r_0^2 l)_A : (\pi r_0^2 l)_B : \frac{4\pi}{3}z_0^3 \tag{5.2}$$

where $N_{2D, A}$, $N_{2D, B}$, and N_{3D} are the ion capacities in the SCIEX, Thermo Finnigan, and QIT instruments, respectively; $(\pi r_0^2 l)_A$, $(\pi r_0^2 l)_B$, and $4\pi z_0^3/3$ are the trapping volumes of the same instruments, respectively. On the basis of values given above and $z_0 = 0.707$ cm, the ratio $N_{2D, A} : N_{2D, B} : N_{3D} = 4.7 : 1.0 : 1$. However, when the volumes of the trapped ion clouds subjected to collisional focusing in each instrument are considered,

$$N_{2D, A} : N_{2D, B} : N_{3D} = (\pi r_{2D}^2 l)_A : (\pi r_{2D}^2 l)_B : \frac{4\pi r_{3D}^3}{3} = l_A : l_B : \frac{4r_{3D}}{3} \tag{5.3}$$

where $r_{2D, A} = r_{2D, B} = r_{3D} = 1.0$ mm based on ion tomography experiments [32] with the QIT, the ratio $N_{2D, A} : N_{2D, B} : N_{3D} = 95 : 22 : 1$. However, when $r_{2D, A}$ and $r_{2D, B}$ are each taken to be one-tenth of the respective values of r_0 and r_{3D} is taken as 60% of the FWHM (1.1 mm) of the radially averaged axial distribution at $z = 0$ [32], the ratio is changed to $N_{2D, A} : N_{2D, B} : N_{3D} = 58 : 13 : 1$. The difference in RF frequencies for the LITs (1 MHz) and the three-dimensional ion trap (0.76 MHz) has been ignored here. This ratio is close to the ratio (18 : 4 : 1) of $l_A : l_B : z_0$, which begs the question as to whether the ion capacity of a LIT is directly proportional to its length. The answer is assuredly positive provided that variation in electrode spacing is negligible. However, the subsequent question is concerned with the efficiency with which ions can be ejected and detected in mass-selective ejection.

5.2.5. Characteristics of Linear Ion Trap Operation

The characteristics of LIT operation examined here are, in turn, trapping efficiency, mass discrimination, ion isolation, ion activation, tandem mass spectrometry, spectral space charge limit, ion ejection, enhanced mass resolution, and sensitivity.

5.2.5.1. Trapping Efficiency   Determination of the efficiency with which ions may be trapped in a two-dimensional quadrupolar device requires measurement both of the ions flowing through the device per unit time and of the ions ejected under mass-independent ejection conditions. The efficiency with which ions may be trapped in a two-dimensional quadrupolar device is dependent upon the q_x value upon entry, ion kinetic energy upon entry, and gas pressure within the device. Assuming no ion losses during axial ejection, mass-independent ion ejection indicates that the efficiency of ion trapping ranges from zero at $q_x = 0.14$ to approximately 29% at $q_x = 0.2$ to ~17% as the LMCO (or limit of stability or $\beta_x = 1$) is approached [26]. The form of the variation of the trapping frequency with the injection value of q_x is similar to that observed for ions created in situ in a QIT [33, 34]. In Figure 5.4, obtained at an injection value of $q_x = 0.40$, it is seen that the trapping efficiency for ions having 5 eV in the laboratory system varied from 58% at a Q_c pressure of 1.0×10^{-4} Torr to ~95%

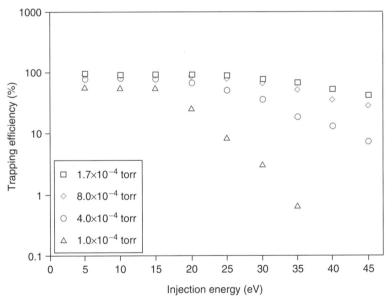

Figure 5.4. Variation of trapping efficiency of the Q_c LIT on ion kinetic energy upon injection and for several pressures within Q_c. A reserpine solution of 100 pg/μL was used to obtain these data. (Reprinted from *Rapid Communications in Mass Spectrometry*, Vol. 16, "A new linear ion trap mass spectrometer," by J. W. Hager, Fig. 8, 512–526 (2000). © John Wiley & Sons Limited. Reproduced with permission.)

at a pressure of 1.7×10^{-3} Torr. Although the trapping efficiency in Q_c is reduced as ion kinetic energy upon entry is increased [28], the data of Figure 5.4 show that the trapping efficiency in Q_c is high for ions with 5–15 eV of kinetic energy over the pressure range 1×10^{-4}–17×10^{-4} Torr, and these data augur well for the performance of LITs. There is evidence that the onset of trapping is pressure dependent and extends to lower q_x values for ions of $m/z > 1000$.

5.2.5.2. Mass Discrimination   In the injection of ions along the z axis into a LIT as shown in Figures 5.2 and 5.3, there is minimal RF field in the z direction, unlike the three-dimensional ion trap. For the application of a single RF amplitude corresponding to a LMCO of 100 Th, a wide mass range from 150 to 2000 Th can be trapped simultaneously with high efficiency. Thus, mass discrimination during the ion injection and ion trajectory stabilization processes is very much reduced in a linear compared to a three-dimensional ion trap.

5.2.5.3. Ion Isolation   Ion isolation has been demonstrated as in the three-dimensional ion trap such that the ^{13}C isotope of the peptide methionine-argenine-phenylalanine-alanine (MRFA) at m/z 525.3 was isolated to $> 95\%$ efficiency [26].

5.2.5.4. Ion Activation   Activation of ions isolated in a LIT is carried out at $q_x = 0.25$, which is a compromise for efficient fragmentation ($\sim 74\%$) and a relatively large fractional (72%) mass range for confinement of product ions.

5.2.5.5. Tandem Mass Spectrometry   Tandem mass spectrometry can be achieved by repeated cycles of isolation and activation using procedures similar to those used for MS^n in the QIT.

5.2.5.6. Spectral Space Charge Limit   When the onset of space charge perturbation is defined as a shift in apparent mass by 0.1 Th, the corresponding ion density is found [28] to be ~ 460 ions/mm^3, which compares well with the value of ~ 400 ions/mm^3 for the three-dimensional ion trap [35]. It is not surprising that the onset of space charge perturbation occurs at a common ion density in quadrupolar devices under similar operating conditions. Hager has described a method [36] for reducing space charge in an LIT; this method is essentially analogous to the automatic gain control used in the QIT.

5.2.5.7. Ion Ejection   Mass-selective ion ejection was observed with an efficiency of 44% relative to the number of ions detected under mass-independent ion ejection conditions, yielding an overall efficiency of 12.7% [26]. In a study employing 5-eV ions of m/z 609 from reserpine, mass-selective axial ejection with an auxiliary AC potential of 480 kHz and 5.1 V_{p-p} yielded an extraction efficiency of 8% at 5.0×10^{-4} Torr to 18% at 1.2×10^{-3} Torr [28]. This result has been interpreted in terms of the length of the extraction region in the vicinity of the exit aperture; 18% of the length of the LIT, 12.7 cm, yields an extraction length of 2.3 cm or $5.5r_0$. The question posed earlier concerning a practical limit to the length of a LIT may have been answered by this result; perhaps the useful length of a two-dimensional ion trap is some 2.3–3 cm only. For axial ion ejection, the ejected ion signal intensity was observed to increase some threefold as the auxiliary AC frequency was increased from 334 kHz ($q_x = 0.451$) to 762 kHz ($q_x = 0.844$) [28].

5.2.5.8. *Enhanced Mass Resolution*
Mass spectra with enhanced mass resolution can be obtained with LITs as for the three-dimensional ion trap [37] by reduction of the mass scanning rate and axial modulation amplitude. When ion signals due to three different charge states of melittin were observed from a LIT, only a single scan was required for a mass spectrum, whereas the same observation from a three-dimensional ion trap necessitated signal averaging: it was estimated that a 10-fold increase in the number of ions was obtained from the linear compared to the three-dimensional device [26]. At a scanning rate of 5 Th/s, a mass resolution ($m/\Delta m$) of ~6000 measured at half maximum was observed for the protonated reserpine molecule [28]. The mass spectrum that shows wide baseline separation of isotopic peaks was recorded with an auxiliary frequency of 450 kHz applied to the Q_c of Figure 5.3.

5.2.5.9. *Sensitivity*
A measured trapping efficiency of 29% [38] indicates that the sensitivity of the LIT should be some six times higher than that of the three-dimensional ion trap, where the efficiency is ~5% [39]. The additional sensitivity translates into lower detection limits. A fivefold improvement in the detection limits of a LCQ Deca three-dimensional ion trap has been demonstrated with a LIT [26]. An observed extraction efficiency of 18% at a pressure of 1.2×10^{-3} Torr in Q_c in combination with the calculated ratio of $N_{2D, A} : N_{3D} = 58 : 1$ indicates that the LIT may be some $0.18 \times 58 = 10$ times more sensitive than a three-dimensional device. When the higher trapping efficiency and ion capacity of the LIT are compared with those of the QIT, it is seen readily that the LIT has the capability to become a highly sensitive scanning mass spectrometer.

5.2.6. Low-Pressure Linear Ion Trap, Q_2

The principal difference between the Q_c LIT and the low-pressure Q_2 LIT is the operating pressure; Q_c is pressurized to 5×10^{-3} Torr and Q_2 has an ambient pressure of ~3×10^{-5} Torr. An auxiliary AC potential is applied in quadrupolar mode and, within the exit fringing field, coupling of radial and axial motion leads to axial mass-selective ejection. The trapping efficiency for the low-pressure Q_2 LIT was determined to be ~45% for ions of m/z 609 having kinetic energies in the range 5–15 eV; the efficiency decreased to ~30% as the ion kinetic energy was increased to 45 eV [28]. For a calculated target thickness of 3.8×10^{-5} Torr·cm, the trapping efficiencies are somewhat more than may be expected and the relative insensitivity of trapping efficiency to ion kinetic energy is advantageous. The gas plume emerging from Q_c at a pressure greater than ambient pressure in Q_2 may enhance trapping efficiency. Mass-selective ejection from Q_2 was found to be ~20% for ions having kinetic energies of 5–40 eV upon entry to Q_c. The variation of trapping efficiency with the q_x value on admission to Q_2 is similar in form to that for Q_c, but the overall efficiency for admission to Q_2 is higher.

Let us compare the product ion mass spectrum of protonated reserpine obtained from the low-pressure Q_2 LIT with those obtained when the apparatus of Figure 5.3 is employed as a standard triple-stage quadrupole mass spectrometer and when Q_c acts as a LIT [28]. The conventional triple-stage quadrupole product ion mass spectrum obtained at a collision energy of 33 eV with a 10-pg/μL solution shows a range of fragment ions to m/z 174. The base peak of m/z 195 was observed at a signal intensity of 1375 counts/s. The corresponding product ion mass spectrum observed from the Q_c LIT with an auxiliary AC frequency of 480 kHz and mass scan rate of 1000 Th/s shows the middle range (m/z 350–450) fragment ions having higher relative signal intensities; the base peak of m/z 195 was observed at a signal intensity of 15,700 counts/s, that is, some 11 times more intense. The corresponding product ion mass spectrum obtained with the low-pressure Q_2 LIT (shown in Figure 5.5) was observed with an auxiliary AC frequency of 816 kHz and mass scan rate of 5200 Th/s; a 100-pg/μL solution was used. The relative fragment ion intensities in the mass spectrum were very similar to those obtained with the Q_c LIT and the base peak, m/z 195, was observed at a signal intensity of 70,000 counts/s; correcting for the different solution concentrations, the base-peak intensity corresponded to some 7000 counts/s for a 10-pg/μL solution. The product ion mass spectrum obtained with the low-pressure Q_2 LIT was some 5 times more intense than that obtained with the standard triple-stage quadrupole mass spectrometer for solutions of equal concentration. The 50%

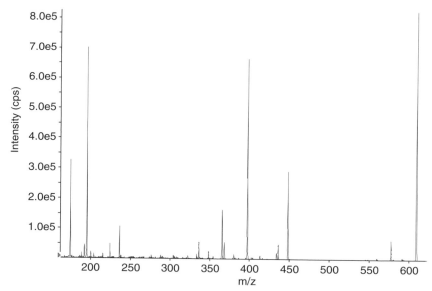

Figure 5.5. Product ion mass spectrum of protonated reserpine, m/z 609, obtained with the Q_2 LIT at mass scan rate of 5200 Th/s. Q_1 was operated with unit resolution; ion kinetic energy was 35 eV; auxiliary AC frequency was 816 kHz. (Reprinted from *Rapid Communications in Mass Spectrometry*, Vol. 16, "A new linear ion trap mass spectrometer," by J. W. Hager, Fig. 17, 512–526 (2000). © John Wiley & Sons Limited. Reproduced with permission.)

peak widths were somewhat mass dependent in that they ranged from ~0.35 Th at m/z 174 to ~0.50 Th at m/z 609. Greater mass resolution could be obtained at lower mass scan rates such that a mass resolution of ~6000 was observed for m/z 609 at a mass scan rate of 100 Th/s.

An interesting characteristic of the low-pressure Q_2 LIT is a dramatic reduction in the susceptibility to space charge compared with the Q_c linear and quadrupole ion traps. Product ion mass spectra obtained at ion fluxes that were varied over a factor of 10^4 (1.93×10^2 and 2.26×10^6 counts/s) exhibited a centroided mass shift of only ~0.02 Th. It is suggested that a greater radial distribution of ions is permitted in the low-pressure Q_2 LIT due to the reduced frequency of ion/neutral collisions.

In a further comparative experiment of the performances of the low-pressure Q_2 LIT and a triple-stage quadrupole mass spectrometer, the mass analyzers were operated at a constant mass scanning rate of 1000 Th/s for the determination of product ion mass spectra of protonated reserpine from a 100-pg/μL solution. The two mass spectra were acquired for the same period of time and were virtually identical with respect to ion relative intensities. However, not only was the low-pressure Q_2 LIT mass spectrum some 16 times more intense than that acquired from the triple-stage instrument, but the mass resolution and the signal/noise ratio of the former instrument were superior to those observed with the latter instrument.

5.3. RECTILINEAR ION TRAP

A rectilinear ion trap (RIT) has been constructed and characterized [40]. An RIT combines the advantages of a LIT with the geometric simplicity of a cylindrical ion trap (CIT), as shown in Figure 5.6. This figure, created by Cooks of Purdue University, is a visual summary of this book. From the original mass filter having hyperbolic electrodes [5], there is the simplification to round rods, then the addition of a DC trapping field (to either hyperbolic or round rods) to affect the LITs discussed above. The geometry of the LITs can be simplified to arrive at the RIT trap discussed here, which in turn is related to the six-electrode cubic structure constructed with six planar sheets of metal proposed by Langmuir et al. [11]. It is of interest that an RIT could be operated as a three-dimensional ion trap by applying the same RF drive potential to all four sides and grounding the end plates. Similarly, the six-electrode cubic structure could be operated as a LIT by the application of V to two opposite sides and $-V$ to the remaining two sides, with the end plates grounded.

From the original QIT having hyperbolic electrodes [5], there is the elongation to the stretched ion trap (see Chapter 3), the application of a rectangular waveform for the digital ion trap (see Chapter 8), and the simplification of geometry to the CIT (see Chapter 6). In turn, the addition of a DC trapping field to the CIT leads to the RIT.

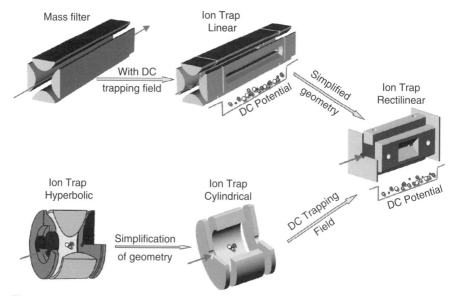

Figure 5.6. Conceptual evolution of RIT; RITs evolved from the LIT and the CIT. (Reprinted by permission of Professor R. G. Cooks.)

5.3.1. RIT Structure

The simple geometry, as shown in Figure 5.7, consists of three electrode pairs (x, y, and z directions). The x, y electrodes are rectangular plates of high aspect ratio (as for quadruple rods) in that the length, 40 mm, is some 4–5 times the width and they are arranged in a rectangular array. Centrally located on each x electrode is a 15-mm $\times$ 1 mm slit through which ions can be ejected onto a detector. There is a rectangular plate at each end of the x, y electrodes; these plates form the z electrodes. Ion trapping is accomplished by the application of an RF potential across the x, y pairs. Normally, DC potentials only are applied to the z electrodes. Thus, an RIT is similar to a LIT wherein the quadrupole rod array has been replaced by a rectangular arrangement of flat plates.

An interesting feature of the RIT is the relative ease with which the essential dimensions of the RIT can be modified to explore variation in RIT performance. That is, the magnitude of the half-distance between the x electrodes (d_x) and that between the y electrodes (d_y) can be varied independently so as to permit exploration of the effects of both physical size and the ratio of $d_x : d_y$; this degree of flexibility for the RIT is similar to that for the CIT. An optimized RIT has been constructed with $d_x = 5.0$ mm and $d_x = 3.8$ mm.

5.3.2. Optimization of the RIT Geometry

Optimization of the RIT geometry consisted of plotting the electric field as a function of d_x, d_y, and $d_x : d_y$ for a number of devices using the program CreatePot, written in

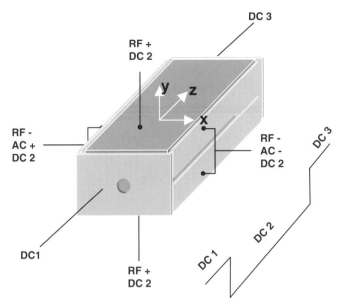

Figure 5.7. Configuration of the RIT and its operational mode. (Reprinted by permission of Professor R. G. Cooks.)

the laboratory of R. G. Cooks, which automatically calls up functionality from Poisson/Superfish [41]. Generally, the fields were similar to those for the QIT and CIT. The same CreatePot program was used to calculate the higher-order fields for the various RIT geometries, as had been carried out for the CIT (Chapter 6) [42]. For the optimized RIT with $d_x = 5.0$ mm and $d_x = 3.8$ mm, quadrupole field coefficient $A_2 = 0.6715$, octopole field coefficient $A_4 = 0.0832$, and dodecapole field coefficient $A_6 = -0.1260$. The positive octopole coefficient is desirable because a negative octopole coefficient can cause ejection delays leading to chemical mass shifts [43].

5.3.3. Stability Diagram

The stability diagram is an important characteristic of a quadrupole device, and the RIT presented an opportunity for the investigation of the variation of the stability diagram for the RIT as a function of the ratio $d_x : d_y$. The method involved was essentially that used by Todd et al. [44]. A selected ion species, of m/z 105, was isolated at several values of the RF drive potential amplitude, V_{p-p}, from which the corresponding values of q_x can be determined. At each q_x value, a DC potential, $\pm U$, was applied and increased until the ion ejection signal disappeared. For a given q_x value, the magnitudes of the $-U$ and U potentials determined the a_x and $-a_x$ boundary values,

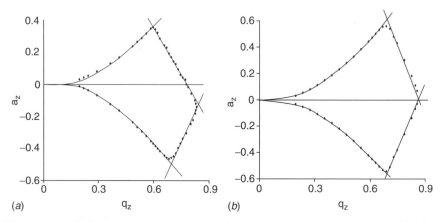

Figure 5.8. Stability diagrams in a_x, q_x space determined experimentally for m/z 105 using two RITs having different geometries: (a) RIT with $d_x = 5.0$ mm, $d_y = 3.8$ mm; (b) $d_x = d_y = 5.0$ mm. (Reprinted by permission of Professor R. G. Cooks.)

respectively, of the stability diagram. The complete stability diagram can then be plotted either in U, V space or as shown in Figure 5.8 in a_x, q_x space by expressing the values of U and V_{p-p} in terms of a_x and q_x. It is shown in Chapter 2 that

$$a_x = \frac{8eU}{mr_0^2 \Omega^2} \qquad \text{[Eq. (2.39)]}$$

and

$$q_x = \frac{-4eV}{mr_0^2 \Omega^2}. \qquad \text{[Eq. (2.40)]}$$

The RIT stability diagram of Figure 5.8b, for which $d_x = d_y = 5.0$ mm, is symmetric about the q_x axis, just as is the stability diagram for the QMF; the upper part of the stability diagram resembles closely that shown in Figure 2.7. The two apexes are observed at $a_x = \pm 0.59$, $q_x = 0.70$; the q_x axis is intersected by the $\beta_x = \beta_y = 1$ stability diagram boundaries at $q_x = 0.86$, which is remarkably close to the corresponding value of $q_x = 0.9$ for the QMF. The RIT stability diagram of Figure 5.8a, for which $d_x = 5.0$ mm, $d_y = 3.8$ mm, is markedly asymmetric and resembles a truncated stability diagram for a QIT, as shown in Figure 2.15. The $\beta_x = \beta_y = 0$ stability diagram boundaries in Figure 5.8a that emerge from the origin are similar to those of Figure 5.8b. The $\beta_x = \beta_y = 1$ stability diagram boundaries have been shifted to lower q_z values and the shift for the $\beta_x = 1$ stability boundary is greater than that for the $\beta_y = 1$ stability boundary, with the result that the former intersects with the q_z axis at $q_z = 0.78$. A result of having $d_x > d_y$ appears to be that the secular frequencies ω_y are shifted to higher values.

5.3.4. RIT Performance

The optimized RIT, when operated in the mass-selective instability mode with resonance excitation at 444 kHz, produced a mass spectrum for perfluorotributylamine showing ion signals from m/z 69 to m/z 502; there was little noise, the valleys between ion signals from ions of low mass/charge ratio reached close to zero, and the peaks were observed with good, though not quantified, mass resolution. A comparative study of mass-selective instability and resonance excitation at 444 kHz and resonant excitation at a lower frequency of 388 kHz (away from the $\beta_x = 1$ stability boundary) was carried out with 1,3-dichlorobenzene. The mass spectrum obtained at 388 kHz showed peak intensities some four times those obtained at 444 kHz. When ions are brought into resonance at 388 kHz, by ramping the RF potential amplitude, ions come into resonance with ω_x before they would come into resonance with ω_y, and so they are ejected preferentially in the x direction. However, the ejection of only one-quarter of the ions at a q_x value close to $q_{x,\,max}$ which is less than $q_{y,\,max}$ is not readily explained by this argument.

The MS^2 capability of the optimized RIT has been demonstrated using acetophenone. The molecular ion of m/z 120 was isolated, then dissociated to form m/z 105. This capability was demonstrated using each of RF/DC isolation and of SWIFT (stored waveform inverse Fourier transform) isolation. The SWIFT method described by Guan and Marshall [45] was calculated using the ion trap simulation program ITSIM research version [46]. Similarly, MS/MS was demonstrated using the primary fragment ion of m/z 105 and its dissociation to form m/z 77. (See also FNF, Section 3.4.3.1.)

The MS/MS/MS capability of the RIT is demonstrated with reference to the mass spectra for acetophenone shown in Figure 5.9. From the electron impact mass spectrum of acetophenone (top frame), the molecular ion of m/z 120 was isolated (second frame) using combined RF and DC potentials. Following a cooling period, the q_z value for molecular ions was adjusted to 0.61 and the ions were excited resonantly with 232 kHz at 500 mV$_{0-p}$. The fragment ion of m/z 105 observed as a product ion of m/z 120 (third frame) was isolated using combined RF and DC potentials. Following a second cooling period, the q_z value for ions of m/z 105 was adjusted to 0.64 and the ions were excited resonantly with 251 kHz at 225 mV$_{0-p}$. The final frame shows the product ion, m/z 77, from m/z 105 as a result of MS/MS/MS.

5.4. STACKED-RING SET

In 1969, Bahr, Gerlich, and Teloy [47] introduced an RF device that consisted of a stack of ring electrodes, like washers, for the storage or transmission of ions. The RF potential applied to each ring was 180° out of phase with that applied to its two adjacent rings such that the potential sign alternated down the stack. The stacked ring could act as an ion pipe or ion guide. Ions traveling at constant speed along the

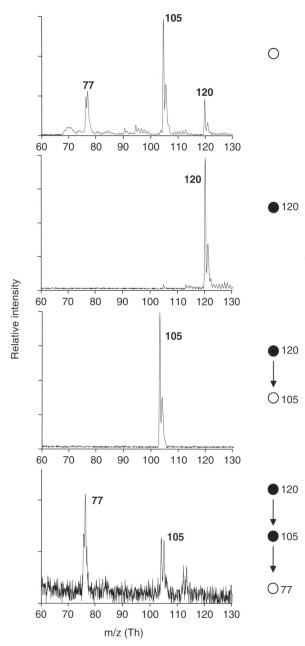

Figure 5.9. MS/MS/MS spectra of acetophenone. Molecular ion m/z 120 was first isolated using RF combined with DC. The RF was ramped to bring m/z 120 to $q_z = 0.61$; resonance excitation applied at 232 kHz, 500 mV$_{0-p}$. Fragment ion m/z 105 was then isolated using RF combined with DC. The RF was then ramped to bring m/z 105 ion to $q_z = 0.64$; resonance excitation applied at 251 kHz, 225 mV$_{0-p}$. (Reprinted by permission of Professor R. G. Cooks.)

axis of the ring stack are subjected to periodic potential oscillations somewhat akin to the time-varying potentials obtained when RF voltages are applied between adjacent rods of a quadrupole, hexapole, or octopole rod array. The first RF-only stacked-ring device was used by Bahr et al. [47] to study rate coefficients of ion/molecule interactions at thermal energies.

5.4.1. Electrostatic Ion Guide

Guan and Marshall reported on the operation of an ion guide using electrostatic potentials rather than RF potentials [48]. The oscillating electric potential of the stacked-ring ion guide focuses ions by exerting a field gradient force on the ions so as to direct ions to the central axis where the field is weakest. The radial pseudopotential well for the ion guide is steep sided (the radial pseudopotential $\propto e^r$), thus producing a broad electric field-free region in the center. In comparison, the radial pseudopotential well for a quadrupole field is less steep at the sides (the radial field is parabolic, that is, $\propto r^2$) with a minimum only at the central axis. The radial pseudopotential well for an octopole field is less steep (the radial field is $\propto r^6$) than that of the stacked-ring ion guide but steeper than that of the quadrupole field; it has a central field-free region that is smaller than that of the radial pseudopotential well. The application of DC voltages rather than high-amplitude RF voltages is convenient for an external ion source Fourier transform ion cyclotron mass spectrometer, for which ion detection is vulnerable to electrical noise in the same frequency range. A disadvantage of the stacked-ring ion guide is the requirement for high ion axial kinetic energy in order to avoid ion trapping in the shallow pseudopotential well between adjacent ring electrodes.

5.4.2. Ion Tunnel

In 2001, Giles and Bateman reported the application of a stacked-ring ion guide, shown in Figure 5.10, as an ion transmission device at intermediate pressures [49]. Each ring electrode of the ion tunnel is 0.5 mm thick and the interelectrode separation is 1 mm. An ion tunnel of length 100 mm will consist of 67 ring electrodes. The RF potential applied to each ring was 180° out of phase with that applied to its two adjacent rings such that the potential sign alternated down the ion tunnel. As discussed above (see the introduction), there is a need for highly efficient ion pipes for the transmission of ions through various pressure regions when using an atmospheric pressure ionization source.

5.4.2.1. Transmission Efficiency The transmission of ions through an ion tunnel was investigated as a function of ambient pressure, RF frequency, mass/charge ratio (i.e., transmission window), and the interelectrode separation, and the observations were compared with those obtained with a hexapole rod assembly. A limiting factor was the breakdown voltage; at a pressure of 3.8 mbars (~3 Torr, given that 1 Torr $\equiv$ 1.33 mbars) and an RF frequency of 0.8 MHz, electrical breakdown occurred at some 385 V_{p-p}. A change of RF frequency to 2.1 MHz did not affect the breakdown limit,

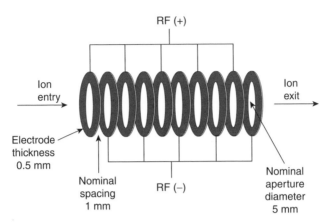

Figure 5.10. Schematic diagram of the stacked-ring ion guide. (From K. Giles, R. H. Bateman, paper presented at the Forty-Ninth Conference of the American Society for Mass Spectrometry and Allied Topics, Chicago, IL, May 27–31, 2001. Reprinted by permission of Waters Corporation.)

which varied almost insignificantly when the RF frequency was increased to 2.1 MHz, when the interelectrode separation was increased to 2.25 mm, and when a hexapole rod assembly was substituted for the ion tunnel. The transmission window was much reduced with an interelectrode separation of 2.25 mm compared with that for an interelectrode separation of 1 mm.

In Figure 5.11 is shown a transmission efficiency comparison of the ion tunnel (IT) and a hexapole rod assembly (Hx) at an RF frequency of 0.8 MHz and at two pressures, approximately 2 and 3 Torr. The comparison is given for four ion species of m/z 152, 472, 964, and 2034. For each ion species at a pressure of 3 Torr, the transmission efficiency of the ion tunnel (IT, 3 Torr) exceeds (100 cf. 60) that of the hexapole rod assembly (Hx, 3 Torr). Transmission efficiency is reduced generally at the lower pressure of 2 Torr, yet the transmission efficiency of the ion tunnel (IT, 2 Torr) remains superior to that of the hexapole rod assembly (Hx, 2 Torr). Thus, in the pressure region of 2–3 Torr, an ion tunnel is preferable to a hexapole rod assembly.

5.4.2.2. Charge State Discrimination A limiting factor in MS/MS can be the inability to identify a precursor ion, such as a molecular ion in an electron impact mass spectrum (see Chapter 7) or a peptide precursor ion (see Chapter 8). While a large proportion of peptide products from a tryptic digest of a protein have two or more charges by which their recognition is enhanced, low levels of peptide ions may be obscured by more intense singly-charged background ions. Wildgoose et al. [50] reported observations of the abundance of doubly- and triply-charged ions relative to that of singly-charged ions following their accumulation in an RF ion tunnel for up to 100 ms at pressures in the range 1.6–2.8 mbars (1.2–2.1 Torr).

A stacked ion guide, or ion tunnel, was employed as an ion storage device and coupled to a quadrupole/collision cell/time-of-flight mass spectrometer. The ion tunnel is shown in Figure 5.10. In a flow system, the ion tunnel confines ions radially and

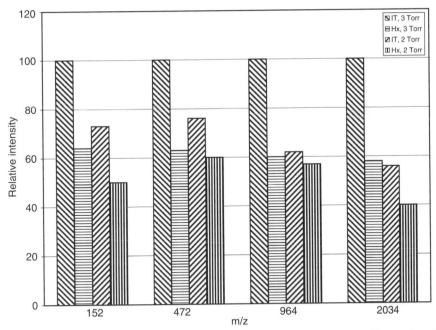

Figure 5.11. Comparison of transmission efficiencies of an ion tunnel (IT) and hexapole rod assembly (Hx) at RF of 0.8 MHz and at two pressures, approximately 2 and 3 Torr. (From K. Giles, R. H. Bateman, paper presented at the Forty-Ninth Conference of the American Society for Mass Spectrometry and Allied Topics, Chicago, IL, May 27–31, 2001. Reprinted by permission of Waters Corporation.)

reduces ion losses during transmission to the mass-selective QMF. When a potential of $+10\,V$ is applied to a gate electrode downstream of the ion tunnel, ions can be stored in the RF guide. Upon removal of the $+10\,V$ potential, ions are extracted for mass analysis in either the QMF or the time-of-flight mass spectrometer. When a mixture of 0.1-ng/μL solutions of gramicidin-S $[M + 2H]^{2+}$ of m/z 571.36 and leucine enkaphalin $[M + H]^+$ of m/z 556.28 was infused continuously from an electrospray source and the pressure in the RF guide was set to 2.8 mbars, mass spectra were obtained with zero storage and with 58 ms of storage in the trapping RF guide. The mass spectra showed a clear enhancement of the ratio of doubly-charged to singly-charged ions after 58 ms of storage. The singly-charged ion intensity was reduced by more than an order of magnitude while that of the doubly-charged ion was reduced by some 15%, leading to overall enhancement of the above ratio.

Upon repeating the experiment with renin substrate $[M + 3H]^{3+}$ of m/z 586.98 in place of gramicidin-S $[M + 2H]^{2+}$ but at a shorter storage time of 18 ms, a similar enhancement of the ratio of triply-charged to singly-charged ions was observed. In Figure 5.12 are shown the normalized intensities of triply-charged renin substrate

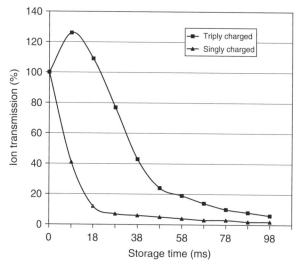

Figure 5.12. Effect of storage time on percentage transmission of triply- and singly-charged ions in an ion tunnel at 2.86 mbars (2.15 Torr). (From J. Wildgoose, J. Hayes, R. Bateman, A. Gilbert, in Proceedings of the Fifty-First Conference of the American Society for Mass Spectrometry and Allied Topics, Montreal, PQ, Canada, June 8–12, 2003. Reprinted by permission of Waters Corporation.)

and singly-charged leucine enkephalin ions. In both cases the singly-charged ion intensity reduces more rapidly than the multiply-charged ion intensity. Charge discrimination has been observed previously in pseudopotential wells; for example, the variations in the ratio of intensities of $Ar^{2+}/Ar^{+\cdot}$ in a QIT [34] bear a striking similarity to the normalized intensities shown in Figure 5.12. Thus with an appropriate storage time and pressure, the enhanced transmission of ions with two or more charges relative to that of singly-charged ions can provide a means of improving detection limits for low-level analysis of proteins.

The time-averaged radial force directed to the center of the RF guide is proportional to z^2/m and is greater for ions of the same mass m with higher charge states z or ions of the same substance with lower mass/charge ratios. Smith and coworkers [51] have reported the radial stratification of ions as a function of mass/charge ratio in a quadrupole ion guide in the presence of buffer gas. Ions of the same substance but with different mass/charge ratios, after sufficient cooling in a quadrupole rod set containing a buffer gas, congregate in separate concentric layers; ions with higher mass/charge ratios occupy layers with larger radii and are more easily lost by collision with the rods.

5.5. CONCLUSIONS

A factor common to the LIT, the RIT, and the IT is that each employs a two-dimensional trapping field for the confinement and/or transmission of ions. The versatilities of these new developments offer new opportunities for the application of mass spectrometry. In addition, each new development appears to offer enhanced efficiencies for ion detection.

There are a number of advantages to be gained from the operation of a quadrupolar ion-trapping device that has been extended longitudinally with respect to a three-dimensional QIT. Not only is the physical size of the ion-trapping volume increased in such a device with concomitant enhancement of ion confinement capability, but the facility with which externally-generated ions can be admitted and confined in the quadrupolar ion-trapping device and ejected subsequently in a mass-selective manner trumpets a novel instrument of both high sensitivity and high mass resolution. In these early days in the development of LITs, it is nevertheless clear that such instruments will find increasing application in the mass spectrometric examination of a wide range of compounds.

The high flexibility of the RIT structure, which is similar to that of the CIT (see Chapter 6), will permit rapid characterization of this device. Finally, the IT appears to offer a modestly superior alternative to multipole rod assemblies for the efficient transmission of ions through regions of relatively high pressure in instruments that are coupled with atmospheric pressure ion sources.

APPENDIX

A new hybrid instrument, the Finnigan LTQ FT™, comprising an LIT coupled with a Fourier transform ion cyclotron resonance (FTICR) mass spectrometer has been announced by Thermo Electron Corporation (http://www.thermo.com/ms). The MS^n capability of the LIT is used concurrently with the high resolution and accurate mass determination capabilities of the ICR on the LC timescale. The LIT and ICR components are combined linearly; ions are ejected axially from the LIT through ion guides into the ICR for an initial low-resolution ($R=25,000$) mass scan. Ions of highest abundance (predominantly doubly-charged ions in the case of peptide analysis) are identified and mass-selected subsequently for MS/MS in the LIT using both radial detectors. Concurrent with MS/MS of the identified doubly-charged ions in the LIT, a high mass resolution ($R=100,000$ at m/z 400) mass spectrum of primary ions is obtained from the ICR. The ICR performs 50 Fourier transform full scans per minute during which the LIT performs 150 data-dependent MS/MS scans. Accurate mass determination is also possible on product ions formed by CID in the LIT; additionally, alternative activation modes, such as IRMPD and ECD, can be applied to mass-selected ions trapped in the FT ICR cell. Sub-femtomole on-column sensitivity is claimed for this unique system

REFERENCES

1. W. M. Brubaker, *Adv. Mass Spectrom.* **4** (1968) 293.
2. W. M. Brubaker, U.S. Patent 3,129,327 (1964).
3. W. M. Brubaker, *J. Vac. Sci. Technol.* **10** (1973) 291.

4. W. Paul, H. P. Reinhard, U. von Zahn, *Z. Phys.* **152** (1958) 143.

5. W. Paul, H. Steinwedel, Apparatus for Separating Charged Particles of Different Specific charges, German Patent 944,900 (1956); U.S. Patent 2,939,952 (June 7, 1960).

6. U. von Zahn, *Z. Phys.* 168 (1962) 129; U. von Zahn, S. Gebauer, W. Paul, paper presented at the Tenth Annual Conference on Mass Spectrometry, New Orleans, 1962.

7. U. von Zahn, *Rev. Sci. Instrum.* **34** (1963) 1.

8. P. H. Dawson, *Quadrupole Mass Spectrometry and Its Applications*, Elsevier, Amsterdam, 1976. Reprinted as an "American Vacuum Society Classic" by the American Institute of Physics (ISBN 1563964554).

9. P. H. Dawson, N. R. Whetten, *Adv. Electron. Electron Phys.* **27** (1969) 59.

10. R. F. Wuerker, H. M. Goldenberg, R. V. Langmuir, *J. Appl. Phys.* **30** (1959) 441.

11. D. B. Langmuir, R. V. Langmuir, H. Shelton, R. F. Wuerker, Containment device, U.S. Patent 3,065,640 (1962).

12. A. F. Haught, D. H. Polk, *Phys. Fluids* **9** (1966) 2047.

13. A. A. Zaritskii, S. D. Zakharov, P. G. Kryukov, *Sov. Phys.-Tech. Phys.* **16** (1971) 174.

14. P. H. Dawson, *Int. J. Mass Spectrom. Ion Phys.* **6** (1971) 33.

15. U. Brinkmann, *Int. J. Mass Spectrom. Ion Phys.* **9** (1972) 161.

16. U. H. W. Brinkmann, U.S. Patent 4,090,075 (1978).

17. J. D. Prestage, G. J. Dick, L. Maleki, *J. Appl. Phys.* **66** (1989) 1013.

18. M. G. Raizen, J. M. Gilligan, J. C. Berquist, W. M. Itano, D. J. Wineland, *J. Mod Opt.* **39** (1992) 233.

19. J. D. Prestage, U.S. Patent 5,420,549 (1995).

20. J. E. Fulford, D. N. Hoa, R. J. Hughes, R. E. March, R. F. Bonner, G. Wong, *J. Vac. Sci. Technol.* **17** (1980) 829.

21. W. Paul, H. P. Reinhard, U. von Zahn, *Z. Phys.* **152** (1958) 143.

22. J. T. Watson, D. Jaoen, H. Mestdagh, C. Rolando, *Int. J. Mass Spectrom. Ion Processes* **93** (1989) 225.

23. L. M. Cousins, B. A. Thomson, Efficient multiple fragmentation (MS)^n using resonance excitation in flow-through RF quadrupole ion guides. Proc. 47th Ann. ASMS Conf. on Mass Spectrometry and Allied Topics, Dallas, TX, June 13–18, 1999, CD

24. R. E. March, R. Javahery, I. Galetich, C. Hao, X. Miao, Resonance excitation in flow-through RF-only ion guides. Proc. 49th Ann. ASMS Conf. on Mass Spectrometry and Allied Topics, Chicago, IL, May 27–31, 2001, CD.

25. J. C. Schwartz, M. W. Senko, J. E. P. Syka, U.S. Patent 5,420,425 (1995).

26. J. C. Schwartz, M. W. Senko, J. E. P. Syka, A two-dimensional quadrupole ion trap mass spectrometer, *J. Am. Soc. Mass Spectrom.* **13** (2002) 659.

27. J. W. Hager, U.S. Patent 6,177,668 (2001).

28. J. W. Hager, A new linear ion trap mass spectrometer, *Rapid Commun. Mass Spectrom.* **16** (2002) 512.

29. R. L. Alfred, F. A. Londry, R. E. March, *Int. J. Mass Spectrom. Ion Processes* **125** (1993) 171.

30. F. A. Londry, J. W. Hager, *J. Am. Soc. Mass Spectrom.* **14** (2003) 1130.

31. J. M. Campbell, B. A. Collings, D. J. Douglas, *Rapid Commun. Mass Spectrom.* **12** (1998) 1463–1474.

32. P. H. Hemberger, N. S. Nogar, J. D. Williams, R. G. Cooks, J. E. P. Syka, *Chem. Phys. Lett.* **191** (1992) 405.

33. P. H. Dawson, *Quadrupole Mass Spectrometry and Its Applications*, Elsevier, Amsterdam, 1976.

34. R. E. March, R. J. Hughes, J. F. J. Todd, *Quadrupole Storage Mass Spectrometry*, Wiley Interscience, New York, 1989.

35. J. C. Schwartz, Do space charge effects limit LC quadrupole ion trap performance? 9th Sanibel Conference on Mass Spectrometry, Sanibel Island, FL, January 25–28, 1997, p. 61.

36. J. W. Hager, U.S. Patent 6,627,876 (2003).

37. J. C. Schwartz, J. E. P. Syka, I. Jardine, *Rapid Commun. Mass Spectrom.* **2** (1991) 198.

38. M. W. Senko, J. C. Schwartz, Trapping efficiency measurements in a 2D ion trap mass spectrometer. Proc. 50th Ann. ASMS Conf. on Mass Spectrometry and Allied Topics, Orlando, FL, June 2–6, 2002, CD.

39. S. T. Quarmby, R. A. Yost, *Int. J. Mass Spectrom.* **190/191** (1999) 81.

40. Z. Ouyang, G. Wu, H. Li, W. R. Plass, J. R. Green, H. Chen, R. G. Cooks, private communication.

41. J. H. Billen, L. M. Young, POISSON/SUPERFISH on PC compatibles, Proc. of the 1993 Particle Accelerator Conference, May 17–20, 1993, Edited by S. T. Corneliussen, IEEE, New York, 1993, pp. 790–792.

42. G. Wu, Z. Ouyang, W. R. Plass, R. G. Cooks, Geometry optimization of cylindrical ion trap. Proc. 50th Ann. ASMS Conf. on Mass Spectrometry and Allied Topics, Orlando, FL, June 2–6, 2002, CD.

43. W. R. Plass, H. Li, R. G. Cooks. *Int. J. Mass Spectrom.* **228** (2003) 237.

44. J. F. J. Todd, R. M. Waldren, R. E. Mather, G. Lawson. *Int. J. Mass Spectrom. Ion Phys.* **28** (1978) 141.

45. S. Guan, A. G. Marshall, *Int. J. Mass Spectrom. Ion Processes* **157/158** (1996) 5.

46. W. R. Plass, The dependence of RF ion trap mass spectrometer performance on electrode geometry and collisional processes, dissertation, Justus-Liebig Universitaet Giessen, Giessen, 2001.

47. R. Bahr, D. Gerlich, E. Teloy, *Verhandl. DPG (VI)* **4** (1969) 343.

48. S. Guan, A. G. Marshall, *J. Am. Soc. Mass Spectrom.* **7** (1996) 101.

49. K. Giles, R. H. Bateman, paper presented at the Forty-Ninth Annual Conference of the American Society for Mass Spectrometry and Allied Topics, Chicago, IL, May 27–31, 2001.

50. J. Wildgoose, J. Hayes, R. Bateman, A. Gilbert, Charge state discrimination by means of ion storage at intermediate pressures. Proc. 51st Ann. ASMS Conf. on Mass Spectrometry and Allied Topics, Montreal, PQ, Canada, June 8–12, 2003, CD.

51. A. Tolmachev, R. Harkewicz, K. Alving, C. Masselon, G. Anderson, V. Rakov, L. Pasa-Tolic, E. Nikolaev, M. Belov, H. Udseth, R. D. Smith, Radial stratification of ions as a function of *m/z* ratio in collisional cooling radiofrequency multipoles used as ion guides with ion traps. Proc. 48th Ann. ASMS Conf. on Mass Spectrometry and Allied Topics, Orlando, FL, June 2–6, 2000, CD.

6

CYLINDRICAL ION TRAP MASS SPECTROMETER

Quadrupole Ion Trap Mass Spectrometry, Second Edition, By Raymond E. March and John F. J. Todd
Copyright © 2005 John Wiley & Sons, Inc.

6.1. INTRODUCTION

The ability to confine charged species to a specified element of space is of great importance for the examination of charged species, the study of their reactions with neutral species and with photons, and the determination of the mass/charge ratio of such species. The CIT has been derived as a simplified version of a Paul trap or QIT to affect such confinement of charged species. Two early reports [1, 2] described the operation with oscillatory fields of ion storage devices fabricated crudely from wire mesh. The fields at the electrodes of such devices are clearly different from those derived from electrodes with continuous surfaces, yet ion storage was observed when mesh electrodes were used. It was surmised that, because the fields at the center of each of the mesh and continuous electrode devices may not differ significantly, similar fields at the center of a device may be obtained by using barrel and planar electrodes to form an ion trap of cylindrical geometry.

Simplicity of fabrication of CITs permits ready investigation of their ion-confining properties as functions of size and geometry. The general direction of research into CITs has been the study of small ion traps, described as miniature ion traps, for the confinement of a single ion or for application as sophisticated tandem mass spectrometers. In a further development, multiple CITs have been mounted in an array. While the operating conditions for a QIT of usual size with $r_0 \approx 1$ cm can be transferred virtually unchanged (except for matching of the impedance of the device to the RF drive potential circuit) to a CIT of the same magnitude, this situation does not hold when the magnitude of the cylindrical device is reduced. In such a case, the drive frequency required for the smaller cylindrical device must be appreciably higher in order to realize a trapping potential well of sufficient magnitude to permit excitation of confined charged species.

6.2. INITIAL STUDIES OF A CYLINDRICAL ION TRAP

In 1962, Langmuir et al. [3] obtained a patent for a CIT that was, to all intents and purposes, an ion trap configured as the inscribed cylinder to a three-dimensional QIT, as shown in Figure 6.1. The equations given in the patent for the electric fields and for ion motion were identical with those for the QIT having hyperbolic electrodes. It is not clear whether a device was fabricated, but no experimental details of the performance of such a device were given. Benilan and Audoin [4] first gave a detailed description of the theory of ion containment in a CIT; their treatment considered the application of an RF potential to each planar end-cap electrode and the cylinder barrel was grounded.

6.2.1. Operation of a CIT

Bonner et al. [5] described the construction of a CIT having $r_1 = 1$ cm and $r_1^2 = 2z_1^2$ and, in so doing, showed that the cylindrical device is easier to construct than a

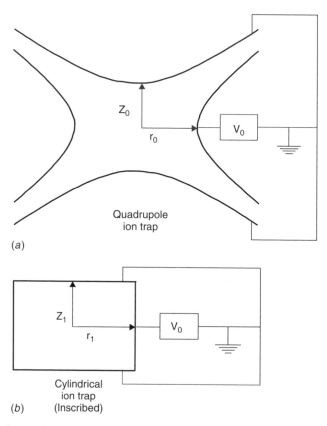

Figure 6.1. Comparison of geometries and operating modes for the three-dimensional quadrupole ion trap (QUISTOR) and a cylindrical ion trap: (*a*) QUISTOR, $r_0^2 = 2z_0^2$; (*b*) "inscribed" cylindrical ion trap, $r_1^2 = 2z_1^2$. (Reprinted from *Quadrupole Storage Mass Spectrometry* by R. E. March, R. J. Hughes, J. F. J. Todd, © Wiley-Interscience 1989. Reprinted with permission of John Wiley & Sons, Inc.)

QIT having hyperbolic electrodes and that variation of the ratio $r_1 : z_1$ is relatively facile. They reported the confinement of ions from the observation of self-chemical ionization of methane and presented a numerical analysis of the potential distribution within the ion trap. They commented further that ion storage was achieved at a lower RF potential than for a QIT and that extraction of stored ions may be more efficient in the cylindrical device. Mather et al. [6] determined stability diagrams for a number of ionic species confined within three ion traps having cylindrical geometry and reported on the first analysis of the mass-selective storage mass spectrum of *n*-hexane. Fulford et al. [7] described the operation of a CIT as a reactor for the study of ion/molecule reactions and reported on the application of resonant ion ejection to the elucidation of ion/molecule reaction mechanisms.

6.2.2. Further Development of the CIT

Much of the early work was reviewed in Chapter 5 of the first edition of *Quadrupole Storage Mass Spectrometry* [8]. Further development of the CIT, except for the theoretical work of Jardino and co-workers [9] and the photodissociation experiments of Mikami et al. [10–12], was stalled until after the commercialization of the QIT. Jardino and co-workers examined the effective potential of a CIT and the spatial symmetry and harmonicity of that potential. They found that the potential is not perfectly spherically symmetrical as in the case of the hyperbolic QIT. Nevertheless, they concluded that greater ion densities can be achieved and a greater total number of ions confined in a CIT than in one possessing quadrupole geometry, as had been reported earlier [5]. Mikami et al. [10–12] used a relatively large CIT ($r_0 = 2$ cm, $z_0 = 1.4$ cm) for two-color laser spectroscopy in conjunction with supersonic molecular beams. Ions generated by multiphoton ionization of chlorobenzene were stored for 10 ms or more, then photodissociated by radiation from a dye laser. Recently, Lee et al. [13] investigated the characteristics of a CIT; they reconfirmed that, contrary to a QIT, the stability space in a CIT is dependent upon the position of an ion.

6.2.3. Miniature Ion-Trapping Devices

From a survey of the literature, it would appear that for an ion trap to be described as miniature the radius of the ring or barrel electrode should be < 10 mm; when the radius is < 1 mm, such traps are described as submillimeter or micro ion traps. Miniature devices have been designed and optimized for specific applications [14, 15].

6.2.3.1. Stored-Ion Spectroscopy

Neuhauser et al. [14], in their study of stored-ion spectroscopy, constructed a submillimeter ion trap because they needed to optimize simultaneously ion density and laser irradiation of the ion cloud. Due to the difficulties of machining such small electrodes, high-accuracy electrodes of spherical form were machined. At the heart of their apparatus was a small ion trap with hemispherical end-cap electrodes and a ring electrode of circular cross section; the latter was slightly deformed elliptically to produce an effective trapping potential for a single Ba^+ ion. The theory of ion motion in such an ion trap has been discussed at some length by O and Schuessler [16]. Note that the deliberate introduction here of a degree of asymmetry to improve performance of the ion trap is similar to the optimization of the rectilinear ion trap, as discussed in Chapter 5, by decreasing the separation of the y plates relative to that of the x plates. An excellent review of the various trapping devices used for stored-ion spectroscopy has been given by Dehmelt [17].

6.2.3.2. Readily Machined Ion Traps

An alternative geometry for a small ion trap that was examined theoretically by Beaty [15] was employed by Wineland and co-workers [18, 19] in their study of laser-cooled Hg^+ ions; the dimensions of their ion trap were $r_0 \cong 445$ μm and $z_0 = 312$ μm. The RF drive frequency was 21 MHz. Beaty

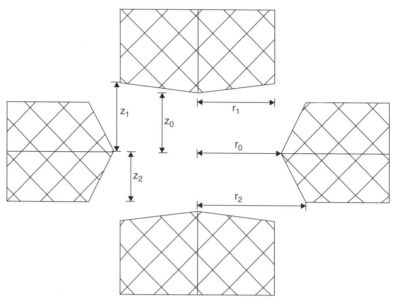

Figure 6.2. Readily machinable and variable electrode structures having the same general topology as ideal electrodes. (Reprinted from "Simple electrodes for quadrupole ion traps," by E. Beaty, *Journal of Applied Physics*, **61** (1987), 2118–2122. Reprinted with permission of the American Institute of Physics.)

considered a geometry in which the electrodes could be machined with simple lathe cuts. A cross-sectional view of the ion trap is shown in Figure 6.2; the end-cap electrodes have cylindrical symmetry with conical boundaries. The dimensions z_0, z_1, z_2, r_1, and r_2 are relative to r_0 and denote the degree of variation of this particular geometry; the magnitude of r_0 defines the size of the ion trap. Using a computational method [20], Beaty obtained values of z_0 through r_2 for which the fourth- and sixth-order anharmonic contributions to the potential vanished, and a close approximation to the ideal (quadrupole) field could be achieved in the central region of the ion trap.

6.3. MINIATURE CYLINDRICAL ION TRAPS (MINI-CITs)

Once the discovery of mass-selective axial ejection had been exploited with the QIT, thoughtful reflection on the application of the same technology to CITs, and particularly to cylindrical miniature ion traps, led to a resurgence of interest in this field. Mass analysis using the mass-selective axial instability scan in a CIT mass spectrometer was described first by Wells et al. [21]. Multiple-stage mass analysis has been demonstrated in a mini-CIT [22]. The performance of the CIT was similar to that of a commercial QIT fabricated with hyperbolic electrodes: sequential product ion scans to MS³ were demonstrated, baseline resolution of adjacent mass/charge ratios was observed, and the range of the CIT was comparable to that of a hypothetical QIT to which the CIT was inscribed.

6.3.1. Driving Force

The principal driving force for the development of CITs has been the increasing importance of field-portable analytical instrumentation [23–25] for in situ analysis of toxic pollutants and detection of chemical warfare (CW) agents [26, 27]. The detection and identification of CW agents is a problem in trace organic analysis of great challenge and relevance. An analytical device appropriate to this challenge must possess the following performance criteria: (i) high sensitivity for low-level detection and quantitation limits, (ii) high specificity (that translates as tandem mass spectrometry to MS^3 for a CIT) for detection and identification of target compounds, (iii) good precision and accuracy (that translates as good mass resolution for a CIT), (iv) high speed of analysis (short scan function and high ramping rate for a CIT), (v) continuous operation in real time and in situ, and (vi) low weight, compact, and battery powered for extended periods. A CIT can be fabricated readily from a polished metal cylinder with an internal radius of a few millimeters and two discs with suitable perforations for admission of electrons or ions and for ejection of ions. Such an ion trap is capable of storing an adequate mass range of charged species [21], can be operated as a mass spectrometer with tandem MS capabilities, and has good sensitivity.

6.3.2. Miniaturization

Once a physically small ion trap has been selected, miniaturization of the necessary power supplies is facilitated by the lower demands of the small trap, particularly the drive RF amplitude, that scale with the size of the CIT. The performance of a CIT depends strongly on its geometry [28]. Patterson et al. [29] have described the construction and performance of a mini-CIT mass spectrometer that had a mass/charge ratio range of ~250 Th and was capable of tandem mass spectrometry. The choice of CIT size was a compromise between increasingly severe relative field faults as the size was decreased and increasing power consumption and weight of the instrument as the size was increased. The compromise was a cylindrical barrel of internal radius r_0 of 2.5 mm and the separation of the end-cap electrode disks was 5.77 mm. A schematic representation of the ion source, ion optics, CIT, and detector is shown in Figure 6.3. The length of the cylindrical barrel was 4.21 mm because the distance between the cylindrical barrel and each end-cap electrode was 0.78 mm, defined by the thickness of delrin spacers. The end-cap electrode spacing was adjusted to 5.77 mm so as to add a small octopolar field (see Chapter 3) to that of the CIT in order to compensate for the field defects that arose due to the perforations of 2 mm diameter in the end-cap electrodes.

To achieve a mass/charge ratio range of ~250 Th with these electrode dimensions, a maximum drive RF voltage amplitude of 800 V_{0-p} oscillating at 2 MHz was employed. For a typical tandem mass spectrometer experiment, these experimental parameters correspond to irradiation of m/z 250 at, say, $q_z = 0.4$ (because 352 V_{0-p} is applied to the barrel electrode of the CIT) to yield a product ion mass spectrum

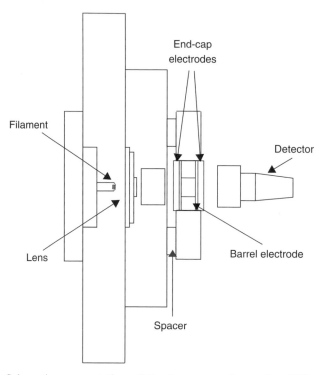

Figure 6.3. Schematic representation of the ion source, ion optics, CIT, and detector. Filament, mounting hardware, and gate electrode are based on those used in Finnigan ITS-40 instrument. (Reprinted by permission from "Miniature cylindrical ion trap mass spectrometer," by G. E. Patterson, A. J. Guymon, L. S. Riter, M. Everly, J. Griep-Raming, B. C. Laughlin, Z. Ouyang, R. G. Cooks, *Analytical Chemistry* **74** (2002), 6145–6153, Fig. 3. Copyright (2002) American Chemical Society.)

from m/z 110 to m/z 250; this range of fragment ions is reasonable. The next factor to consider is the pseudopotential well depth for m/z 250 at $q_z = 0.4$ in order to determine whether the pseudopotential well is sufficiently deep to permit resonance excitation leading to CID. The trapping depth $\bar{D}_z$ of the pseudopotential well (see Chapter 3) is given as

$$\bar{D}_z \approx \tfrac{1}{8} V_{0\text{-p}} q_z \qquad \text{[Eq. (3.18)]}$$

such that $\bar{D}_z \approx 352\,V_{0\text{-p}} \times 0.4/8 = 17.6\,\text{eV}$, that is, more than sufficient for CID.

6.3.3. Portable Miniature CIT

The mini-CIT mass analyzer described by Patterson et al. [29] accounts for <1% of the total volume and weight of the instrument and thus the major miniaturization effort was directed to the ancillary components. The weight of the version 5 mini-CIT [29] was 55 kg, including the computer and battery pack, and measured $\sim$45 $\times$ 60 $\times$ 71 cm; a smaller version 7 instrument of 16 kg and measuring 28 $\times$ 70 $\times$ 18 cm has been constructed also. Some 20 kg of version 5 was due to the system battery.

6.3.3.1. *Vacuum System*

Vacuum systems are normally large and heavy and consume a lot of power. However, in a mini-CIT, the oscillation amplitude of an ion at a given secular frequency is smaller than in an ion trap of standard size. As a result, not only do ions experience fewer collisions with buffer gas per RF cycle at a given pressure but these collisions occur with a diminished collision energy. Therefore, it is possible to operate a mini-CIT at a higher pressure than is normal and the pumping requirements are reduced. Turbomolecular pumps were chosen to evacuate the system because they can operate in any orientation to yield the required pumping speed and ultimate vacuum. The necessary secondary backing pumps used were KNF Neuberger (Trenton, NJ) four-stage diaphragm pumps because they are oil free, can be used in any orientation, and are available in small packages. This type of pump, which is capable of providing $\sim$400 mTorr at a flow rate of 13 L/min, is 14 $\times$ 14 $\times$ 30 cm and consumes $\sim$15 W of power in the absence of sample. Once again, a standard vacuum manifold was used; in this case, the manifold was taken from a Finnigan (ThermoFinnigan, San Jose, CA) ITS-40 ion trap mass spectrometer because it had appropriate flange connections and sufficient gas and electrical feedthroughs.

6.3.3.2. *Ionization Source*

Electron impact (EI) ionization was employed for the investigation of volatile compounds and a standard EI filament assembly from an ITS-40 instrument was used. The filament power supply was a 24-V variable-current power unit that provided $\sim$4 A of current and had trim pots for control of the bias voltage and emission current. The entire circuit, including power unit and test points, is enclosed in a 5 $\times$ 5 $\times$ 1-cm volume. A gate electrode provides a means for arresting the electron flow into the CIT during ion cooling and mass analysis.

6.3.3.3. *Detector System*

The higher operating pressure of the mini-CIT than that of the QIT is advantageous for CIT operation but is a cause for concern with respect to the possible shortening of the lifetimes of the filament, vacuum pumps, and detector. A channel electron multiplier detector from K and M Electronics (West Springfield, MA) was chosen because it is small (3.6 $\times$ 0.6 cm) and can operate at a pressure of 7 $\times$ 10^{-4} Torr. Initially [29], the detector was mounted axially, as shown in Figure 6.3; later, in version 7, an off-axis mounting was used. Because the detector output is below the noise floor of standard digitization circuits, the output must be fed into a preamplier. The unit selected was a Keithley Instruments (Cleveland, OH) model 427 current amplifier; this instrument has variable gain control from 10^4 to 10^{11} that permits the monitoring of very high level events such as multiplier gating and very low level events such as ion signals of low abundance. The variable time constant of the preamplifier permitted a choice between peak width and sensitivity.

6.3.3.4. Waveform Generation   The five waveforms used in the operation of the mini-CIT (and for almost every other type of QIT) are the drive RF potential for the confinement of ions, the RF amplitude modulation potential, an arbitrary waveform for the isolation of selected ion species in the ion trap, an AC potential for ion excitation in tandem mass spectrometry (MS/MS and MS/MS/MS) experiments, and a single frequency potential for resonant ejection of ions in mass-selective axial ejection. Each of these potentials is independent of the other potentials and must be generated and applied separately. The drive RF potential applied to the ring or barrel electrode is obtained from a crystal oscillator circuit operated at a single frequency of 2000 MHz. The amplification circuit is frequency tuned to a single frequency. The RF amplitude modulation potential for modulation of the drive RF potential is obtained from an arbitrary waveform.

The remaining potentials are applied normally to the end-cap electrodes. The waveform used for ion isolation is calculated using the SWIFT (stored waveform inverse Fourier transform) method described by Guan and Marshall [30]. The waveform is calculated such that ranges of ion species can be ejected resonantly but the frequencies corresponding to the secular frequencies of ions to remain in the ion trap are set to zero. The waveform used for ion excitation leading to CID can be varied but is set normally to correspond to $q_z \cong 0.4$, as discussed above. Typical AC amplitudes used in the mini-CIT are $\sim 2\,V_{0\text{-}p}$. The application of a single frequency potential for resonant ejection (or axial modulation) of ions in mass-selective axial ejection is to eject ions rapidly and to improve mass resolution [31]. The potential is applied in dipolar mode to the end-cap electrodes and corresponds to some value of q_z on the q_z axis; all three electrodes are held at the same DC level so that a_z for all ions is zero. The frequency used for resonant ejection was an octopole resonance that caused rapid ejection of ions with the production of narrow peaks in the mass spectrum [32]. The choice of the ejection q_z value determines directly the degree of mass range extension [33]. See also FNF, Section 3.4.3.1.

In a typical experiment, the SWIFT potential is applied following a cooling period once ionization has been terminated; then a further cooling period is followed by application of the CID excitation potential. The SWIFT pulses are calculated using the ITSIM [34] ion trajectory simulation program. The card used for waveform generation for the end-cap electrodes has 16 MB of memory so that it may store all three of the waveforms. The arbitrary waveform is split into two wires and one of the potentials is inverted so as to provide dipolar excitation. The waveform circuits can be floated with respect to ground in order to reduce ion kinetic energy where necessary, such as when a glow discharge source is used.

6.3.3.5. Control Software   Virtually all of the electronics required custom-software control. The user interface was also custom designed. Many of the required components were purchased from National Instruments Corporation, and so it was possible to take advantage of National Instruments' proprietary programming language, LabVIEW, for which they provide support. Because LabVIEW drivers are available for all of the components purchased from National Instruments, much of the basic software/firmware was prewritten.

The user interface was designed to run within the LabVIEW program window and the data are displayed automatically as the average of a user-selected number of scans.

The instrument panels are addressed in three subwindows that permit control of the waveforms and potentials applied to all three electrodes as well as the ionization parameters and data acquisition. The subwindow by which the barrel electrode waveforms are controlled permits user-defined timing sequences and application of user-selected waveforms independent of the sequences employed in the other two subwindows. For example, the software has a built-in ramp function available for the ion ejection event; when the RF amplitude at the start and stop points of the RF ramp is inputted by the user, the corresponding ramp length and ramp rate are obtained. Such independence permits much higher flexibility than is available in commercial ion trap control software.

In Figure 6.4 is shown a typical timing diagram or scan function by which an EI mass spectrum may be obtained. A scan function is a diagram showing the temporal relationships of the various potentials applied during the course of an ion trap experiment. The duration of application of each of the potentials is indicated along with the duration of each cooling period. The SWIFT waveform is indicated schematically as a square wave with an interposed slot that corresponds to the secular frequency of the ion species to be isolated. The CID waveform is shown as a single-frequency sine

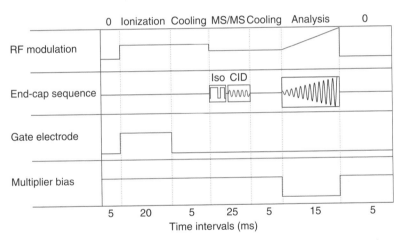

Figure 6.4. Timing diagram or scan function for miniature CIT. Gate electrode is opened concurrently with application of drive RF potential at $t = 5$ ms. Gate electrode permits electrons to enter CIT during 20 ms for ionization of compound present at low pressure and closed again. Amplitude of drive RF potential determines LMCO of ion trap. At conclusion of cooling period of 5 ms to permit ion ensemble to form ion cloud, drive RF amplitude is lowered (normally) so as to establish q_z value of ~0.4 for ion isolation and CID. Ion isolation waveform is applied at $t = 30$ ms for ~10 ms; then after 2–3 ms, ion activation waveform is applied for CID. Following further cooling period of 5 ms, analytical ramp of drive RF commences. Also shown in scan function is multiplier bias that ensures operation of multiplier during mass analysis period only. (Reprinted by permission from "Miniature cylindrical ion trap mass spectrometer," by G. E. Patterson, A. J. Guymon, L. S. Riter, M. Everly, J. Griep-Raming, B. C. Laughlin, Z. Ouyang, R. G. Cooks, *Analytical Chemistry* **74** (2002), 6145–6153, Fig. 5. Copyright (2002) American Chemical Society.)

wave. The analysis waveform is shown as a fixed-frequency sine wave increasing in amplitude at a constant rate.

6.3.4. Mini-CIT System Performance

This examination of system performance is limited to verification of the linearity of the mass calibration scale; investigation of the efficiencies of ion isolation, ion excitation for CID (MS/MS), and MS/MS/MS; and determination of the experimental conditions that either promote or diminish mass resolution.

6.3.4.1. Mass Calibration   In Figure 6.5 are shown two mass spectra obtained with mini-CITs using perfluorotri-*n*-butylamine (PFTBA or FC-43), a mass calibrant used

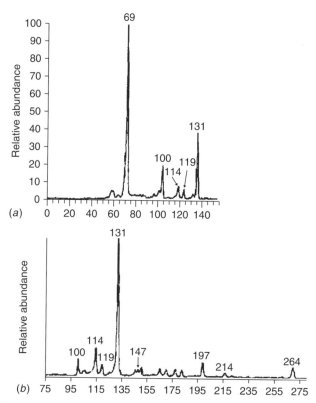

Figure 6.5. Mass spectra obtained from PFTBA with two miniature CITs: (*a*) lower mass range obtained with version 5 miniature CIT, correlation coefficient (R^2) of 0.9999 calculated for this mass spectrum; (*b*) higher range mass spectrum, also of PFTBA, obtained with version 7 miniature CIT. (Reprinted by permission from "Miniature cylindrical ion trap mass spectrometer," by G. E. Patterson, A. J. Guymon, L. S. Riter, M. Everly, J. Griep-Raming, B. C. Laughlin, Z. Ouyang, R. G. Cooks, *Analytical Chemistry* **74** (2002), 6145–6153, Fig. 7. Copyright (2002) American Chemical Society.)

commonly in mass spectrometry. The low-range mass spectrum in Figure 6.5a was obtained with version 5 mini-CIT using a scan function similar to that shown in Figure 6.4 with the isolation and CID periods omitted. The indicated pressure of PFTBA was 4×10^{-6} Torr. The drive RF amplitude established a LMCO of ~48 Th initially and, during the analysis period, it was ramped to ~250 Th in 15 ms, giving a ramp rate of ~13,500 Th/s. The resonance ejection frequency was 700 kHz at an amplitude of ~5 V_{0-p}. The ion signal intensity data were imported to an Excel spreadsheet program for calculation of the linearity of the calibration. The expected peaks in this mass range were observed together with the expected base peak of m/z 69 (CF_3^+). The mass calibrations were linear with $R^2 = 0.9999$. The higher range mass spectrum of PFTBA in Figure 6.5b was obtained using the smaller version 7 mini-CIT. Again, the expected peaks in this mass range were observed, particularly the base peak, m/z 131 ($C_3F_5^+$), in addition to some background peaks, for example, m/z 197.

6.3.4.2. Tandem Mass Spectrometry

In Table 6.1 are shown the results from some representative experiments in which mass spectra were obtained and, subsequently, product ion mass spectra were obtained by MS/MS and MS/MS/MS [29]. Ion species are identified by their mass/charge ratio; the intensity of each peak is given in parentheses and in units of volts so that the absolute performance of the instrument may be appreciated. In the MS column are listed the components of single-stage mass spectra obtained from p-nitrotoluene, acetophenone, methyl salicylate, and dimethyl methyl phosphonate (DMMP) in the absence of bath/buffer or collision gas. DMMP is an important precursor and an impurity in G nerve agents (cholinesterase inhibitors) of the organophosphate family that includes Tabun, Sarin, Soman, and VW. It is a widely used simulant for phosphorus ester CW agents and has been examined previously by ion trap mass spectrometry [35, 36].

The major peaks observed in the single-stage mass spectrum from p-nitrotoluene are shown in the MS column; the base peak is m/z 137 and the observed peak intensity was 4.8 V. Also shown in the MS column are the single-stage mass spectra for acetophenone, methyl salicylate, and DMMP. The mass spectra are in good agreement with published data, except for that of DMMP. The mass spectrum observed here is similar to that observed from a QIT by Riter et al. [37] and shows the same ions as given in the National Institute of Standards and Technology (NIST) mass spectrum [38]; however, the base peak in the mini-CIT mass spectrum was observed, in the absence of collision gas, to be the molecular ion, m/z 124. In the presence of helium collision gas, the base peak became the fragment ion m/z 109 while the ion of m/z 94 is the base peak in the NIST mass spectrum.

The third and fourth columns show data from MS/MS experiments. The column labeled MS/MS Isolation gives the peak heights of ions after isolation with the SWIFT waveform but prior to CID. Normally, in an MS/MS experiment, one does not observe ion signals corresponding to the isolated ion (prior to CID) because such an observation requires ejection of the isolated ion species from the CIT. Once the isolated ions have been ejected, no ions remain for CID! Thus, the results shown in

TABLE 6.1. Single-Stage MS, MS/MS, and MS/MS/MS Spectra of *p*-Nitrotoluene, Acetophenone, Methyl Salicylate, and DMMP[a]

Compound	MS[b]	MS/MS		MS/MS/MS	
		Isolation[b]	Dissociation[c]	Isolation[b]	Dissociation[c]
p-Nitrotoluene	137 (4.8)	137 (4.6)	137 (2.0)	121 (0.5)	121 (0.5)
		121 (0.4)	121 (2.0)		107 (0.08)
			107 (0.2)		91 (0.1)
			91 (0.3)		
	121 (1.1)				
	107 (1.4)				
	91 (1.2)				
Acetophenone	120 (1.3)	120 (1.0)	120 (0.4)	105 (0.6)	105 (0.4)
		105 (0.2)	105 (0.7)		77 (0.06)
	105 (2.0)		77 (0.06)		
	77 (1.2)				
Methyl salicylate	152 (1.4)	152 (1.1)	152 (0.2)	152 (0.06)	N/A
			120 (0.7)	120 (0.6)	
			92 (0.1)		
	120 (0.9)				
	92 (0.7)				
DMMP	124 (1.0)	124 (1.7)[c]	124 (0.7)	N/A	N/A
		79 (0.2)	94 (0.1)		
			79 (0.5)		
	109 (0.2)				
	94 (0.4)				
	79 (0.8)				

Source: Reprinted by permission from "Miniature cylindrical ion trap mass spectrometer," by G. E. Patterson, A. J. Guymon, L. S. Riter, M. Everly, J. Griep-Raming, B. C. Laughlin, Z. Ouyang, R. G. Cooks, *Analytical Chemistry* **74** (2002), 6145–6153, Table 1. Copyright (2002) American Chemical Society.

[a]All data were obtained from pure samples of the compounds. In all cases, the data are given as mass/charge ratio with the peak height given in volts in parentheses.
[b]No bath or buffer or collision gas.
[c]Helium was used as the bath/buffer/collision gas.

the third and fourth columns were obtained from two experiments with each compound. For each of the four compounds examined, only the ion of highest mass/charge ratio was subjected to isolation. For *p*-nitrotoluene, the *m/z* 137 ion has an intensity of 4.6 V following isolation. Thus, in this case, the isolation procedure using 0.04 V amplitude, 270–300 kHz, and 4 ms isolation time has an efficiency of 96%. Immediately below 137 (4.6) is shown the intensity (0.4 V) of *m/z* 121 that remained in the mini-CIT along with *m/z* 137 after isolation. The signal intensity of *m/z* 121 constitutes a percentage residual ion abundance of 8%. The *m/z* 120 ion from acetophenone shows an isolation efficiency of 77% and percentage residual ion abundance of 17%. The *m/z* 152 ion from methyl salicylate shows an isolation efficiency of 79% and percentage residual ion abundance of 0%. These results are very encouraging.

In the column labeled MS/MS Dissociation are presented the components of the product ion mass spectrum of each of the ions isolated previously. For example, CID of m/z 137 from p-nitrotoluene yields a residual ion signal intensity of 2.0 V for m/z 137 and intensities of 2.0, 0.2, and 0.3 V for m/z 121, 107, and 91, respectively. Thus the attenuation of the molecular ion of m/z 137 is 56% (corresponding to a dissociation efficiency of 44%). The total ion intensity of the components of the product ion mass spectrum of m/z 137 is 4.5 V so that, when compared with the total ion intensity (including that for m/z 121) at completion of the isolation stage, 5 V, it yields an overall ion retention efficiency during CID of 90%. The conditions employed for CID of p-nitrotoluene were excitation at 0.05 V at 270 kHz for 51 ms. Similar dissociation and ion retention efficiencies were observed for the other compounds.

The fifth and sixth columns show data from MS/MS/MS experiments that permit evaluation of the efficiencies of the isolation and CID procedures and retention of total ion charge. For example, in the case of p-nitrotoluene, the intensity of m/z 121 in the MS/MS Dissociation column is 2.0 V and in the MS/MS/MS Isolation column is 0.5 V. Thus the isolation efficiency here is 25%. Collision-induced dissociation of m/z 121 with 0.003 V at 300 kHz for 51 ms does not change the intensity of m/z 121, as shown in the sixth (and last) column, thus yielding zero dissociation efficiency; however, two fragment ions are observed at low intensity, m/z 107 at 0.08 V and m/z 91 at 0.1 V, signifying an ion retention in excess of 100%. Acetophenone provides an almost ideal example of an MS/MS/MS experiment. From the fourth and fifth columns, it is seen that the isolation efficiency for m/z 105 is 86% and, from the fifth and sixth columns, the dissociation efficiency is 33% while the ion retention efficiency is 77%. The MS^3 product ion mass spectrum shows unchanged m/z 105 and a single fragment ion of m/z 77. The overall efficiency of the formation of m/z 77, through two repeated processes of isolation and CID, from m/z 120 is 5%. With complete dissociation of m/z 120 and m/z 105, which is possible with ion traps, the overall efficiency could be increased to 19%; this MS^3 achievement is remarkable for a mini-CIT.

6.3.4.3. *Limit of Detection and Mass Resolution*   The limit of detection for the CW simulant methyl salicylate was estimated as 1 pg. The mass spectrum obtained from methyl salicylate vapor was recorded over 1 s and yielded a single peak corresponding to the molecular cation at m/z 152; the signal/noise ratio was 7:1.

6.4. MEMBRANE INTRODUCTION MASS SPECTROMETRY

The application of a permeable membrane for the introduction of volatile compounds to a mass spectrometer has been combined with a mini-CIT for air analysis and for aqueous solution analysis [39]. Miniaturized membrane inlet systems were designed to the scale of a mini-CIT, as shown in Figure 6.6, so as to reproduce

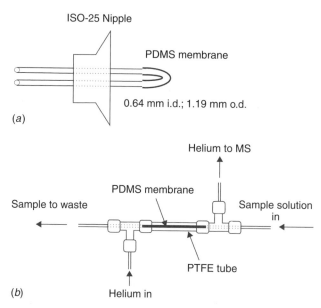

Figure 6.6. Miniature membrane inlet systems fabricated to fit a mini-CIT mass spectrometer: (*a*) internal MIMS system used for air analysis; (*b*) external MIMS system used for water analysis. In each case, the PDMS membrane is in a tubular configuration. (Reprinted by permission from "Analytical performance of a miniature cylindrical ion trap mass spectrometer," by L. S. Riter, Y. Peng, R. J. Noll, G. E. Patterson, T. Aggerholm, R. G. Cooks, *Analytical Chemistry* **74** (2002), 6154–6162, Fig. 1. Copyright (2002) American Chemical Society.)

many of the features of a direct-insertion membrane probe [40]. The miniature membrane introduction mass spectrometry (MIMS) system shown in Figure 6.6*a* was fabricated from two stainless tubes [1.65 mm outside diameter (o.d.), 1.0 mm internal diameter (i.d.)] of length 8 cm with 2 cm exposed to the vacuum; the tubes were inserted through holes 6 mm apart drilled into a blank stub and welded in place. The tubes were connected inside the vacuum by a loop of poly(dimethylsiloxane), PDMS, of length 4 cm with a surface area of 0.75 cm². External air was passed through the steel tubing and PDMS membrane; volatile compounds in the air sample permeated the membrane and passed into the mini-CIT where they could be ionized and analyzed. In Figure 6.7 is shown a mass spectrum of air sampled online while a cough drop (active ingredient menthol) was held 3 cm from the membrane inlet of the mini-CIT mass spectrometer. The labeled peaks in Figure 6.7 are characteristic of menthol.

A second MIMS inlet system described previously [41] was coupled to a mini-CIT via a vacuum feedthrough. A cross-linked PDMS tube 3.5 cm in length, 0.64 mm i.d., and 1.19 mm o.d. was used as the permeable membrane. Aqueous solutions were passed through the permeation tube in the direction shown in Figure 6.6*b*. Helium was flowed countercurrently to the solution direction to sweep the permeating compounds from the inlet to the mini-CIT for mass analysis.

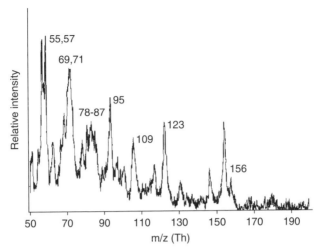

Figure 6.7. Mass spectrum of air sampled online while a cough drop (active ingredient menthol) is held 3 cm from membrane inlet of mini-CIT mass spectrometer. The labeled peaks are characteristic of menthol. (Reprinted by permission from "Analytical performance of a miniature cylindrical ion trap mass spectrometer," by L. S. Riter, Y. Peng, R. J. Noll, G. E. Patterson, T. Aggerholm, R. G. Cooks, *Analytical Chemistry* **74** (2002), 6154–6162, Fig. 4. Copyright (2002) American Chemical Society.)

6.5. MINIATURE CYLINDRICAL ION TRAP ARRAY

In some of those areas where mass spectrometry has become the analytical technique of choice, there exist needs to screen large numbers of samples as quickly as possible. The demands for high throughput occur in combinatorial synthesis [42], in proteomics where there is a need to monitor proteins and peptide mixtures [43], in industrial process monitoring [44], and in the emerging field of metabolomics [45]. One solution to the problem of high throughput is to use a multiplexed inlet system with a multiplexed mass spectrometer having an equal number of parallel sample channels. To this end, the simultaneous trapping of ion clouds in miniature ion traps, with each ion trap of a different size and where the size dictates a specific mass trapping range, has been proposed [28]. A high-throughput CIT array mass spectrometer has been developed and tested [46–48], where each mini-CIT is of the same size.

In this array, there are four mini-CITs having barrel electrodes of 2.5 mm i.d., as discussed above. Each CIT has its own inlet system and filament assembly together with a small (2-mm-diameter) electron multiplier, all within a common vacuum manifold and with a single set of control electronics. An array of identical mini-CITs has been characterized previously as a single-channel mass spectrometer [49]. In addition, sample ions are generated externally to the mini-CIT and transported through an Einzel lens into each mini-CIT, as shown in Figure 6.8 for two of the four parallel

References pp. 208–210.

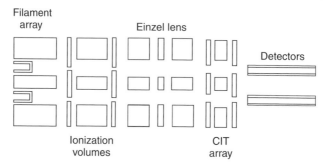

Figure 6.8. Schematic representation of two of the four parallel analysis channels in the 2×2 array of mini-CITs. (Reprinted by permission from "High-throughput miniature cylindrical ion trap array mass spectrometer," by A. M. Tabert, J. Griep-Raming, A. J. Guymon, R. G. Cooks, *Analytical Chemistry* **75** (2003), 5656–5664, Fig. 2. Copyright (2003) American Chemical Society.)

channels. The instrument provides a mass/charge ratio range of approximately m/z 50–500. The average peak width is 0.3 Da/charge, yielding a mass resolution $(m/\Delta m)$ of 1000 at m/z 300. Because each channel has its own inlet system, the compounds examined in the four channels may differ from each other. In Figure 6.9 are displayed the results of a high-throughput experiment [48] in which three of the four channels were operated simultaneously; acetophenone, bromobenzene, and 1,3-dichlorobenzene were introduced into channels 1, 2, and 3, respectively. Channel 4 was idle. Helium pressure was $\sim 1 \times 10^{-4}$ Torr, and resonance ejection was employed with an amplitude of 5 V_{p-p} at 700 kHz. The mass spectra required 100 ms of ionization and the total emission current was $\sim 20 \mu A$. The successful operation of this array of mini-CITs is a remarkable achievement.

6.6. FIELD APPLICATIONS OF MINI-CITs

A mini-CIT is currently under development as a portable explosives' detector mass spectrometer for use in the field [50, 51]. The instrument employs a Teledyne Discovery 2 ion trap mass spectrometer that has been modified by the substitution of a mini-CIT for the usual ion trap and by augmentation of the RF drive frequency to 2.1 MHz. The mini-CIT is similar to that described by Badman and Cooks [52] in that $r_0 = 2.5$ mm and $z_0 = 2.35$ mm. Resonant ejection of ions is accomplished with a forced asymmetric trajectory (FAST) signal. The current mass range of the instrument is 40–550 Th, which is best scanned in two sections, 50–250 and 250–550 Th. When combined with a gas chromatograph, the dynamic range of the mini-CIT was determined for a fixed ionization period by measurement of the area under the chromatographic peak for a range of concentrations of o-nitrotoluene; a linear response was observed for the range 500–10 ng. In Figure 6.10 is a total ion current chromatographic trace obtained with the mini-CIT and showing separation and identification of each of 12 explosives and related compounds injected as a mixture

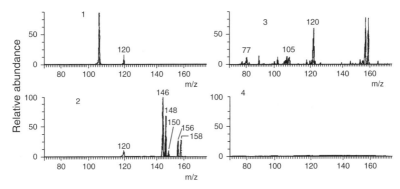

Figure 6.9. Mass spectra obtained from a high-throughput experiment during which three channels of the four-channel array instrument were used to analyze three different compounds simultaneously: (1) acetophenone; (2) bromobenzene; (3) 1,3-dichlorobenzene; (4) idle. (Reprinted by permission from "High-throughput miniature cylindrical ion trap array mass spectrometer," by A. M. Tabert, J. Griep-Raming, A. J. Guymon, R. G. Cooks, *Analytical Chemistry* **75** (2003), 5656–5664, Fig. 7. Copyright (2003) American Chemical Society.)

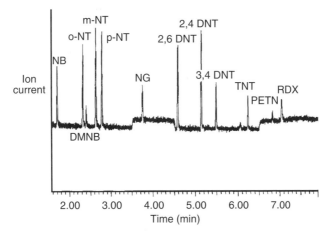

Figure 6.10. Total ion current trace showing peaks due to 12 explosive compounds and related compounds separated by gas chromatography. Each peak represents the total ion charge of mass spectra obtained from a cylindrical ion trap. Compounds are as follows: NB, nitrobenzene; *o*-NT, *ortho*-nitrotoluene; DMNB, diaminonitrobenzene; *m*-NT, *meta*-nitrotoluene; *p*-NT, *para*-nitrotoluene; NG, nitroglycerine; 2,6-DNT, 2,6-dinitrotoluene; 2,4-DNT, 2,4-dinitrotoluene; 3,4-DNT, 3,4-dinitrotoluene; TNT, trinitrotoluene; PETN, pentaerythritol-tetranitrate; RDX. (From S. N. Cairns, J. Murrell N. J. Alcock; S. R. Dixon, P. A. Cartwright, M. D. Brookes, D. M. Groves, presented at the Sixteenth International Mass Spectrometry Conference, Edinburgh, UK, 2003, Poster 08064. © British Crown copyright, DSTL, Published with the permission of the Controller of Her Majesty's Stationery Office.)

in solution onto the gas chromatographic column. The tandem mass spectrometry efficiency, defined as the ratio of fragment ion to precursor ion signal intensities, of 30–40% has been achieved using n-butylbenzene. The mini-CIT has a variable scan rate from 1000 to 20000 amu/s.

6.7. MICRO ION TRAPS

The development of micro ion traps, where the critical dimensions are in the sub-millimeter range, is becoming an area of great interest. In the achievement of specific objectives, such as the study of a single ion or the utilization of micro ion traps with microfabricated electrospray ion sources, efforts must be made to optimize the field within the device to the task in hand.

6.7.1. Single-Ion Study

A micro-CIT has been designed for laser interrogation of a single ion of Ca^+ in a metrology application [53]. Here, strong confinement was required in order to access the Lamb–Dicke regime; for an ion trap wherein the barrel electrode radius is of the order of 1 mm, this could be achieved only with a high confinement or drive frequency. The micro-CIT used here [53] is of particular interest because it has no apparent end-cap electrodes and it uses two pointed positioning electrodes, as shown in Figure 6.11; the positional electrodes bring about small corrections in the storing potential. The micro-CIT has an open Paul–Straubel structure [54–56] for easy visualization of the confined charged particles and access for the laser beams. The barrel electrode has $r_1 = 0.7$ mm and an overall height of $2z_1 = 0.85$ mm. The two compensation electrodes are flat rings, 13 mm in diameter, with centered openings of diameter 10.5 mm covered by a fine mesh. Each of these electrodes is 5.5 mm from the trap center and they screen the trapping volume from stray electrical fields. For a confining potential of $700\,V_{rms}$ at a frequency of 11.6 MHz, a potential well depth of 9.2 eV is realized for $q_r = 0.1$.

6.7.2. Optimization of Micro Ion Traps

Recent reports of microfabricated electrospray ion sources enhance the possibilities for combining electrospray ionization with miniature mass spectrometers [57, 58], though such a combination has not been reported thus far. Several geometries for miniature ion traps are amenable to microchip construction methods, for example, the planar sandwich of three electrodes used by Brewer et al. [59] and the multiple-electrode structure of Wang and Wanczek [60]. The effects of ion trap geometry have been examined by Hartung and Avedisian [61]. Kornienko et al. [26] have explored the possibility of performing mass spectrometry with submillimeter ion traps of cylindrical geometry. In the selection of the geometry of an ion-trapping device, the principal criterion is the generation of a quadrupole field at the device center when one or more RF potentials are applied to the electrodes. The fraction of the applied potential from which the field is generated is, however, geometry dependent. For a CIT, Kornienko et al. [27] have calculated both the ratio of octopole to quadrupole

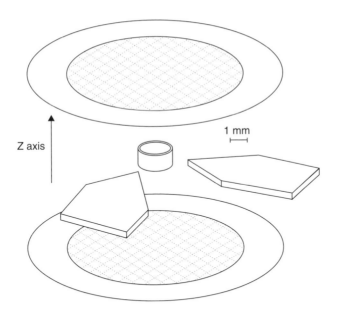

Figure 6.11. View of a miniature CIT showing the two circular compensation electrodes with large apertures covered by fine mesh and two pointed positioning electrodes. (Reprinted from "Characterization of a miniature Paul–Straubel trap," by C. Champenois, M. Knoop, M. Herbane, M. Houssain, T. Kaing, M. Vedel, F. Vedel, *European Physics Journal* **D15** (2001), 105–112, Fig. 1. Reprinted with permission of Springer-Verlag GmbH.)

coefficients ($r_0^2 C_4/C_2$) and the coefficient of the quadrupole contribution to the potential ($r_0^2 C_2/V_{ring}$) versus z_0/r_0, as shown in Figure 6.12. For values of z_0/r_0 less than 0.8, the quadrupole component is a larger fraction of the applied potential than that for an ideal quadrupole ion trap having the same value of r_0.

Multipole fields of order higher than quadrupolar can have significant effects on the operation of ion traps [62]. Wells et al. [21] showed that the octopolar field of a CIT changes sign as z_0/r_0 is varied; thus it may be possible to choose an appropriate CIT geometry such that the influence of the octopolar component is optimized. From Figure 6.12 it is seen that, for the z component of the octopolar field to be in the same direction as the quadrupolar field (i.e., $r_0^2 C_4/C_2 \rightarrow$ positive), the value of z_0/r_0 should be greater than 0.83. Under this condition, ion secular frequency will increase with trajectory amplitude and mass resolution will be enhanced. For their submillimeter CIT, Kornienko et al. [27] chose $r_0 = 0.5$ mm and $2z_0 = 1.0$ mm and employed an RF potential oscillating at 5.80 MHz; they achieved a mass range of 30–400 Da.

6.8. CONCLUSIONS

Much progress has been achieved recently in the development of CITs as field-portable analytical instruments. The major problems associated with miniaturization

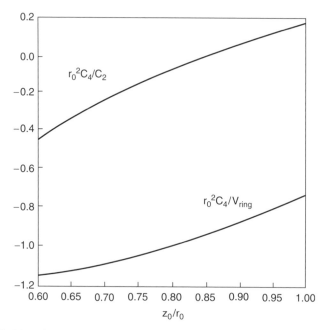

Figure 6.12. Plot of the ratio of the octopole to quadrupole coefficients (upper curve) and coefficient of the quadrupole contribution to the potential, $r_0^2 C_2/V_{ring}$ versus z_0/r_0 (lower curve). (Reprinted from *Rapid Communications in Mass Spectrometry*, Vol. 13, "Micro ion trap mass spectrometry," by O. Kornienko, P. T. A. Reilly, W. B. Whitten, J. M. Ramsey, Fig. 1, 50–53 (1999). © John Wiley & Sons Limited. Reproduced with permission.)

of power supplies and vacuum pumps have been recognized and are being surmounted at this time. While the in situ analysis of trace organic toxic pollutants and the detection of chemical warfare agents using mini-CITs are problems of great challenge and relevance, perhaps the greatest advances involving CITs will lie in the applications of micro ion traps in combination with microfabricated electrospray ion sources.

REFERENCES

1. P. H. Dawson, J. Hedman, N. R. Whetten, A simple mass spectrometer, *Rev. Sci. Instrum.* **40** (1969) 1444–1450.

2. G. Lawson, R. F. Bonner, J. F. J. Todd, The quadrupole ion store (QUISTOR) as a novel source for a mass spectrometer, *J. Phys. E* **6** (1973) 357–362.

3. D. B. Langmuir, R. V. Langmuir, H. Shelton, R. F. Wuerker, U.S. Patent 3, 065, 640 (1962).

4. M.-N. Benilan, C. Audoin, *Int. J. Mass Spectrom. Ion Phys.* **11** (1973) 357.

5. R. F. Bonner, J. E. Fulford, R. E. March, G. F. Hamilton, *Int. J. Mass Spectrom. Ion Phys.* **24** (1977) 255.

6. R. E. Mather, R. M. Waldren, J. F. J. Todd, R. E. March, *Int. J. Mass Spectrom. Ion Phys.* **33** (1980) 201.

7. J. E. Fulford, R. E. March, R. E. Mather, J. F. J. Todd, R. M. Waldren, *Can. J. Spectrosc.* **25** (1980) 85.

8. R. E. March, R. J. Hughes, J. F. J. Todd, *Quadrupole Storage Mass Spectrometry*, Chemical Analysis Series, Vol. 102, Wiley, New York, 1989.

9. H. Lagadec, C. Meis, M. Jardino, *Int. J. Mass Spectrom. Ion Processes* **85** (1988) 287.

10. N. Mikami, Y. Miyata, S. Sato, S. Toshiki, *Chem. Phys. Lett.* **166** (1990) 470.

11. N. Mikami, S. Sato, M. Ishigaki, *Chem. Phys. Lett.* **180** (1991) 431.

12. N. Mikami, S. Sato, M. Ishigaki, *Chem. Phys. Lett.* **202** (1993) 431.

13. W.-W. Lee, C.-H. Oh, P.-S. Koo, M. Yang, K. Song, *Int. J. Mass Spectrom.* **230** (2003) 25.

14. W. Neuhauser, M. Hohenstatt, P. E. Toschek, H. Dehmelt, *Phys. Rev. Lett.* **12** (1978) 233.

15. E. Beaty, Simple electrodes for quadrupole ion traps, *J. Appl. Phys.* **61** (1987) 2118–2122.

16. C.-S. O, H. A. Schuessler, Ion storage in a radio-frequency trap with semi-spherical electrodes, *Int. J. Mass Spectrom. Ion Phys.* **35** (1980) 305–317.

17. H. G. Dehmelt, Stored ion spectroscopy, in F. T. Arecchi, F. Strumia, H. Walther (Eds.), *Advances in Laser Spectroscopy*, Plenum, New York, 1983, p. 153.

18. J. C. Bergquist, D. J. Wineland, W. M. Itano, H. Hemmati, H. U. Daniel, G. Leuchs, *Phys. Rev. Lett.* **55**(15) (1985) 1567.

19. D. J. Wineland, J. C. Bergquist, W. M. Itano, J. J. Bollinger, C. H. Manney, *Phys. Rev. Lett.* **59** (1987) 2935–2938.

20. E. C. Beaty, Calculated electrostatic properties of ion traps, *Phys. Rev. A Gen. Phys.* **33**(6) (1986) 3645–3655.

21. J. M. Wells, E. R. Badman, R. G. Cooks, *Anal. Chem.* **70** (1998) 438.

22. T. F. Meaker, B. C. Lynn, Tandem mass spectometry in a miniature cylindrical ion trap. Proc. 49th Ann. ASMS Conf. on Mass Spectrometry and Allied Topics, Chicago, IL, 2001, CD.

23. V. Lopez-Avila, H. H. Hill, *Anal. Chem.* **69** (1997) R269.

24. S. D. Richardson, *Anal. Chem.* **74** (2002) 2719.

25. R. A Ketola, T. Koliaho, M. E. Cisper, T. M. Allen, *J. Mass Spectrom.* **37** (2002) 457.

26. O. Kornienko, P. T. A. Reilly, W. B. Whitten, J. M. Ramsey, *Rapid Commun. Mass Spectrom.* **13** (1999) 50.

27. O. Kornienko, P. Reilly, W. Whitten, M. Ramsey, *Rev. Sci. Instrum.* **70** (1999) 3907.

28. E. R. Badman, R. C. Johnson, W. R. Plass, R. G. Cooks, *Anal. Chem.* **70** (1998) 4896.

29. G. E. Patterson, A. J. Guymon, L. S. Riter, M. Everly, J. Griep-Raming, B. C. Laughlin, Z. Ouyang, R. G. Cooks, *Anal. Chem.* **74** (2002) 6145.

30. S. Guan, A. G. Marshall, *Int. J. Mass Spectrom. Ion Processes* **157/158** (1996) 5.

31. D. E. Goeringer, W. B. Whitten, J. M. Ramsey, G. L. Glish, *Anal. Chem.* **64** (1992) 1434.

32. J. Franzen, R. H. Gabling, M. Schubert, Y. Wang, Nonlinear ion traps, in R. E. March and J. F. J. Todd (Eds.), *Practical Aspects of Ion Trap Mass Spectrometry*, Vol. I, CRC Press, Boca Raton, FL, 1995, Chapter 3, pp. 49–167.

33. R. E. Kaiser, R. G. Cooks, J. Moss, P. H. Hemberger, *Rapid Commun. Mass Spectrom.* **3** (1989) 50.

34. H. A. Bui, R. G. Cooks, *J. Mass Spectrom.* **33** (1998) 297.

35. S. M. Gordon, P. J. Kenny, J. D. Pleil, *Rapid Commun. Mass Spectrom.* **10** (1996) 1038.

36. M. E. Cisper, P. H. Hemberger, *Rapid Commun. Mass Spectrom.* **11** (1997) 1449.

37. L. S. Riter, Z. Takats, R. G. Cooks, *Analyst* **126** (2001) 1980.

38. S. E. Stein, *IR and Mass Spectra*. In NIST Chemistry WebBook, W. G. Mallard and P. J. Lindstrom (Eds.), NIST Standard Reference Database Number 69, National Institute of Standards and Technology, Gaithersburg, MD (February) 2000, 1,2-Ethanediol (http://webbook.nist.gov).

39. L. S. Riter, Y. Peng, R. J. Noll, G. E. Patterson, T. Aggerholm, R. G. Cooks, *Anal. Chem.* **74** (2002) 6154.

40. M. E. Bier, T. Kotiaho, R. G. Cooks, *Anal. Chim. Acta* **231** (1990) 175.

41. L. S. Riter, L. Charles, R. G. Cooks, *Rapid Commun. Mass Spectrom.* **15** (2001) 2290.

42. C. Enjalbal, J. Martinez, J. Aubagnac, *Mass Spectrom. Rev.* **19** (2000) 139.

43. R. Aebersold, D. R. Goodlett, *Chem. Rev.* **101** (2001) 269.

44. J. Workman, K. E. Creasy, S. Doherty, L. Bond, M. Koch, A. Ullman, D. J. Veltkamp, *Anal. Chem.* **73** (2001) 2705.

45. M. Yanagida, *J. Chromatogr. B* **771** (2002) 89.

46. R. G. Cooks, B. C. Laughlin, A. G. Talbert, A. J. Guymon, Z. Ouyang, J. Griep-Raming, High throughput mass spectrometry using cylindrical ion tray arrays, Pittsburg Conference on Analytical Chemistry and Applied Spectroscopy, New Orleans, LA, March 21, 2002.

47. R. G. Cooks, R. J. Noll, A. Makarov, W. Plass, Z. Ouyang, Frontiers in quadrupole ion traps. Proc. 51st Ann. ASMS Conf. on Mass Spectrometry and Allied Topics, Montreal, PQ, Canada, June 8–12, 2003, CD.

48. A. M. Tabert, J. Griep-Raming, A. J. Guymon, R. G. Cooks, *Anal. Chem.* **75** (2003) 5656.

49. E. R. Badman, R. G. Cooks, *Anal. Chem.* **72** (2000) 3291.

50. S. N. Cairns, J. Murrell, N. J. Alcock, S. R. Dixon, P. A. Cartwright, M. D. Brookes, D. M. Groves, Characterization of a cylindrical ion trap mass spectrometer, 16th International Mass Spectrometry Conference, Edinburgh, UK, August 31–September 5, 2003, DSTL (Defence Science & Technology Laboratory) Poster 08064.

51. S. N. Cairns, N. J. Alcock, S. R. Dixon, P. A. Cartwright, J. Murrell, M. D. Brookes, D. M. Groves, Development of a cylindrical ion trap mass spectrometer for explosive trace detection, 16th International Mass Spectrometry Conference, Edinburgh, UK, August 31–September 5, 2003, DSTL (Defence Science & Technology Laboratory) Poster 08065.

52. E. R. Badman, R. G. Cooks, *J. Mass Spectrom.* **6** (2000) 659.

53. C. Champenois, M. Knoop, M. Herbane, M. Houssin, T. Kaing, M. Vedel, F. Vedel, *Eur. Phys. J. D* **15** (2001) 105.

54. N. Yu, W. Nagourney, H. Dehmelt, *J. Appl. Phys.* **65** (1991) 3779.

55. C. Schrama, E. Peik, W. Smith, H. Walther, *Opt. Commun.* **101** (1993) 32.

56. H. Straubel, *Naturwissenschaften* **18** (1955) 506.

57. Q. Xue, F. Foret, Y. M. Dunayevskiy, P. M. Zavracky, N. E. McGruer, B. L. Karger, *Anal. Chem.* **69** (1997) 426.

58. R. S. Ramsey, J. M. Ramsey, *Anal. Chem.* **69** (1997) 1174.

59. R. G. Brewer, R. G. DeVoe, R. Kallenbach, *Phys. Rev. A* **46** (1992) R6781.

60. Y. Wang, K. P. Wanczek, *J. Chem. Phys.* **98** (1993) 2647.

61. W. H. Hartung, C. T. Avedisian, *Proc. Roy. Soc. Lond. A* **437** (1992) 237.

62. J. D. Williams, K. A. Cox, R. G. Cooks, S. A. McLuckey, K. J. Hart, D. E. Goeringer, *Anal. Chem.* **66** (1994) 725.

7

GAS CHROMATOGRAPHY/ MASS SPECTROMETRY

Quadrupole Ion Trap Mass Spectrometry, Second Edition, By Raymond E. March and John F. J. Todd
Copyright © 2005 John Wiley & Sons, Inc.

7.1. INTRODUCTION

The first commercial ion trap mass spectrometer was the ITD700 instrument that was described by the manufacturer as an "ion trap detector" for a gas chromatograph. Under computer control, the ITD700 was capable of acquiring full-scan mass spectra over a mass range of $\sim$20–650 Th, with a scan rate of 5555 Th/s [1]. In 1983, the ion trap detector was presented by Finnigan MAT Corporation of San Jose, California, at the Pittsburgh Conference on Analytical Chemistry, held in Atlantic City [2]. Data obtained with the ion trap detector were presented subsequently at the Thirty-First Annual Conference of the American Society for Mass Spectrometry [3] held in Boston, Massachusetts, and at the Thirteenth Annual Meeting of the British Mass Spectrometry Society [4]. A publication [5] during the same year described the application of new ion trap technology as an economical detector for gas chromatography.

Prior to 1983, it was not possible to purchase an operating QIT. All of the research carried out with the QIT during three decades from the time of the original public disclosure [6] until 1983 was accomplished with home-made devices fashioned in industrial and university workshops. After 1983, research involving home-made devices continued for about 7–10 years only. The commercialization of the QIT has been discussed in Chapter 3 and has been described in detail elsewhere [7]. Although there had been some early thoughts about the viability of a commercial analytical instrument based upon ion trap technology [8], the linkage between university research and an industrial, entrepreneurial corporation was made at the Twenty-Seventh Annual Conference of the American Society for Mass Spectrometry held in Seattle, Washington, in 1979. George Stafford of Finnigan MAT Corporation attended a lecture given by Ray March on the characterization and application of the QIT [9]; among the applications discussed in this lecture were the use of the QIT as a detector for a gas chromatograph [10] and the use of resonant ion ejection for the study of ion/molecule reactions in a QIT [11]. Stafford became interested in ion traps and, after some reflection, had an idea that represented an entirely new strategy for the use of the QIT in mass analysis.

7.2. GAS CHROMATOGRAPHY

Gas chromatography is used widely for compound identification in the analysis of mixtures of volatile organic compounds from a variety of chemical classes, such as acids, bases, alcohols, ketones, esters, hydrocarbons, and halogenated compounds. Conventionally, identification of a specific compound by GC is based on the ability of one or more chromatographic columns to resolve chromatographically one or more components of interest and, using one of a number of detectors, to measure accurately the retention time and signal intensity of each component. While many advances have been made in GC in recent years with the advent of capillary columns, coelution of components continues to require additional and separate gas chromatographic analysis.

7.2.1. Gas Chromatography/Mass Spectrometry

The combination of mass spectrometry with gas chromatography, GC/MS, is clearly advantageous and superior to conventional GC alone. Unencumbered by the technical difficulties and financial challenges encountered in trying to realize a functioning GC/MS instrument, let us consider the advantages to be gained [12].

7.2.1.1. Information Theory The concepts of information theory have been applied by Fetterolf and Yost [13] to tandem mass spectrometry and can be extended to include GC also. Because the range of each experimentally-variable parameter and the ion signal intensity is limited in time, space, and magnitude, these ranges may be represented by a finite numerical value that is known as the *informing power*, P_{inf}. The informing power of an analytical procedure, as defined by Kaiser [14], can be expressed in terms of "binary digits," or bits, and is given by

$$P_{inf} = \sum_{i=1}^{n} \log_2 S_i \tag{7.1}$$

where n is the number of parameters or quantities to be determined, such as elution time or mass number, and S_i is the number of measurable steps for a given quantity. If m is the variable parameter and δm is the smallest distinguishable increment in m, then the number of steps of m is given by $m/\delta m$, which in mass spectrometry is the mass resolution $R(m)$. By substituting into Eq. (7.1) and allowing for the variation in the parameter m, the summation can be replaced with an integral evaluated from m_a to m_b, as in

$$P_{inf} = \int_{m_a}^{m_b} R(m) \log_2 S(m) \frac{dm}{m}. \tag{7.2}$$

However, when the resolution $R(m)$ is constant and $S(m)$ is fixed by the detection system, Eq. (7.2) may be simplified to yield

$$P_{inf} = R(m) \log_2 S(m) \ln\left(\frac{m_b}{m_a}\right). \tag{7.3}$$

References pp. 246–249.

The informing power is a figure of merit for an instrument. The informing power may be increased by improving any of the three terms in Eq. (7.3), say the mass resolution or the mass range $(m_a - m_b)$, and by the addition of another resolution element such as a gas chromatograph or a second mass spectrometer. In tandem mass spectrometry, a second mass analyzer enters as a resolvable parameter. Following the derivation of Eq. (7.2), Fitzgerald and Winefordner [15] have shown that for two resolvable parameters

$$P_{\text{inf}} = \int_{y_a}^{y_b} \int_{x_a}^{x_b} R(x)R(y) \log_2 S(x, y) \frac{dx}{x} \frac{dy}{y}. \tag{7.4}$$

7.2.1.2. Informing Power in Mass Spectrometry
In the cases of the QIT and the QMF, it is the minimum resolution element δm that is constant rather than the resolution $R(m)$. For unit mass resolution, $\delta m = 1$. The informing power may then be expressed as

$$P_{\text{inf}} = \frac{1}{\delta m} \log_2 S(m) \int_{m_a}^{m_b} dm \tag{7.5}$$

such that

$$P_{\text{inf}} = 1 \times \log_2 S(m) \times (m_b - m_a). \tag{7.6}$$

For a QIT with a mass range of 10–650 Th and an ion intensity range of 2^{12} bits, $P_{\text{inf}} = 1 \times 12 \times 640 = 7.7 \times 10^3$ bits.

For a QIT combined with GC, Eq. (7.4) becomes

$$P_{\text{inf}} = \frac{1}{\delta m} \log_2 S(m) \int_{m_a}^{m_b} dm \, R(g) \int_{g_a}^{g_b} \frac{dg}{g} \tag{7.7}$$

where $R(g)$ is the constant resolution of a capillary GC column with 1.0×10^5 theoretical plates (N) and the period of analysis is 1 h; there is a delay of 3 min to permit elution of solvent. The parameter $R(g)$, defined as the ratio of retention time t_r to GC peak width t_w, is given by $(N/5.54)^{1/2}$ so that P_{inf} for GC/MS is given by

$$P_{\text{inf}} = 7.7 \times 10^3 \left(\frac{10^5}{5.54} \right)^{1/2} \ln \left(\frac{3600}{180} \right) = 3.1 \times 10^6 \text{ bits} \tag{7.8}$$

so the informing power of the QIT (7.7×10^3 bits) has been increased by a factor of 402 by coupling with a gas chromatograph to yield an informing power of 3.1×10^6 bits. The converse holds also that the informing power of a gas chromatograph (4.02×10^2 bits) will be increased by a factor of ~ 7700 by coupling with a QIT such as the ITD.

7.2.1.3. Informing Power in Tandem Mass Spectrometry
For a GC/MS/MS instrument, the informing power is simply the product of the informing power of a GC/MS instrument and that of a second mass analyzer. Therefore P_{inf} for GC/MS/MS is $3.1 \times 10^6 \times 10^3 \approx 3 \times 10^9$. Note that the informing power of the second mass analyzer is only $\approx 13\%$ of the first mass analyzer due to (i) the reduction in the ion intensity range $S(m)$ of ions issuing from the first analyzer and (ii) the

reduction in mass of a fragment ion with respect to its precursor ion mass. An ion species transmitted through the first mass analyzer and into the second mass analyzer cannot exhibit in the second mass analyzer an ion signal intensity greater than the original ion signal intensity.

In tandem mass spectrometry, experimental variables associated with CID such as resonant frequency, amplitude, and duration of resonant excitation are other potential resolution elements. In addition to these measurable experimental parameters, other variables that are procedural in nature, such as chemical ionization, selection of positive or negative ions, and the choice of reactive collisions, can be used to increase the informing power of tandem mass spectrometry. However, the inability to scan these variables continuously makes it difficult to express these effects on informing power.

7.2.1.4. The State of the GC Market

In 1987, it was estimated that, of approximately 220,000 gas chromatographs in North America, less than 5% were coupled to a mass spectrometer. There were several reasons for this situation. First, the pressure of the carrier gas at the outlet of a GC column is some orders of magnitude higher than the normal operating pressure of a mass spectrometer; various coupling devices combined with additional pumping were required to effect the GC/MS combination. Second, mass spectrometers were relatively more complex instruments than were gas chromatographs and required a knowledgeable operator; thus upgrading a gas chromatograph to a gas chromatograph/mass spectrometer could require a change of personnel. Third, the generation of one to three mass spectra per second created a major problem in data handling. Fourth, the capital cost of a mass spectrometer was considerably greater than that of a gas chromatograph; therefore to upgrade a gas chromatograph to a gas chromatograph/mass spectrometer constituted an appreciable investment.

In 1983, the ion trap detector was marketed at a significantly lower cost than the prevailing costs for mass spectrometers and offered such simplicity of operation as to obviate the requirement for personnel with long-term mass spectrometric experience.

7.2.1.5. Carrier Gas

In conventional GC, a common carrier gas that sweeps volatile compounds through the GC column is helium and, when a gas chromatograph is coupled to a mass spectrometer, much of the helium carrier gas must be pumped away; such is not the case for the ion trap detector. As discussed in Chapter 3, not only does helium play a critical role in the ion trap detector in that it serves to focus ions at the center of the ion trap but optimal ion focusing occurs at an ambient pressure of helium of $\sim 10^{-3}$ Torr. A pressure of this magnitude would be anathema in normal mass spectrometry where the required pressure range is a factor of $\sim 10^{-3}$–10^{-4} lower in order to preserve collision-free conditions for gaseous ions during flight in a magnetic or electrostatic field. A heated transfer line coupled the gas chromatograph to the ion trap via an open-split interface. The effluent from the gas chromatograph flowed through this interface and into an incompletely sealed QIT. The resulting pressure of helium in the ion trap was close to the optimal magnitude for collisional focusing in the ion trap; additional helium was added through the open-split interface for optimization of the helium pressure. The exit of the GC column protrudes slightly into the electrode structure of the QIT so that helium will flush background gases out of the ion trap. The pumping requirements for the ITD were modest.

7.2.2. The ITD Instrument

The ITD as developed by Kelley, Stafford, and Stevens [16] is shown schematically in Figure 7.1. Shown is the electrode assembly consisting of a ring electrode and two end-cap electrodes in cross section as well as the dimensions r_0 and z_0 and their relationship. The spacing between each end-cap electrode and the ring electrode is determined by the thickness of an interposed flat-ring ceramic spacer, as shown in Figure 7.1. A further function of the ceramic spacers is to seal incompletely the interior of the ion trap; a degree of sealing is necessary to maintain the required pressure of helium while the incompleteness of the seal allows the GC effluent to escape from the ion trap interior. An RF drive potential is applied to the ring electrode while the end-cap electrodes are grounded. An electron gun, consisting of a filament and gating lens, is shown above the upper end-cap electrode. The upper end-cap electrode has a small perforation through which the gated electron beam passes. The lower end-cap electrode has several perforations in a pepper-pot formation through which ions are ejected onto the electron multiplier. Later, the lower end-cap electrode had but a single perforation. Ion signals are amplified and stored. A microcomputer provides instrument control and data acquisition.

The ITD is compact and was designed to couple with any commercially available capillary gas chromatograph and to interface with an IBM PC data system for control and storage of data. A wide variety of software was included with the ITD that

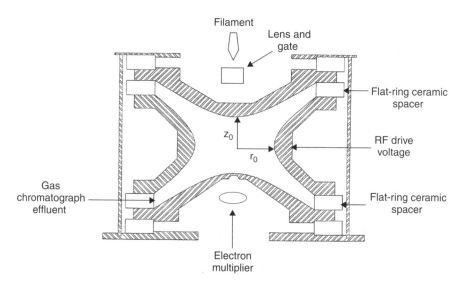

Figure 7.1. Schematic diagram of ion trap detector.

permitted spectral library searches, quantitation, plotting of chromatograms, mass spectra, and control of the ITD.

7.2.2.1. Mass Analysis by Mass-Selective Instability

The mass-selective instability mode of operation of the QITs, which revolutionized the use and application of these devices, has been discussed in Chapter 3. The mass-selective instability technique of Kelley, Stafford, and Stevens [16] relied on external detection of mass-selective, axial ejection of ions whose trajectories had become unstable. The simplicity, speed, and mass resolution of the mass-selective instability technique coupled with the structural simplicity of the QIT permitted the application of mass spectrometric techniques over an extraordinarily wide range of fields [17].

In general terms the new mode of operation was as follows: DC and RF voltages U and $V \cos \Omega t$ are applied to the three-dimensional quadrupole electrode structure such that ions over the entire mass/charge range of interest could be trapped within the field imposed by the electrodes. Normally, $U = 0\,V$ and the RF potential is applied to the ring electrode. Ions are created within the quadrupole field by a short burst of electrons from the gated filament, where the gating action is achieved by the imposition of appropriate DC potentials to the lens in Figure 7.1. After a brief storage time, the amplitude of the RF potential is increased at the rate of $55,500\,V_{0\text{-}p}/s$ such that, for example, the mass range of 50–250 Th is ejected in ~37 ms. When the amplitude of the RF potential is increased, the working point q_z on the q_z axis of each ion species increases until it attains a value close to 0.908, whereupon the trajectories of ions of consecutive values of m/z become unstable as ion axial excursions exceed the dimensions of the ion trap. These ions pass out of the trapping field through one or more perforations in the exit end-cap electrode and impinge on a detector, such as an electron multiplier. The detected ion current signal intensity as a function of time corresponds to a mass spectrum of the ions that were trapped initially.

7.2.2.2. Mode of Operation

In Figure 7.2 is shown a scan function that depicts the sequence used to perform mass analysis. A scan function is a visual representation of the temporal variation of DC and RF potentials applied to electrodes of the ion trap assembly and the ion signals detected. In the simplest variant of this new method, the ring electrode is driven at an initial RF voltage V_i and at a fixed frequency such that all ions in the m/z range of interest may be trapped within the imposed quadrupole field. No DC voltage is applied between the ring and end-cap electrodes ($U = 0$) so that the confining field is purely oscillatory, that is, RF only. With this arrangement, the locus of all possible working points (a_z, q_z) maps directly onto the q_z axis on the stability diagram (Figures 1.8 and 2.5). A number of ion trap parameters for the ITD in the mass-selective axial instability mode are given in Table 7.1.

The initial ion trap operating RF voltage V_i and frequency are chosen such that all ions of interest have specific masses greater than the LMCO (see Chapter 2); usually, the LMCO is set at m/z 50. With the ion trap electrodes maintained at this initial voltage and frequency, the filament is switched on and a positive voltage applied to the gate electrode to permit the electron beam to enter the quadrupole field region. After a

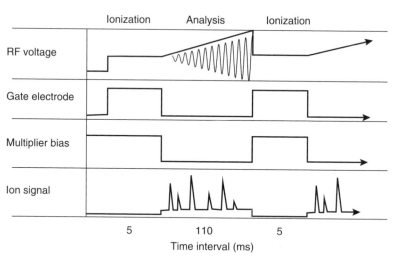

Figure 7.2. Operation of QIT in mass-selective mode. Scan function showing sequence of operations for obtaining electron impact mass spectrum with QIT in mass-selective axial instability mode. (Copyright 1985 Finnigan Corporation. All rights reserved. Reprinted with permission. Figure 6.3 from *Quadrupole Storage Mass Spectrometry*, by R. E. March, R. J. Hughes, J. F. J. Todd, © Wiley-Interscience, 1989.)

TABLE 7.1. ITD Parameters for Mass-Selective Axial Instability Mode

Radio frequency, 1.0 MHz
Radius of ring electrode (r_0), 10.00 mm
Half-separation of end-cap electrodes (z_0), 7.83 mm
Ring electrode hyperbolic profile with $r_0 = 10.00$ mm
End-cap electrode hyperbolic profile with $z_0 = 7.07$ mm
Ionization time, 1.0 ms
RF voltage, 6 kV (zero to peak)
Equivalent RF voltage scan speed, 5555 Th/s
Helium pressure, 0.1 Pa
Electron multiplier gain, 1×10^5
Mass range, 20–650 Th

time interval, typically 1 ms, the electron beam is switched off, and ionization within the trapping field ceases. It is during the ionization period that molecules eluting from a gas chromatograph will be ionized. In Figure 7.2 the ionization period is shown as 5 ms; this period is too long when the density of molecules eluting from a gas chromatograph is moderately high and too many ions are formed. In these circumstances, the duration of the ionization must be reduced to the typical value. The problem of matching the duration of the ionization period to the molecular density was overcome by the introduction of automatic gain control (AGC), as discussed in Chapter 3.

Ion species created in the trapping field region and whose values of m/z are less than the LMCO are lost within a few RF cycles (1 μs/cycle). Ions created in the trapping field region and having trajectories that are so large as to cause the ion to impinge on the electrodes are lost also, usually within a few hundred RF cycles. Therefore, several hundred RF cycles after termination of ionization, few such ions are leaving the trapping field and striking the detector behind the lower end-cap electrode in Figure 7.1. The multiplier is "switched off" during the ionization and ion-settling processes.

For the ions remaining in the trapping field the next step, as shown in Figure 7.2, is to ramp the magnitude of the trapping field potential amplitude $V \cos \Omega t$ to V_f and, simultaneously, to bias negatively the multiplier, that is, the detector is "switched on" so that positive ions may reach the detector. The ramping of the RF potential amplitude and the switching of the detector are depicted schematically in Figure 7.2. As the applied RF voltage is increased from V_i to V_f, at a rate of approximately 185 μs/Th, the lower limit of the range of masses that may be trapped, that is, the LMCO, is increased proportionally:

$$\left(\frac{m}{z}\right)_{\mathrm{LMCO}_f} = \frac{(m/z)_{\mathrm{LMCO}_i} V_f}{V_i} . \tag{7.9}$$

Hence, as the applied RF voltage amplitude increases, stored ions develop unstable trajectories in order of increasing value of m/z. During this voltage sweep, trajectory instability develops only in the axial direction of motion. Because ions have been focused collisionally as a result of ion/helium collisions to the center of the ion trap, axial trajectory instability causes the ions to move along the axis of cylindrical symmetry. A significant fraction (but <50%) of the ions pass through the perforation in the exit end-cap electrode and strike the detector.

The sweep rate of the RF voltage was chosen so that ions of consecutive values of m/z develop trajectory instability at the rate at which ions depart from the trapping field region. The time–intensity profile of the signal detected at the electron multiplier corresponds to a resolved mass spectrum of the ions stored originally within the trapping field, as shown in Figure 7.2. When the mass range of interest is extensive, say a range of 595 Th, the total scan time is 110 ms, as shown in Figure 7.2. In a mass scan from m/z 55 to m/z 650, any m/z 650 ions will have spent at least 110 ms in the trapping field and, during this period, ion/molecule reactions may occur that modify the mass spectrum. Subdividing the mass range into four smaller mass ranges and adding the resulting mass spectra can reduce the effects of such modification (see Ref. 12, p. 145). Segmented scans were introduced to overcome the problems of excessive ionization during a fixed ionization period of 1 ms; during each scan segment, the RF level could be optimized so that the mass spectra obtained would resemble more closely the standard library EI mass spectra. Later, the duration of ionization was optimized using AGC (see Chapter 3), rather than the RF level [18]. Eventually, ion trap performance was improved to the point where a single scan acquisition could be made.

7.2.2.3. A Remarkable Achievement

The observation of a mass spectrum in this manner was a remarkable achievement, yet the ITD was an even more remarkable instrument. Direct-current power supplies were incorporated into the construction of the ITD, presumably for operation of the ITD away from the q_z axis, yet they were not accessible in normal operation. However, it was relatively simple to incorporate a further degree of computer control that permitted use of the DC power supplies and to add an auxiliary supplementary potential that could be applied to one or both end-cap electrodes. In this manner, resonant excitation (see Chapter 3) of ions could be demonstrated in a Finnigan MAT model 800 ITD and mass range extension achieved by resonant ion ejection at a value of q_z less than 0.908 [19–21].

7.2.3. Ion Trap Mass Spectrometer, ITMS

In the 1980s, Finnigan introduced the ITMS as a multipurpose research instrument to encourage the development and characterization of new ion trap capabilities. The ion trap electrode assembly, electronics, and GC/MS transfer-line interface were similar to the Finnigan ITD700 and ITD800 instruments. The ion trap assembly was housed in a relatively large rectangular multiport UHV vacuum system that permitted additional hardware. Hardware supplied with the ITMS included a probe lock, a heated solids probe, and a programmable DC power supply and frequency synthesizer. Examples of the high-performance features studied using the ITMS include tandem mass spectrometry [22, 23], ion injection [24, 25], chemical ionization [26, 27], mass range extension [28, 29], and high-mass resolution [30, 31]. Although some GC/MS/MS studies with an ITMS have appeared [22, 32–35], the ITMS was not ideally suited for such studies because the AGC and automatic reaction control (ARC, see Chapter 3) routines had not been integrated into the scan editor software that is used to create MS/MS scan programs.

7.2.4. SATURN Model I Ion Trap Detector

The Varian SATURN model I ion trap detector was virtually identical to the Finnigan ITS40 instrument, even to the inclusion of inaccessible DC power supplies; one important addition was the incorporation of a waveform generator for the application of AC potentials to the end-cap electrodes. Modification of the standard form of computer control of this model permitted MS^n, variation of the mass scan rate over six orders of magnitude, and the observation of enhanced mass resolutions of 0.8×10^7 and 1.2×10^7 for m/z 414 and m/z 614, respectively [36]. These examples are given to illustrate the enormous capability of the physical ITD instrument. It is relatively unimportant that the normal mode of operation was limited by the manufacturer's software; as instrument development proceeded, the range of options available to customers increased also and upgrading of instruments was possible.

7.2.5. SATURN Model 4000 GC/MS System

The commercial success of the QIT as a mass detector for a gas chromatograph has permitted the continued development of this instrument. These developments consist of the operation of the ion trap as a tandem mass spectrometer in conjunction with positive-ion chemical ionization, including selective-reagent chemical ionization, negative-ion chemical ionization, selected-ion monitoring, injection of externally-generated ions into the ion trap, multiresonance ion ejection techniques, and variable mass resolution. It is beyond the scope and purpose of this book to present an exhaustive comparison of ion trap instruments that are available commercially; rather we focus attention on one particular instrument as being somewhat representative of the choice of instruments available and illustrative of the developments accomplished recently. Here we focus on the SATURN model 4000 GC/MS system. For further information on the Varian SATURN model 4000 GC/MS system, the reader is directed to www.varianinc.com. For information on the GC/MS instruments offered by Finnigan, the reader is directed to www.finnigan.com and www.thermo.com/ms_chrom and to www.agilent.com for information on the GC/MS instruments offered by Agilent.

The ion trap in the Varian SATURN model 4000 is slightly smaller than in earlier models. The radius of the inscribed circle tangential to the inner surface of the ring electrode, r_0, is 7.07 mm, as compared to 10 mm for the ion trap detector; the separation of each end-cap electrode from the center of the ion trap, z_0, is 7.82 mm. This value for z_0 is virtually unchanged from that in the original Finnigan (and Varian) stretched ion trap. The three electrodes are of hyperbolic geometry and Eqs. (2.57) and (2.58) are the equations for the ring electrode and for the end-cap electrode surfaces, respectively. The slopes of the asymptotes of the ring electrode and of the end-cap electrodes are given by Eqs. (2.47) and (2.48) such that the asymptotes have an angle of 35.26° with respect to the radial plane of the ion trap. Because the asymptotes bisect the gaps between adjacent electrodes, the quality of the quadrupolar potential will be maintained at greater distances from the origin (see Section 2.3.1).

Thus the surfaces of the electrodes of the Varian ion trap are defined by their respective r_0 and z_0 values and by Eqs. (2.57) and (2.58). The cross section of the Varian ion trap is similar to that shown in Figure 2.12b, wherein the asymptotes and the dimensions r_0 and z_0 are identified. It should be noted that for the ion trap cross-section shown in Figure 2.12b the surfaces of the electrodes are defined by the r_0 value and by Eqs. (2.57) and (2.58); however the z_0 value is constrained by Eq. (2.77) such that z_0 is equal to $r_0/\sqrt{2}$. The cross section of the Varian ion trap is similar also to the ion trap cross section depicted with light lines in Figure 3.10. Note also that the surfaces of the electrodes shown in the ion trap cross section depicted with light lines in Figure 3.10 have been calculated using Eqs. (2.57) and (2.58) with $z_0 = 0.70711 \times 110.6\% = 0.78206$ unit and $r_0 = 1$ unit. The geometry of the commercial ion trap electrodes has been modified so as to introduce an octopole component into the trapping field to enhance mass resolution [37] (see Section 8.6.1). Higher-order fields can be obtained by increasing the separation between the end-cap electrodes while maintaining ideal hyperbolic surfaces [38].

The RF drive potential is applied at a frequency of 1.05 MHz and the mass/charge ratio range of the ion trap is to m/z 1000. The ramp rate of the RF drive potential amplitude is adjustable so that some variation in mass resolution can be achieved; the standard rate for mass scanning is 5000 Th/s and the maximum scan rate is

10,000 Th/s. Automatic gain control is employed to control the ion density within the ion trap. The principal developments in the SATURN model 4000 GC/MS system concern the location and type of ion creation and the method of ion ejection; each is discussed in turn.

7.2.5.1. Ion Creation

The original method of ionization used in the SATURN model I ion trap detector, that is, electron impact within the ion trap (described as in situ EI or internal EI), is retained in the SATURN model 4000 instrument. Electron impact ionization of the eluent from a gas chromatograph permits the observation of both the mass spectrum (GC/MS) of the eluting compound and a product ion mass spectrum (GC/MS/MS) of a selected ion species. With the addition of an external ion source, three further modes of ionization can be employed; these modes are external electron impact ionization, positive-ion chemical ionization (PICI), and negative-ion chemical ionization (NICI). In each case, ionization in the external ion source is followed by injection of ions through an Einzel lens into the ion trap whereupon the ion trap can be operated in both GC/MS and GC/MS/MS modes. Thus the additional external ion source permits greater flexibility among ionization modes and permits both GC/MS and GC/MS/MS operation with each mode.

The PICI mode allows the selection of a specific reactant ion for chemical ionization from a chemical ionization reagent. The specific reactant ion may be a primary ion or a secondary ion (formed by ion/molecule reaction) that is selected according to either the proton affinity of the associated Brønsted base relative to that of a target molecule or the binding energy of the specific reactant ion to a target molecule (see Section 7.2.6.1). Injection of the specific reactant ion into the ion trap is followed by either protonation of the target molecule or formation of an adduct ion with the target molecule. In each case, the specificity of the reaction of the specific reactant ion leads to high sensitivity for the detection and identification of the target molecule. Specific negative reactant ions are formed by electron capture within the external ion source and are injected into the ion trap subsequently. The most common mode of ionization by a specific negative reactant ion is proton abstraction from a target molecule M to form $M^{-\cdot}$. The external ion source permits the utilization of pulsed ionization that can lead to enhancement of sensitivity.

The entire range of ionization modes and scanning modes can be activated during a single run. This degree of flexibility within a single run is illustrated in Figure 7.3, which shows the use of EI and CI combined with a mass scan (MS), a product ion mass scan (MS/MS), and selected-ion storage (SIS).

7.2.5.2. Ion Ejection

A new method for ion ejection is employed that takes advantage of an asymmetric trapping field to improve the mass scanning performance of the ion trap [39–42] (see Section 3.3.2.3). In the normal operation of an ion trap, a buffer gas such as helium is used to dampen the ion trajectories to the center of the ion trap (see Section 3.4) so that the ion cloud is focused there. When a supplementary oscillating potential is applied to excite the ions (and, in the limit, to eject them), it is imperative that the excitation field has a finite strength at the ion cloud location in order to displace the ions from the center of the ion trap. For example, parametric resonant excitation by a supplementary quadrupolar field causes ion amplitudes to

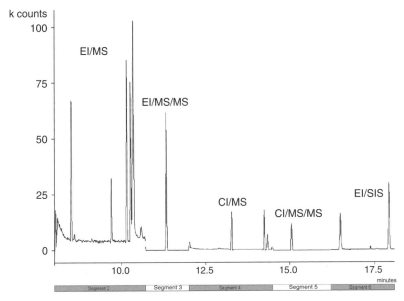

Figure 7.3. Total ion current obtained from single gas chromatographic run during which the ion trap detector was operated in five different mass spectrometric modes: segment 2, EI ionization followed by mass scan; segment 3, EI ionization followed by product ion mass spectrum; segment 4, chemical ionization followed by mass scan; segment 5, chemical ionization followed by product ion mass spectrum; segment 6, EI ionization followed by selected ion storage. (Copyright 2004 Varian Inc. All right reserved. Reprinted with permission. Figure "Flexibility within a single run" from Varian Inc. gcms4000.dpf.)

increase axially when the ion frequency is one-half of the supplementary quadrupole frequency [43]. However, parametric resonant excitation, which has been investigated theoretically [43, 44], is ineffectual because of the vanishing strength of the supplementary field at the geometric or mechanical center of the ion trap. Yet the radial pseudopotential well for a quadrupole field is of parabolic form, that is, $\propto r^2$ (see Section 5.4.1) and so assumes a nonzero value a short distance from the center of the ion trap. Thus, to take advantage of parametric resonant excitation in an efficient manner, the ion cloud must first be moved away from the geometric center of the ion trap.

An asymmetric trapping field is generated by the application of an alternating potential out of phase to each end-cap electrode and at the same frequency as that of the RF drive potential applied to the ring electrode. This trapping field dipole component causes the center of the trapping field to move away from the geometric center toward one of the end-cap electrodes. The addition of the dipole component causes the center of the trapping field to move toward the end-cap electrode that has the trapping field dipole component in phase with the RF drive potential applied to the ring electrode. The cross section of the Varian ion trap is shown in Figure 7.4; the trapping

References pp. 246–249.

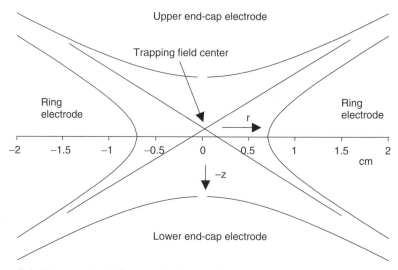

Figure 7.4. Cross-sectional diagram of electrodes in the Varian SATURN 4000 QIT showing trapping field center displaced by 15% from geometric center. Trapping field asymptotes only are shown in this figure and trapping field center is located, as indicated, at intersection of these asymptotes ($r_0 = 7.07$ mm, $z_0 = 7.82$ mm). Asymmetric trapping field is produced by application of -1.15 V on lower end-cap electrode, -0.85 V on upper end-cap electrode, and $+1.0$ V on ring electrode. (Reprinted from the *International Journal of Mass Spectrometry*, Vol. 190/191, M. Splendore, E. Marquette, J. Oppenheimer, C. Huston, G. Wells, "A new ion ejection method employing an asymmetric trapping field to improve the mass scanning performance of an electrodynamic ion trap," Fig. 1, pp. 129–143 (1999), with permission from Elsevier.)

field dipole component on the upper end-cap electrode is in phase with the RF drive potential, and so the trapping field center is displaced toward the upper end-cap electrode. The trapping field center, indicated by a sloping arrow in Figure 7.4, is located at the intersection of the field asymptotes shown in Figure 7.4; the geometric asymptotes that are not shown in this figure intersect at the zero value of the radial scale. The radial direction in Figure 7.4 is indicated by a horizontal arrow (r) and the axial direction in the lower half of the ion trap is indicated by a vertical arrow ($-z$). Not only does the ion cloud move in the same direction as does the trapping field center, that is, toward the upper end-cap electrode, but resonant ejection of the components of the ion cloud through the upper end-cap electrode is now favored. Because the ions are ejected exclusively through the upper end-cap electrode and strike a detector behind the electrode, the number of ions detected is twice the number detected when the trapping and geometric centers coincide.

A second-order effect of the application of the trapping field dipole component is the superimposition of a substantial hexapole field [25, 45]. The resulting multipole trapping field has a nonlinear resonance at $\beta_z = \frac{2}{3}$ [46–48]. Because the ions are already displaced from the geometric center of the ion trap by the asymmetric trapping field, the hexapole resonance has a finite value at the new location of the ion cloud.

Similarly, at this location, a parametric resonance attributed to a supplementary quadrupole field will also have a nonzero value. Finally, the addition of a supplementary dipole field at this location will cause dipolar resonant excitation. All three fields have nonzero values at the operating point of $\beta_z = \frac{2}{3}$ and, therefore, a triple-resonance condition exists. An ion that is moved to the displaced trapping field center and has an operating point of $\beta_z = \frac{2}{3}$ will be in resonance with and will absorb power nonlinearly from three fields simultaneously (see Sections 3.3.2.2 and 8.6.1). The amplitude of ion axial motion also increases nonlinearly and the ion is ejected rapidly from the ion trap with improved mass resolution. Due to the rapidity of ion ejection, there are fewer collisions of ions with buffer gas hence, mass resolution is enhanced for ions that do not have a low dissociation threshold.

7.2.6. Chemical Ionization

It was discussed above that ion/molecule reactions can occur in the QIT that modify the mass spectrum. These modifications may be the results of reactions between primary ions and their neutral precursor. Because such reactions occur, it is possible to choose a type of ion/molecule reaction that will be advantageous in the examination of the effluent from a gas chromatograph. For example, a particularly useful type of ion/molecule reaction is CI where a specified reagent ion will transfer a proton to a molecule that has a higher proton affinity than does the deprotonated reagent ion. In early studies with the QUISTOR–quadrupole mass filter combination, it was demonstrated that analysis by CI could be performed easily [49]. Upon ionization and storage in the QUISTOR, a $400:1$ mixture of CH_4 and CH_3OH yielded a dominant peak due to protonated methanol $(CH_3OH_2)^+$ formed by proton transfer from CH_5^+ to CH_3OH.

7.2.6.1. Chemical Ionization Mass Spectral Mode

A CI reagent gas, for example, CH_4, is introduced at a pressure of 1.3 mPa. Parenthetically, the application of CI in other types of mass spectrometers at that time required a much higher pressure of CH_4, around 1 Torr, which required an additional specific pumping system to maintain the normal low pressure for collision-free flight of ions. The mean residence time of primary ions in the ion source was ~1 μs. The much lower pressure of CH_4 used in the ITD was sufficient because the reaction time was longer. The extent of reaction, that is, of proton transfer, is dependent on the product of reagent pressure and reaction time. The product of reagent pressure and reaction time in other mass spectrometers (133 Pa $\times$ 1 μs) is equal to that (1 mPa $\times$ 100 ms) in a QIT; therefore the extent of reaction in each instrument is virtually identical.

The procedure for acquiring an EI mass spectrum in the presence of CI reagent gas is as illustrated in Figure 7.2 and by the identical sequence A, B, and C in Figure 7.5. The magnitude of the RF potential amplitude indicated by A permits the storage of sample ions of interest during the ionization period. The RF level is then increased rapidly, B, to the level appropriate for the selected low mass at which the scan is to

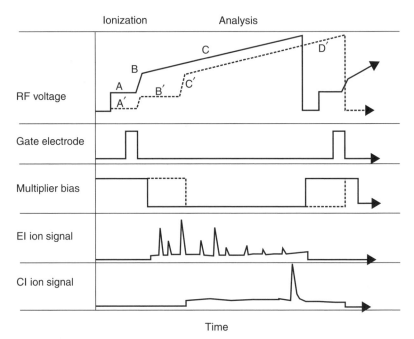

Figure 7.5. Operation of the ion trap in the mass-selective axial instability mode in the presence of chemical ionization reagent gas. Sequence of operations for obtaining either electron impact or chemical ionization mass spectra. Stages *A–C* and *A'–D'* are discussed in text. Note that the multiplier bias is synchronized with the analytical scan in each case. (Copyright 1985 Finnigan Corporation. All rights reserved. Reprinted with permission. Figure 6.8 from *Quadrupole Storage Mass Spectrometry*, by R. E. March, R. J. Hughes, J. F. J. Todd, © Wiley-Interscience, 1989.)

commence. Sequence *C* represents the scan of the RF voltage during which an EI mass spectrum such as that shown in Figure 7.6 is acquired.

In Figure 7.6*a* is shown an EI mass spectrum of nicotine, molecular weight (MW) 162, obtained with the ITD using the sequence shown in Figure 7.2. The base peak is the ion species of *m/z* 84, and there is extensive fragmentation. The addition of methane reagent gas does not alter the EI mass spectrum within the mass range shown in Figure 7.6*a* when the sequence shown in Figure 7.2 is followed. Electron ionization mass spectra may be obtained thus in the presence of CI reagent gas provided the initial RF voltage amplitude establishes an LMCO in excess of the molecular ion of the CI reagent gas.

To obtain a CI mass spectrum, that is, to operate the ion trap in the CI spectral mode, the sequence of operations *A'*, *B'*, *C'*, and *D'* as shown in Figure 7.5 is followed. The RF amplitude is set initially at *A'* (a value lower than *A* and therefore a lower LMCO) and held constant for a period of 10–20 ms so as to permit the formation and storage of primary ions of the CI reagent. Primary ions, for example, $CH_4^{+\cdot}$, react with methane to form secondary ions:

$$CH_4^{+\cdot} + CH_4 \rightarrow CH_5^+ + CH_3^{\cdot} . \tag{7.10}$$

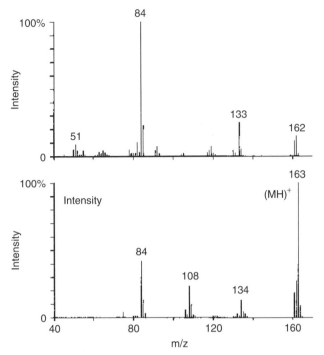

Figure 7.6. Mass spectra derived from nicotine: (*a*) EI mass spectrum acquired in presence of methane reagent gas; (*b*) methane chemical ionization mass spectrum. (Copyright 1985 Finnigan Corporation. All rights reserved. Reprinted with permission. Figure 6.9 from *Quadrupole Storage Mass Spectrometry*, by R. E. March, R. J. Hughes, J. F. J. Todd, © Wiley-Interscience, 1989.)

The RF amplitude is increased to B' so as to eject low-mass primary ions and to permit CH_5^+, a secondary reagent ion, to react by proton transfer with neutral sample molecules, M:

$$CH_5^+ + M \rightarrow MH^+ + CH_4. \qquad (7.11)$$

Rapid ramping of the RF amplitude C' to the voltage corresponding to the selected low-mass limit m_i for the CI mass spectrum is followed by a normal RF voltage ramp D' to the end of the mass range of interest, m_f. The upper limit for a CI mass spectrum is usually greater than that for an EI mass spectrum to allow for addition products, such as $[M + CH_4]H^+$.

The time sequence shown in Figure 7.5 includes an extended pause in the sequence of events that permits the reaction of CH_5^+ with neutral nicotine to proceed. The methane CI mass spectrum obtained thus is shown in Figure 7.6*b*. The base peak is due to protonated nicotine of *m/z* 163. While the accompanying degree of fragmentation is

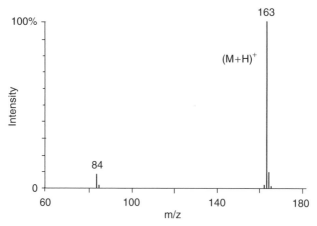

Figure 7.7. Chemical ionization mass spectrum of nicotine using ammonia as reagent gas. (Copyright 1985 Finnigan Corporation. All rights reserved. Reprinted with permission. Figure 6.10 from *Quadrupole Storage Mass Spectrometry*, by R. E. March, R. J. Hughes, J. F. J. Todd, © Wiley-Interscience, 1989.)

reduced, it is sufficient for the extraction of structural information of the nicotine precursor. Under these conditions, the molecular weight of the neutral molecule M is obtained readily from the mass/charge ratio of the base peak, which is protonated nicotine, that is, MH^+, together with some information on the structure of M. The fragmentation of nicotine shown in Figure 7.6*b* arises from the exothermicity of the proton transfer process from CH_5^+ to nicotine. The proton affinity of nicotine is appreciably greater than that of methane, and reaction exothermicity leads to fragmentation of the product species in which a new bond is formed [50].

When methane is replaced by ammonia as the CI reagent, the extent of fragmentation is reduced appreciably as there is now only a small energy difference in the proton affinities of ammonia and nicotine and reaction exothermicity is low. This effect is shown in Figure 7.7, where about 82% of the total charge now resides in the MH^+ species of *m/z* 163. The dearth of fragmentation yields little information of structural value but, as virtually the entire charge resides in one highly characteristic species, ammonia CI is a sensitive method for the quantitative determination of nicotine.

7.2.6.2. Chemical Ionization with Specific Reagent Ions

A wide variety of reagent ions of the type AH^+ can be used as CI agents. For example, a problem arose in the determination of two coeluting polychlorinated biphenyl (PCB) compounds, congener 77 (3,3′,4,4′-tetrachlorobiphenyl) and congener 110 (2,3,3′,4′,6-pentachlorobiphenyl), where congener 77 is much more toxic than is congener 110 but, in environmental samples, congener 110 is present at greater concentration. In an attempt to determine the amount of each congener by CI, several CI reagent ions were examined; these reagents included $C_4H_9^+$ from isobutene, $C_3H_5^+$ from each of methane and ethylene, and $C_2H_5^+$ from methane. Reagent $C_2H_5^+$ is obtained from methane by the reaction

$$CH_3^+ + CH_4 \rightarrow C_2H_5^+ + H_2. \qquad (7.12)$$

Only the reagent ion $C_2H_5^+$ obtained from methane forms $[M+H]^+$ ions with each congener, and the similarity of CI efficiencies of the two congeners permits observation of $[M+H]^+$ ions from each congener with almost equal facility [51].

To isolate reagent ions such as $C_2H_5^+$, a SIS waveform with a frequency notch window must be applied after the reaction period during which reagent ions $C_2H_5^+$ were formed. The SIS is a multifrequency waveform with a single notch (containing no frequencies) centered on the fundamental secular frequency of the selected reagent ion. The frequency notch was $\pm 2\,kHz$ from the secular frequency and the SIS waveform was applied at an amplitude of $10\,V_{p-p}$ for 1 ms. This waveform causes the ejection of all ions with secular frequencies outside the frequency notch, leaving only the selected CI reagent ion species in the ion trap to react subsequently with PCB molecules eluting from the GC column. This technique for the isolation of $C_2H_5^+$ reagent ions was applied also to a study of CI of 60 PCBs [52]. A review of the wide variety of CI reagent ion systems has been given elsewhere [50].

7.2.7. Tandem Mass Spectrometry

When two mass spectrometers are coupled together, the whole is greater than the sum of the two parts. J. J. Thomson [53] demonstrated clearly the veracity of this statement when he built a special instrument in which a beam of positive ions passed successively between the poles of two perpendicular magnets. In addition to the field-free region preceding both magnets, a second field-free region between the magnets presented a novel opportunity for a mass-selected ion to undergo a collision with background gas; charged products of this encounter could then be mass selected in the field of the second magnet. Tandem mass spectrometry came about thus, yet some 60 years were to pass before it was applied to the analysis of mixtures [54] and as a probe for structure determination of ions in the gas phase [55]. In tandem mass spectrometry, which was given the acronym MS/MS [56, 57], the achievable sensitivity in terms of signal/noise ratio is enhanced in comparison with single-stage mass spectrometry, or MS. While the ion signal intensity in MS/MS is reduced both in the fragmentation stage, that is, in CID following the first stage of mass selectivity, and in transmission through the second stage of mass selectivity, the reduction in noise level is greater. Thus in MS/MS the separate roles of electronic noise and chemical noise are more clearly differentiated. The considerable reduction in chemical noise as a direct result of the introduction of a second independent resolving element brings about a lower detection limit for the tandem mass spectrometric configuration despite the accompanying loss of ion signal intensity.

The first comprehensive review of the state of the art of tandem mass spectrometry appeared in 1983, concurrently with the announcement of the Finnigan MAT ion trap detector [2–5]. This review [58], to which approximately 50 of the leading practitioners contributed, describes the general types of MS/MS applications that may be pursued with the instrumentation available at the time of writing. The first MS/MS studies using a commercial ion trap system (the ITMS) were reported in 1987 by Louris et al. [59].

7.2.8. Scan Function for Tandem Mass Spectrometry

In GC coupled with single-stage MS, the cycle of ionization, mass analysis, and emptying of the ion trap as shown in the scan function of Figure 7.2 is run repetitively each second or each fraction of a second depending on the mass range of the mass analysis scan. The many EI mass spectra obtained thus are stored for examination later. It is possible to sum the numbers of ions of each species in each mass spectrum to obtain a total ion charge and to plot this sum against elution time to yield a total ion current (TIC) for the GC/MS run. A TIC plot is an informative presentation of the data, and appropriate software permits the recall of the mass spectrum for any given elution time in the TIC plot or a plot of the temporal variation of a given ion species can be obtained. All of this type of information can be obtained from a single, simple scan function.

In Thomson's MS/MS experiments, an ion was mass selected in the first magnetic field and dissociated by a collision with background gas, and the product ions of this encounter were mass analyzed by the field of the second magnet, yet the coupling of such an MS/MS procedure to GC is not trivial. It is relatively facile to create a single scan function for MS/MS and to apply this scan function to the effluent from a GC, as shown in Figure 6.4; however, the ion species isolated in the first mass-selective stage is fixed within this scan function. In Chapter 8 there is an example of the recognition (by appropriate software) of doubly-charged ions of a peptide and automatic selection of such ions for examination by MS/MS. Recognition of these ions is based on the dominance of their relative signal intensity.

In GC/MS/MS, where primary ions are formed by electron impact and the principal candidate for examination by MS/MS is the molecular ion, one cannot rely on recognition based upon molecular ion signal intensity because it is frequently low. In a QIT where the effluent from a gas chromatograph is directed into the trapping volume, ions are created by electron impact and confined. Once the ions have been allowed to cool, the next task is to isolate a selected ion species; once the species has been isolated, the remaining ions are subjected to resonant excitation so as to dissociate all of the trapped ions and to trap as many as possible of the fragment ions formed. The stored fragment ions, or product ions, are mass analyzed using the normal mass-selective axial instability scan. Let us examine this situation using a specific example of the determination of dioxins and furans.

7.3. TANDEM MASS SPECTROMETRIC DETERMINATION OF DIOXINS AND FURANS

This brief description of the development of the QIT for tandem mass spectrometry is illustrated by reference to the determination of dioxins and furans. The power and limitation of the ion trap using but a single scan function are discussed. Determination of dioxins and furans requires the utilization of many scan functions not only for isolation of molecular ions from the various congener groups, tetrachloro-, pentachloro-, and so on, but also for the isolation of coeluting isotopically-labeled compounds added to the original sample. The demand for multiple scan functions was met through rapid development of appropriate software. Much of the new software that came online in the 1990s could be loaded and used on QIT instruments with little modification, if any, of ion trap hardware. The original ITD was well constructed and could be shown to be extremely versatile given appropriate software.

At the point in this discussion when the stage is reached that the ion trap functions as a tandem mass spectrometer, attention will be directed to the tuning of the instrument, the isolation of mass-selected ion species, and resonant excitation of the isolated ion species. This chapter will then conclude with a comparison of the performance of the QIT with those of a triple-stage quadrupole (TSQ) tandem mass spectrometer and a high-mass-resolution tandem sector instrument of reverse geometry. The comparison is made with respect to dioxins.

The molecular ion cluster for, say, 2,3,7,8-tetrachorodibenzo-p-dioxin (2,3,7, 8-T$_4$CDD) is composed of several peaks due to the presence of naturally-occurring ^{13}C and ^{37}Cl isotopes. The molecular ion of greatest ion signal intensity is m/z 322, $[C_{12}H_4Cl_3^{37}ClO_2]^{+\cdot}$. Upon CID, this ion species loses the radical COCl$\cdot$ to form a product ion of m/z 259 as shown in

$$[C_{12}H_4Cl_3^{37}ClO_2]^{+\cdot} \rightarrow [C_{11}H_4Cl_2^{37}ClO]^{+} + COCl\cdot. \qquad (7.13)$$
$$(m/z\ 322) \qquad\qquad (m/z\ 259)$$

There are 15 isomeric T$_4$CDDs and in each molecular ion cluster the ion of greatest ion signal intensity is m/z 322. In the absence of retention time data, it is not possible to distinguish between the 15 isomeric T$_4$CDDs. Initially with a modified ITD, it was possible to use *only one scan function* for the determination of a single congener group (i.e., tetra) in a single chromatographic run. Any congener group could be selected but, once selected, that congener group was the only group examined with the scan function. In 1994, a rapid screening technique carried out with a QIT operated tandem mass spectrometrically (MS/MS) was reported for the detection and quantitation of T$_4$CDDs [60, 61]. One microliter of an extract from a clam was injected onto a DB-5 fused silica capillary column (J&W Scientific, Folsom, CA) and the single scan function appropriate to Eq. (7.13) was used repeatedly throughout the GC run.

In Figure 7.8 is shown a comparison of two chromatograms of TCDD isomers from a clam extract; the chromatograms were obtained by a QIT and by high-resolution mass spectrometry (HRMS). The upper chromatogram shows the original ion signals for m/z 259 obtained by MS/MS with the QIT where m/z 322 was isolated initially; the lower chromatogram (HRMS) shows the original ion signals as obtained in the single-ion monitoring (SIM) mode for [M+2]$^{+\cdot}$ molecular ions of m/z 321.8936. The variations in relative peak signal intensities of 12 T$_4$CDDs in each chromatogram are seen to be remarkably similar; this observation was quite extraordinary considering

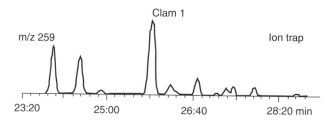

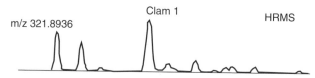

Figure 7.8. Two chromatograms for tetrachlorodibenzo-*p*-dioxins obtained from clam extract. Upper chromatogram shows original ion signals for *m/z* 259 obtained by MS/MS with QIT; lower chromatogram (HRMS) shows original ion signals as obtained in SIM mode for [M+2]$^{+\cdot}$ molecular ions of *m/z* 321.8936. Variations in relative abundances of T$_4$CDDs in each chromatogram are seen to be remarkably similar. Chromatographic retention times (min) correspond to ion trap data. (Reprinted from *Organic Mass Spectrometry*, Vol. 29, "Rapid screening technique for tetrachlorodibenzo-*p*-dioxins in complex environmental matrices by gas chromatography/tandem mass spectrometry with an ion trap detector," by J. B. Plomley, C. J. Koester, R. E. March, upper half on Figure 6, pp. 372–381 (1994). © John Wiley & Sons Limited. Reproduced with permission.)

the significant differences between the two types of instrument and, particularly, with respect to the CID process in the QIT. From inspection of Figure 7.8(upper), it could be concluded either that the available CID energy exceeded the highest activation energy for loss of COCl$^{\cdot}$ from among the congeners or that the T$_4$CDDs have a common activation energy for loss of COCl$^{\cdot}$ from each T$_4$CDD congener. The latter conclusion is favored because, in practice, it has been found that when the CID conditions are tuned for a given congener within a group, for example, 1,2,3,4-T$_4$CDD, these optimized CID conditions are entirely suitable for the remaining congeners in that group. This observation made possible the use of a single QIT scan function (with constant resonant excitation conditions) for the CID of each congener group.

A further restriction imposed by the software at that time (a single scan function only) was the inability to perform quantitation with internal standards that coeluted chromatographically with their native analytes. Software advances overcame these restrictions such that it is possible now to deconvolute mass spectra generated from analytes that coelute chromatographically [62, 63]. When operated in the MS/MS mode, the ion trap is now capable of multiple-reaction monitoring (MRM) of product ions from tetra- to octa-polychlorodibenzo-*p*-dioxins (PCDDs) and polychlorodibenzofurans (PCDFs) in a single chromatographic acquisition [64]. An MS/MS method for the ultratrace detection and quantitation of the tetra- to octa-PCDDs/PCDFs with isotopic dilution techniques has now been developed [65]. The sensitivity of the ion trap MS/MS technique at that time (500 fg/µL instrumental detection limit with a signal/noise ratio of 5 : 1) was shown to be comparable to that of TSQ.

7.3.1. Tuning of the Mass Spectrometer

For the detection of extremely low concentrations (e.g., ≤500 fg/µL) of PCDDs and PCDFs, the optimization of all instrumental parameters is important. Martinez and Cooks reported [66] that ion signal strength in the TSQ is affected by the nature of the collision gas (e.g., He, Ar), the collision gas pressure (the number of collisions that the precursor ion undergoes within the collision chamber, or target gas thickness), the collision energy (the duration of the interaction between the precursor ion and collision gas), electron energy (proportional to the initial internal energy of the precursor ion), the potential of the third quadrupole with respect to that of the second quadrupole, the design of the collision cell (RF voltage and the restrictive interquadrupole aperture of the second quadrupole), and the type of detector. For the QIT, ion signal strength is affected by the nature and pressure of the collision gas, the RF potential during CID, and the supplementary RF potential amplitude, frequency, and duration of application.

Perfluorotributylamine (PFTBA) is used commonly for the optimization of MS/MS parameters when analyzing organic compounds and is introduced at a low partial pressure into the ion source of a HRMS to obtain "lock" mass/charge ratios. The fragmentation of PFTBA under CID conditions does not parallel the behavior of all analytes because, according to the quasi-equilibrium theory [67], the pattern and degree of fragmentation of the precursor ion are dependent on its internal energy. Excitation by collision can form a series of precursor ions with a distribution of internal energies. Kenttämaa and Cooks [68] concluded that, by using breakdown graphs, parameters such as collision energy and collision gas pressure have significant effects on precursor ion internal energy and, therefore, its pattern of dissociation. In principle, parameters that affect precursor ion internal energy can be set to direct fragmentation toward the desired fragmentation, such as the loss of COCl· (or COCl· + 2 COCl·) in the case of PCDDs and PCDFs. Catlow et al. [69] have shown that the optimum collision energy and collision gas pressure for one reaction will almost certainly not be the optimum values for another reaction of that or any other precursor ion. This observation implies that the optimization of a particular fragmentation reaction using the analyte of interest is critical in order to obtain the maximum possible ion signal strength.

7.3.2. Ionization

The scan function employed for the MS/MS determination of T_4CDDs ionized by electron impact is shown in Figure 7.9, wherein the abscissa represents time and the ordinate represents the amplitudes of the voltages. Neither axis is to scale. Voltage

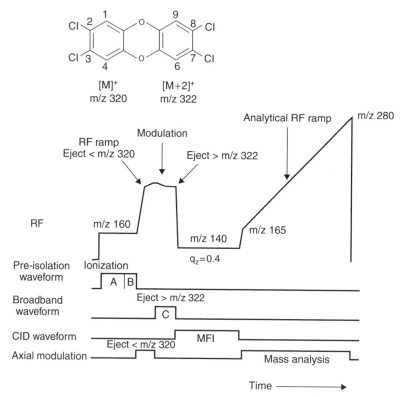

Figure 7.9. Schematic representation of the scan function for MS/MS determination of T₄CDD. (Reprinted from the *International Journal of Mass Spectrometry and Ion Processes*, Vol. 165/166, M. Splendore, J. B. Plomley, R. E. March, R. S. Mercer, "Tandem mass spectrometric determination of polychlorodibenzo-*p*-dioxins and polychlorodibenzofurans in a quadrupole ion trap using multi-frequency resonant excitation," Fig. 1, pp. 595–609 (1997), with permission from Elsevier.)

amplitudes vary from 1–2 V to ~3000 V_{0-p}. The durations of application of potentials are given in the text except for that of the analytical RF ramp; a ramp from *m/z* 165 to *m/z* 280 is accomplished in 21.5 ms. The LMCO value was set to *m/z* 160 during the ionization period *A*. A total ion number target was set for the AGC algorithm; with a filament emission current of 50 μA, the maximum ionization time employed was 20 ms.

7.3.3. Isolation of Mass-Selected Ion Species

Supplementary alternating voltages or waveforms were applied to the end-cap electrodes in dipolar fashion [60]; such waveforms are employed for ion isolation, ion excitation, and axial modulation. Precursor ion isolation is performed in two stages, corresponding to coarse isolation and fine isolation. A coarse-isolation waveform is imposed for ejection of all ions except those selected; the waveform consists of multiple frequencies that cover the range 3.7–513.5 kHz with a 1-kHz notch that corresponds to the secular frequency of the molecular ions to be isolated (see Figure 6.4). This waveform is imposed during ionization (A) and prolonged after the cessation of ionization during period B so as to ensure ejection of unwanted ions. For T_4CDD shown in Figure 7.9, the notch is centered at 174.5 kHz for m/z 320 and m/z 322 with a q_z value of about 0.45. The amplitude of the coarse-isolation waveform is 20 V_{0-p} for all groups (except for hepta-chlorocongeners, where the amplitude is 30 V_{0-p}).

Fine isolation is achieved by ramping the RF amplitude until the LMCO is just less than m/z 320, at which point ions of lower mass/charge ratios are ejected; ion ejection is facilitated by the concurrent application of an axial modulation with an amplitude of 3 V_{0-p}. The RF amplitude is decreased slightly, and ions with $m/z > 322$ are ejected upon application of a 5-ms broadband waveform of amplitude 30 V_{0-p} (C). The strategy for high mass ejection is rationalized on the basis that, as q_z increases, the difference in $\omega_{z,0}$ between successive masses increases. Therefore, the risk of precursor ion ejection by waveform C is minimized at large q_z values.

Once the isolation of the selected ion species (m/z 320 and m/z 322 for T_4CDD in Figure 7.9) is completed, the RF amplitude is reduced to obtain a q_z value of 0.4 for the selected ion with the higher mass/charge ratio. For T_4CDD, a q_z value of 0.4 for m/z 322 corresponds to a LMCO value of m/z 140.

7.3.4. Resonant Excitation of Isolated Ion Species

In all of the scan functions used in this study, CID was carried out at a q_z value of 0.4 for the isolated species (or one of the isolated species). At this q_z value, an adequate range of product ions can be stored and the same waveforms, modified as needed, can be used for CID of a wide range of isolated ion species. There are four possible modes by which CID can be effected: (i) single-frequency irradiation (SFI), (ii) multifrequency irradiation (MFI), (iii) secular-frequency modulation (SFM), and (iv) nonresonant excitation [70]. Single-frequency irradiation is performed at a fixed value of q_z by the application of a supplementary AC signal across the end-cap electrodes in a dipolar fashion. It is the simplest form of resonant excitation and is similar to axial modulation in that a single frequency is employed. Secular-frequency modulation involves small increments and reductions of the RF drive potential amplitude applied to the ring electrode such that ions move into and out of resonance with an applied single-frequency waveform. Because modulation of the RF amplitude introduces small changes in q_z and corresponding changes in β_z, the net result is a sweep of frequency $\omega_{z,0}$ over a narrow range (e.g., 1 or 2 kHz). Multifrequency irradiation involves the application between the end-cap electrodes of a waveform that consists of several frequency components while the value of the trapping parameter q_z is held constant. Modes (ii)–(iv) are less labor intensive during CID tuning because the task of empirically matching an applied single-frequency sinusoidal waveform to the secular frequency of an ion is obviated. Additionally, any shifts in ion secular frequency due to space charge effects, the stretched geometry of the ion trap, the imprecision of the waveform frequency calibration procedure, and ion axial excursion from the ion trap center are compensated for with modes (ii)–(iv).

The MFI waveform employed for CID is composed of 13, 15, or 17 frequency components spaced at intervals of 500 Hz and covering a band of frequencies in the range 6–8 kHz. A constant MFI waveform amplitude can be applied for each group of native and labeled isomers because, as shown in Figure 7.8, the dissociation of each congener group has a common energy of activation. The amplitudes were, for example, 2.45 V_{0-p} for T_4CDD, 2.55 V_{0-p} for T_4CDF, 2.65 V_{0-p} for P_5CDD and H_6CDD, and 3.00 V_{0-p} for O_8CDF; all had a bandwidth of 6 kHz. In each scan function, the MFI band is centered on 153.6 kHz, corresponding to a q_z value of 0.4. Under these conditions, the MFI waveforms were sufficient to dissociate fully the isolated molecular ions in 10 ms, whereas fragment ion capture efficiencies were comparable to those reported previously [71].

7.3.5. Analytical RF Ramp

The preselected retention time windows were 20–27 min for T_4CDF and T_4CDD and 27–34 min for P_5CDF and P_5CDD. Following CID, specified mass ranges were scanned for fragment ions from each chlorocongener. The mass ranges were 165–280 Th for T_4CDF and T_4CDD and 195–315 Th for P_5CDF and P_5CDD. Within these mass ranges, the product ions of all fragmentation channels, save that of chlorine atom loss, could be monitored. The analytical ramp was scanned at 5555 Th/s; axial modulation was carried out with amplitude of 3 V_{0-p} and at a frequency of 485 kHz. The electron multiplier was biased at a voltage of ~1800 V to provide an ion signal gain of 10^5.

7.4. COMPARISON OF THREE MASS SPECTROMETRIC METHODS

A comparison is presented of the performances of three mass spectrometers of high specificity in the determination of dioxin/furan congeners. The three instruments used in this study were a triple-sector EBE (E, electrostatic; B, magnetic) mass spectrometer operated at HRMS, a QIT mass spectrometer, and a TSQ mass spectrometer. The QIT and TSQ instruments were operated in tandem mass spectrometric mode. A mixture of tetra- to octa-chlorodibenzo-p-dioxins (T_4-O_8CDD) containing in all seven dioxin congeners was used for much of this study. The factors considered in this comparison were the tuning of each instrument, the preparation and comparison of calibration curves, the 2,3,7,8-T_4CDD detection limit for each instrument, ion signals due to H_6CDDs obtained with each instrument from two real samples (air and pyrolyzed polychlorinated phenols), average relative response factors, and ionization cross sections. For each dioxin congener, the response factor is expressed relative to that for the O_8CDD congener while the electron impact ionization cross section is expressed relative to that for the T_4CDD congener. The relative ionization cross sections for T_4-O_8CDD from HRMS and the QIT and for T_4-P_5CDD from the TSQ are in good agreement and show an overall decrease of some 10–20% with increasing degree of chlorine substitution; the variation among three H_6CDD congeners is identical in each case. With the TSQ, lower relative ionization cross sections for H_6-O_8CDD are ascribed to mass-dependent fragment ion scattering in the RF-only collision cell [72].

7.4.1. Instruments

For the determinations of dioxins/furans by high-resolution gas chromatography (HRGC)/HRMS, a VG Autospec (Vacuum Generators, Altringham, UK) triple-sector instrument of EBE geometry linked to the gas chromatograph by a direct capillary interface was used and was operated at a resolving power of 10 000 (10% valley). The gas chromatograph was a Hewlett-Packard 5890-II equipped with a splitless injection system and temperature programming. An OPUS data system was used for the collecting, recording, and storing of all MS data. For the determinations of dioxins/furans by TSQ, a Finnigan MAT TSQ 70 triple-stage quadrupole mass spectrometer (Finnigan MAT, San Jose, CA) linked to the gas chromatograph via a direct capillary interface was used. The gas chromatograph was a Varian 3400 equipped with a splitless injection system and temperature programming. The collecting, recording, and storing of MS data used an ICIS II data system. The resolution of the first and third quadrupoles was set to unit mass resolution. The second QMF (RF only) was used to perform CID of the two mass-selected molecular ions isolated consecutively in the first QMF [73]. Argon was used as the collision gas. For the determinations of dioxins/furans by QIT, a Varian Saturn 3D GC/MS/MS instrument equipped with a waveform generator and linked to a gas chromatograph via a direct capillary interface was used. The gas chromatograph was a Varian 3400 equipped with a splitless injection system and temperature programming. The Saturn software version 5.2 that was used for data acquisition is compatible with the multiple-scan-function software Ion Trap Toolkit for MS/MS 1.0 (Varian Chromatography Systems, Walnut Creek, CA). Ion Trap Toolkit software permits the determination of approximately 200 scan functions that can be recalled in a specified time sequence.

7.4.2. Operational Conditions

Multiple-frequency resonant excitation in the presence of helium buffer gas was used to perform CID of mass-selected or isolated molecular ions [74, 75]. All three gas chromatographs had capillary columns of fused silica, 60 m in length, 0.25 mm i.d., J&W, DB-5 stationary phase, and 0.25 μm film thickness. (DB-5 is a trademark of J&W Scientific, now part of Agilent Technologies, http://www.chem.agilent.com.)

The electron energy was 35 eV in HRMS, 22–30 eV in TSQ, and between 50 and 100 eV in QIT. Because the QIT operates with a time-varying voltage, the resultant distribution of electron energies depends on the RF phase upon entry of the electrons. Calculations have suggested an average electron energy of about 50 eV in the QIT when operated at a LMCO of 20–30 Th [76]. In this work, the LMCO varied from 128 to 183 Th such that a linear extrapolation of the above calculations would suggest electron energies appreciably higher than 50 eV. The lifetime of electrons of energy >100 eV in the QIT is relatively short, and the ionization cross section has begun to decrease for such electrons; thus, their contribution to ionization will be minor. Because the observed mass spectra do not differ significantly from those obtained with 70-eV electrons, it is reasonable to assume that the distribution of electron energies does not exceed 50–100 eV. The source temperature was 280°C in HRMS and 245°C in TSQ, and the manifold temperature in the QIT was 240°C.

7.4.3. Product Ions Monitored

In Table 7.2 are listed the ions monitored in each mass spectrometric method for T_4CDD and P_5CDD; $M^{+\cdot}$ is the molecular ion with all ^{35}Cl atoms, and $[M + 2]^{+\cdot}$ is the molecular ion with a single ^{37}Cl atom. In each method, the two most abundant molecular ions were isolated and/or selected. In HRMS, the ions selected were the ions detected. In TSQ, the monitored fragment ions were those that resulted from the loss of $COCl\cdot$ upon CID of the mass-selected ions. In QIT, the monitored fragment ions were those that resulted from the loss of $COCl\cdot$ and $2COCl\cdot$ upon CID of the mass-selected ions. For HRMS and TSQ, the approved methods called for the observation of ions at two mass/charge ratios; for QIT, there is no approved method and thus additional ion species were monitored to increase sensitivity and selectivity. Selectivity is enhanced when a confirmatory ion is monitored with the requirement that its signal

TABLE 7.2. Ion Species Monitored from T_4CDDs and P_5CDDs in Each Mass Spectrometric Method

HRMS (MS)		TSQ (MS/MS)	QIT (MS/MS)[a]	
		T_4CDD		
$[M]^{+\cdot} + [M+2]^{+\cdot}$		$[M-COCl\cdot]^+$	$[M-COCl\cdot]^+ + [M+2-CO^{37}Cl\cdot]^+$	m/z 257
m/z 320	m/z 322	$+ [M+2-CO^{37}Cl\cdot]^+$	$[M+2-COCl\cdot]^+$	m/z 259
		m/z 257	$[M-2(COCl\cdot)]^{+\cdot}$	
		$[M+2-COCl\cdot]^+$	$+ [M+2-C_2O_2Cl^{37}Cl]^{+\cdot}$	m/z 194
		m/z 259	$[M+2-2(COCl\cdot)]^{+\cdot}$	m/z 196
		P_5CDD		
$[M]^{+\cdot} + [M+2]^{+\cdot}$		$[M-COCl\cdot]^+$	$[M-COCl\cdot]^+ + [M+2-CO^{37}Cl\cdot]^+$	m/z 291
m/z 354	m/z 356	$+[M+2-CO^{37}Cl\cdot]^+$	$[M+2-COCl\cdot]^+$ m/z 293	
		m/z 291	$[M-2(COCl\cdot)]^{+\cdot}$	
		$[M+2-COCl\cdot]^+$	$+ [M+2-C_2O_2Cl^{37}Cl]^{+\cdot}$	m/z 228
		m/z 293	$[M+2-2(COCl\cdot)]^{+\cdot}$	m/z 230

[a]In QIT, the $[M]^{+\cdot}$ and $[M+2]^{+\cdot}$ ions of, for example, T_4CDD are isolated along with the $[M+1]^{+\cdot}$ ion, C_{11} $^{13}CH_4O_2Cl_4^{+\cdot}$ which, on dissociation by loss of $^{13}COCl\cdot$, yields m/z 257. Because m/z 257 is one of the mass/charge ratios that are monitored and there is a probability of 1 in 12 that $^{13}COCl\cdot$ will be lost from the $[M+1]^{+\cdot}$ ion, the calculated fraction of the molecular ion cluster isolated includes a fractional contribution from the $[M+1]^{+\cdot}$ ion.

intensity, relative to that of a selected ion, is observed to fall within a narrow specific range. The average signal ratio of the detected fragment ion signals that resulted from the loss of COCl· and 2COCl· for each dioxin investigated is approximately 2 : 1 in favor of the loss of COCl·. However, because the fragmentation channel that involves the loss of 2COCl· is used in HRMS and TSQ for confirmation, it was decided to include the consideration of this channel in the tuning procedure for QIT. A standard solution that contained 1000 pg of each of six dioxin congeners (2,3,7,8-T_4CDD, 1,2,3,7,8-P_5CDD, 1,2,3,4,7,8-H_6CDD, 1,2,3,6,7,8-H_6CDD, 1,2,3,7,8,9-H_6CDD, and 1,2,3,4,6,7,8-H_7CDD) and 2000 pg of O_8CDD was used for all three methods. The furan congener, for which specimen calibration plots obtained with HRMS, QIT, and TSQ are presented, was 2,3,4,7,8-P_5CDF.

7.4.4. Calibration

In each case, the instrument was mass calibrated with PFTBA. The TSQ was first mass calibrated in the Q_1 MS mode, in the Q_3 MS mode, and finally in the MS/MS mode. Once mass calibration was complete, a tetrachloro-dioxin/furan congener on a direct-insertion probe was introduced into the ion source and used for tuning. The precursor ion is first optimized in the Q_1 MS mode, and the instrument is switched to the MS/MS mode and collision gas is allowed to enter the collision quadrupole. For a given fixed collision gas pressure, the ion collision energy is varied and the fragment ion signal intensities monitored. From these data, a breakdown graph of fractional ion abundance versus ion energy similar to that shown in Figure 7.10a can be constructed. It is seen that, at a collision energy in the vicinity of 25 eV, the ion current due to the [M–COCl·]$^+$ species (▲) attains a maximum relative value (recall that it is the total ion current at each collision energy that is plotted on the ordinate of this figure). Additional graphs can be constructed for other collision gas pressures to obtain the optimum collision gas pressure and ion energy for a selected fragment ion channel. At the optimum collision gas pressure ($\sim 3 \times 10^{-3}$ Torr) for fragmentation of 2,3,7,8-T_4CDD, the collision energies were set to optimize the ion signal strength for fragment ions that arise from loss of COCl· for the 2,3,7,8-substituted congeners in each group; these values ranged between 18 and 27 eV in the laboratory frame. Chromatograms obtained with PFTBA tuning and dioxin/furan congener tuning, which are shown elsewhere [77], show clearly that the signal/noise ratio for dioxin/furan fragment ions is much improved with specific dioxin/furan congener tuning.

7.4.5. Resonant Excitation

For the QIT, the pressure of the helium collision gas is optimized ($\sim 10^{-3}$ Torr) with respect to the peak widths of the PFTBA ions used as mass markers. Resonant excitation was carried out at a fixed value of the q_z trapping parameter ($q_z = 0.4$) for the isolated ion species of higher mass/charge ratio. The waveform employed for CID

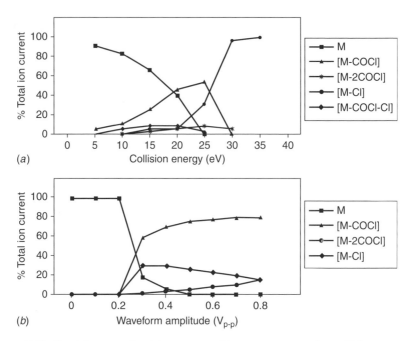

Figure 7.10. Breakdown graphs: (*a*) percent total ion current versus ion collision energy, as used for TSQ tuning; (*b*) percent total ion current versus supplementary RF waveform amplitude, as used for tuning QIT. (Reprinted from the *International Journal of Mass Spectrometry*, Vol. 197, R. E. March, M. Splendore, E. J. Reiner, R. S. Mercer, J. B. Plomley, D. S. Waddell, K. A. MacPherson, "A comparison of three mass spectrometric methods for the determination of dioxins/furans," Fig. 3, pp. 283–297 (2000), with permission from Elsevier.)

with MFI is discussed above [65]. A constant MFI waveform amplitude was applied for 10 ms for each group of congeners; the amplitude varied from 2.65 to 3.10 V_{0-p} for the dioxins, and for the 2,3,4,7,8-P_5CDF, it was 3.60 V_{0-p}. The optimized conditions for each congener group were obtained from breakdown graphs of fractional ion abundance versus supplementary RF waveform amplitude at a fixed duration of irradiation, similar to that shown in Figure 7.10*b*. The scale of the abscissa in Figure 7.10*b* corresponds to the amplitude for each frequency component of the MFI waveform. The suffixes 0-p and p-p refer to the amplitude of a sinusoidal waveform and correspond to zero to peak and peak to peak, respectively. Note the similarity in Figures 7.8*a* and *b* in that both plots show the higher threshold required for observation of the 2COCl⁻ loss channel relative to the threshold for loss of COCl⁻. Note also that the plots differ with respect to the behavior of the COCl⁻ loss channel (▲) at high collision energy and high waveform amplitude; in TSQ (Figure 7.10*a*), the nascent fragment ion formed by loss of COCl⁻ undergoes further fragmentation as a result of either internal excitation or collisions, whereas in QIT (Figure 7.10*b*), the same fragment ion is neither internally excited nor resonantly excited and reaches a plateau in relative abundance at a high waveform amplitude.

7.4.6. Comparisons of Performances

The performances are compared with respect to ion signals at low concentration, detection of H_6CDD in real samples, and ionization cross sections.

7.4.6.1. Ion Signals at Low Concentration Examples of the ion signals obtained with each instrument using gas chromatographic separation for low concentrations of 2,3,7,8-T_4CDD are shown in Figure 7.11. Note that the top trace (HRMS) was obtained with an amount 50 times the detection limit; the middle trace (TSQ) was obtained with an amount some 7 times the detection limit; and the bottom trace (QIT) was obtained with an amount some 5 times the detection limit.

7.4.6.2. Real Samples Two examples are given of the determination by HRMS, TSQ, and QIT of H_6CDDs in real samples; the first example is a sample of ambient air (Figure 7.12), while the second is a sample obtained following the pyrolysis of polychlorinated phenols and is shown in Figure 7.13; in each figure, the last three congeners to elute are 2,3,7,8-tetra chloro-containing congeners. The agreement among the chromatograms in Figure 7.12 with respect to peak relative signal intensities and peak resolution is quite good with the exception of the first peak; here, the QIT shows some tailing of the peak. In Figure 7.13, again agreement is good with respect to peak relative signal intensities (note that the peak relative signal intensities in Figure 7.13 differ from those of Figure 7.12), but the QIT has failed, on this occasion, to resolve the peak that is centered at ~16.88 and is resolved by HRMS and TSQ.

7.4.7. Ionization Cross Sections

For HRMS, the observed ion signal intensity per picogram of material injected for each chlorocongener, $(A_{EI})_{cong}$, is related directly to the electron impact ionization cross section, σ_{cong}, as given by

$$(A_{EI})_{cong} = I_e L (F_{iso})_{cong} \sigma_{cong} N_{cong} \alpha \tag{7.14}$$

where $(A_{EI})_{cong}$, expressed as an area, is the sum of the ion signal intensities of the two mass-selected molecular ions for each chlorocongener as reported in Table 7.2, I_e is the electron beam intensity, L is the ionization path length, $(F_{iso})_{cong}$ is the sum of the fractional abundances of the two mass-selected ions, N_{cong} is the number of molecules per picogram of congener in the HRMS ion source, and α is a fraction corresponding to the ratio of the number of ions detected to the number of ions formed in the ion source. To a first approximation, α can be assumed to be constant over the mass range examined, though this assumption is possibly an oversimplification for HRMS and undoubtedly an oversimplification of a complex process for the TSQ, *vide infra*. To effect a comparison among the HRMS and TSQ beam methods with the QIT pulsed method, each $(A_{EI})_{cong}$ value was normalized to that for T_4CDD for each of the three

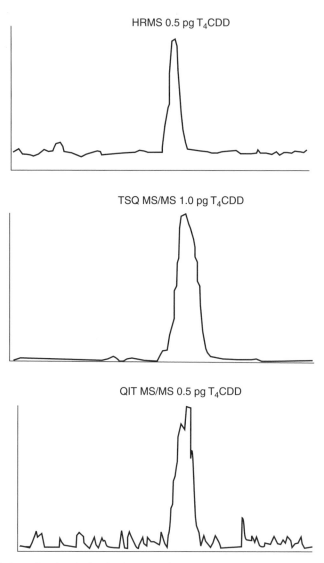

Figure 7.11. Ion signals obtained with each instrument for low concentrations of 2,3,7, 8-T$_4$CDD: (*a*) HRMS, 0.5 pg injected in 1 µL, signal intensity sum due to *m/z* 320 and 322 shown; (*b*) TSQ, 1.0 pg injected in 1 µL, signal intensity sum due to *m/z* 257 and 259 shown; (*c*) QIT, 0.5 pg injected in 1 µL, signal intensity sum due to *m/z* 257, 259, 194, and 196 shown. Signal/noise ratios for HRMS and QIT are comparable and lower than that for TSQ (as expected for higher concentration injected in TSQ) (Reprinted from the *International Journal of Mass Spectrometry*, Vol. 197, R. E. March, M. Splendore, E. J. Reiner, R. S. Mercer, J. B. Plomley, D. S. Waddell, K. A. MacPherson, "A comparison of three mass spectrometric methods for the determination of dioxins/furans," Fig. 4, pp. 283–297 (2000), with permission from Elsevier.)

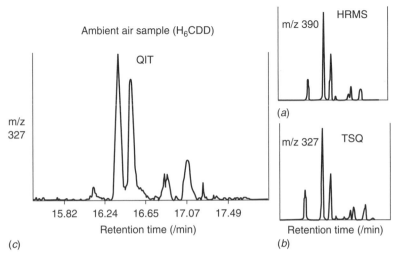

Figure 7.12. Chromatographs obtained by HRMS, TSQ, and QIT showing presence of several H_6CDDs in ambient air. Note that one ion species only is monitored in each case. (Reprinted from the *International Journal of Mass Spectrometry*, Vol. 197, R. E. March, M. Splendore, E. J. Reiner, R. S. Mercer, J. B. Plomley, D. S. Waddell, K. A. MacPherson, "A comparison of three mass spectrometric methods for the determination of dioxins/furans," Fig. 5, pp. 283–297 (2000), with permission from Elsevier.)

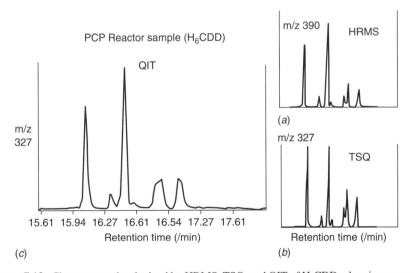

Figure 7.13. Chromatographs obtained by HRMS, TSQ, and QIT of H_6CDDs showing presence of several H_6CDDs in a sample obtained following pyrolysis of polychlorinated phenols. Note that one ion species only is monitored in each case. (Reprinted from the *International Journal of Mass Spectrometry*, Vol. 197, R. E. March, M. Splendore, E. J. Reiner, R. S. Mercer, J. B. Plomley, D. S. Waddell, K. A. MacPherson, "A comparison of three mass spectrometric methods for the determination of dioxins/furans," Fig. 6, pp. 283–297 (2000), with permission from Elsevier.)

methods; the resulting relative ionization cross section ($\sigma_{cong}/\sigma_{T4CDD}$) for each dioxin congener examined was calculated according to

$$\left(\frac{\sigma_{cong}}{\sigma_{T4CDD}}\right)_{HRMS} = \frac{(A_{EI})_{cong}\,(F_{iso})_{T4CDD}\,N_{T4CDD}}{(A_{EI})_{T4CDD}\,(F_{iso})_{cong}\,N_{cong}}. \tag{7.15}$$

Similar expressions can be obtained for $(\sigma_{cong}/\sigma_{T4CDD})_{TSQ}$ and $(\sigma_{cong}/\sigma_{T4CDD})_{QIT}$. It should be noted that the signal intensity (A_{EI}) is expressed in counts per picogram and N_{cong} was calculated using the expression

$$N_{cong} = \frac{weight \times N_{Avogadro}}{MW_{cong} \times Volume} = \frac{1 \times 10^{-12}(g) \times 6.022 \times 10^{23}(molecules/mol)}{MW_{cong}\,(g/mol) \times 22{,}414\,(cm^3)}. \tag{7.16}$$

For the tandem mass spectrometric methods TSQ and QIT, the fragment ion signal intensities $(A_{CID})_{cong}$ are equal to the product of the observed ion signal intensity for each chlorocongener, $(A_{EI})_{cong}$, and the CID efficiency for each chlorocongener, $(\eta_{CID})_{cong}$, as shown in the equation

$$(A_{CID})_{cong} = I_e \times L \times (F_{iso})_{cong} \times \sigma_{cong} \times N_{cong} \times \eta_{(CID)cong} \times \alpha \tag{7.17}$$

where $(\eta_{CID})_{cong}$ is defined as the ratio of the detected fragment ion signal intensities for each congener to the ion signal intensities of the mass-selected ions prior to CID.

Each $(A_{CID})_{cong}$ value was normalized to that for T_4CDD and the following ratio was computed for each dioxin congener:

$$\left(\frac{\sigma_{cong}}{\sigma_{T4CDD}}\right)_{MS/MS} = \frac{(A_{CID})_{cong} \times (F_{iso})_{T4CDD} \times (\eta_{CID})_{T4CDD} \times N_{T4CDD}}{(A_{CID})_{T4CDD} \times (F_{iso})_{cong} \times (\eta_{CID})_{cong} \times N_{cong}} \tag{7.18}$$

where MS/MS refers to both TSQ and QIT. Equation (7.18) is an expression for the relative ionization cross section, $(\sigma_{cong}/\sigma_{T4CDD})_{MS/MS}$, for each dioxin congener and for each MS/MS method.

The relative performances of the three mass spectrometric methods can be compared on the basis of congener-specific relative ionization cross sections, $\sigma_{cong}/\sigma_{T4CDD}$, obtained by HRMS, QIT, and TSQ. In Figure 7.14 are plotted the values of $\sigma_{cong}/\sigma_{T4CDD}$ for each dioxin congener. On the abscissa in Figure 7.14 are shown the congeners examined; the lines have been drawn merely to facilitate recognition of the datum points obtained with each method.

As may be expected, the data obtained by the three methods are generally quite similar except for the marked decrease with TSQ in passing from P_5CDD to the congeners of higher degree of chlorine substitution due possibly to a decrease in α. For HRMS and QIT there is a decrease of some 10–20% in the normalized ionization cross section as the degree of chlorination increases, since there is but a small increase of the correlated ionization energy; theoretical calculation [62] has shown an increase of the ionization energy of about 100 meV as the degree of chlorine

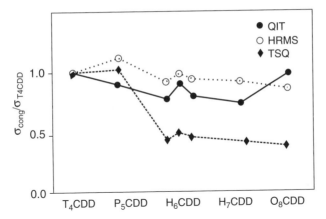

Figure 7.14. Relative ionization cross section ($\sigma_{cong}/\sigma_{T4CDD}$) for seven dioxin congeners obtained from (•) 27 QIT determinations, (○) 24 HRMS determinations, and (♦) 34 TSQ determinations. (Reprinted from the *International Journal of Mass Spectrometry*, Vol. 197, R. E. March, M. Splendore, E. J. Reiner, R. S. Mercer, J. B. Plomley, D. S. Waddell, K. A. MacPherson, "A comparison of three mass spectrometric methods for the determination of dioxins/furans," Fig. 9, pp. 283–297 (2000), with permission from Elsevier.)

substitution increases. Despite the enormous differences between the HRMS, TSQ, and QIT instruments, the ion signals observed in each instrument can be related directly to the ionization efficiency, or cross section, for each congener in the ion sources of the three instruments. The close agreement among HRMS and QIT values of $\sigma_{cong}/\sigma_{T4CDD}$ among the three H_6CDD congeners is of interest, particularly since the same trend is exhibited by TSQ.

7.5. CONCLUSIONS

Gas chromatography coupled with the QIT or TSQ for tandem mass spectrometry or with a single-stage high-resolution mass spectrometer affords a method of high specificity and high sensitivity for the determination of compounds of interest. The determination of dioxins/furans by GC/MS/MS constitutes a severe test of the capabilities of the QIT combined with GC.

The QIT has been compared with two other mass spectrometers with respect to their performance as mass detectors coupled with GC for the determination of dioxins/furans. While the HRMS detection limit for T_4CDD is lower than that of TSQ and QIT, there is evidence that all interferences are not eliminated by high mass resolution alone; thus there is also a need for instruments that achieve high specificity by tandem mass spectrometric operation. The relative ionization cross sections for T_4-O_8CDD from both HRMS and QIT and for T_4-P_5CDD from TSQ are quite close; in addition, the variation among three H_6CDD congeners is identical for the three methods. For TSQ, the values of the relative cross sections for H_6CDD, H_7CDD, and O_8CDD congeners are somewhat lower than those obtained for HRMS and QIT, and this behavior may be explained in terms of mass-dependent fragment ion scattering in the RF-only collision cell.

REFERENCES

1. R. G. Cooks, A. L. Rockwood, *Rapid Commun. Mass Spectrom.* **5** (1991) 93.

2. S. Borman, New Gas Chromatographic Detectors, *Anal. Chem.* **55** (1983) 726A.

3. G. C. Stafford, P. E. Kelley, W. E. Reynolds, J. E. P. Syka, Recent improvements in ion trap technology. Proc. 31st Ann. ASMS Conf. on Mass Spectrometry and Allied Topics, Boston, MA, May 8–13, 1983, pp. 48–49.

4. G. C. Stafford, P. E. Kelley, J. E. P. Syka, W. E. Reynolds, J. F. J. Todd, The ion trap mass spectrometer—A breakthrough in performance. 13th Annual Meeting of the British Mass Spectrometry Society, Warwick, UK, 1983, pp. 18–20.

5. G. C. Stafford, P. E. Kelley, D. C. Bradford, *Am. Lab.*, June 1983, pp. 51–57.

6. W. Paul, H. Steinwedel, Apparatus for separating charged particles of different specific charge, German Patent 944,900 (1956); U.S. Patent 2,939,952 (June 7, 1960).

7. J. E. P. Syka, Commercialization of the quadrupole ion trap. In R. E. March and J. F. J. Todd (Eds.), *Practical Aspects of Ion Trap Mass Spectrometry*, Vol. I, Chapter 4, CRC Press, Boca Raton, FL, 1995. J. E. P. Syka, The geometry of the Finnigan ion trap: History and theory, 9th Asilomar Conference on Mass Spectrometry, Asilomar Conference Grounds, Pacific Grove, CA, September 1992.

8. R. E. Finnigan and J. F. J. Todd, private communication, 1970.

9. M. A. Armitage, J. E. Fulford, R. J. Hughes, R. E. March, Quadrupole ion store characterization and application. Proc. 27th Ann. ASMS Conf. on Mass Spectrometry and Allied Topics, Seattle, WA, June 3–8, 1979, pp. 449–450.

10. M. A. Armitage, Applications of quadrupole ion storage mass spectrometry, M.Sc. Thesis, Trent University, Peterborough, ON, Canada, 1979.

11. M. A. Armitage, J. E. Fulford, D. N. Hoa, R. J. Hughes, R. E. March, *Can. J. Chem.* **57** (1979) 2108.

12. N. A. Yates, M. M. Booth, J. L. Stephenson, Jr., R. A. Yost, Practical ion trap technology: GC/MS and GC/MS/MS, in R. E. March and J. F. J. Todd (Eds.), *Practical Aspects of Ion Trap Mass Spectrometry*, Vol. III, CRC Press, Boca Raton, FL, 1995, Chapter 4.

13. D. D. Fetterolf, R. A. Yost, *Int. J. Mass Spectrom. Ion Processes* **62** (1984) 33.

14. H. Kaiser, *Anal. Chim. Acta* **33B** (1978) 551.

15. J. J. Fitzgerald, J. D. Winefordner, *Rev. Anal. Chem.* **2** (1975) 299.

16. P. E. Kelley, G. C. Stafford, Jr., D. R. Stevens, U.S. Patent 4,540,884 (September 10, 1985); Canadian Patent 1,207,918 (July 15, 1986).

17. G. C. Stafford, Jr., P. E. Kelley, J. E. P. Syka, W. E. Reynolds, J. F. J. Todd, *Int. J. Mass Spectrom. Ion Processes* **60** (1984) 85.

18. J. F. J. Todd, Ion trap theory, design, and operation, in R. E. March and J. F. J. Todd (Eds.), *Practical Aspects of Ion Trap Mass Spectrometry*, Vol. III, CRC Press, Boca Raton, FL, 1995, Chapter 1, p. 16.

19. X. Wang, D. K. Bohme, R. E. March, A commercial ion trap (ITD) modified by monopolar excitation for resonant ejection and extended mass range. Proc. 40th ASMS Conf. on Mass Spectrometry and Allied Topics, Washington, DC, May 31–June 5, 1992, p. 230.

20. X. Wang, D. K. Bohme, R. E. March, *Can. J. Appl. Spectros.* **38** (1993) 55.

21. X. Wang, H. Becker, A. C. Hopkinson, R. E. March, L. T. Scott, D. K. Bohme, *Int. J. Mass Spectrom. Ion Processes* **161** (1997) 69.

22. R. J. Strife, P. E. Kelley, M. Weber-Grabau, *Rapid Commun. Mass Spectrom.* **2** (1988) 105.

23. R. E. Kaiser, R. G. Cooks, J. E. P. Syka, G. C. Stafford, *Rapid Commun. Mass Spectrom.* **4** (1990) 30.

24. J. N. Louris, J. W. Amy, T. Y. Ridley, R. G. Cooks, *Int. J. Mass Spectrom. Ion Processes* **88** (1989) 97.

25. J. D. Williams, H.-P. Reiser, R. E. Kaiser, R. G. Cooks, *Int. J. Mass Spectrom. Ion Processes* **108** (1991) 199.

26. J. S. Brodbelt, J. N. Louris, R. G. Cooks, *Anal. Chem.* **59** (1987) 1278.

27. H. M. Fales, E. A. Sokoloski, L. K. Pannell, P. Quan Long, D. L. Klayman, A. J. Lin, A. Brossi, J. A. Kelley, *Anal. Chem.* **62** (1990) 2494.

28. R. E. Kaiser, R. G. Cooks, J. Moss, P. H. Hemberger, *Rapid Commun. Mass Spectrom.* **3** (1989) 50.

29. R. E. Kaiser, J. N. Louris, J. W. Amy, R. G. Cooks, *Rapid Commun. Mass Spectrom.* **3** (1989) 225.

30. J. C. Schwartz, J. E. P. Syka, I. Jardine, *J. Am. Soc. Mass Spectrom.* **2** (1991) 198.

31. D. E. Goeringer, W. B. Whitten, J. M. Ramsey, S. A. McLuckey, G. L. Glish, *Anal. Chem.* **64** (1992) 1434.

32. R. J. Strife, J. R. Simms, *Anal. Chem.* **61** (1989) 2316.

33. J. V. Johnson, R. A. Yost, P. E. Kelley, D. C. Bradford, *Anal. Chem.* **62** (1990) 2162.

34. C. S. Creaser, M. R. S. McCoustra, K. E. O'Neil, *Org. Mass Spectrom.* **26** (1991) 335.

35. R. J. Strife, J. R. Simms, *J. Am. Soc. Mass Spectrom.* **3** (1992) 372.

36. F. A. Londry, G. J. Wells, R. E. March, *Rapid Commun. Mass Spectrom.* **7** (1993) 43.

37. J. Franzen, R.-H. Gabling, M. Schubert, Y. Wang, Nonlinear ion traps, in R. E. March and J. F. J. Todd (Eds.), *Practical Aspects of Ion Trap Mass Spectrometry*, Vol. I, CRC Press, Boca Raton, FL, 1995, Chapter 3, pp. 49–167.

38. J. N. Louris, J. Schwartz, G. C. Stafford, J. Syka, D. Taylor, The Paul ion trap mass selective instability scan: Trap geometry and resolution. Proc. 40th ASMS Conf. on Mass Spectrometry and Allied Topics, Washington, DC, May 31–June 5, 1992, p. 1003.

39. M. Wang, E. G. Marquette, U.S. Patent 5,291,017 (1994).

40. G. Wells, Mass scanning in an asymmetric trapping field. Proc. 44th Ann. ASMS Conf. on the Mass Spectrometry and Allied Topics, Portland, OR, May 12–16, 1996, p. 126.

41. G. J. Wells, M. Wang, E. G. Marquette, U.S. Patent 5,714,755 (1998).

42. M. Splendore, E. Marquette, J. Oppenheimer, C. Huston, G. Wells, A new ion ejection method employing an asymmetric trapping field to improve the mass scanning performance of an electrodynamic ion trap, *Int. J. Mass Spectrom.* **190/191** (1999) 129–143.

43. D. B. Langmuir, R. V. Langmuir, H. Shelton, R. F. Wuerker, U.S. Patent 3,065,640 (1962).

44. R. L. Alfred, F. A. Londry, R. E. March, *Int. J. Mass Spectrom. Ion Processes* **125** (1993) 171–185.

45. R. K. Julian, Ph.D. Thesis, Purdue University, West Lafayette, IN, 1993.

46. J. Franzen, U.S. Patent 5,170,054 (1992).

47. J. Franzen, *Int. J. Mass Spectrom. Ion Processes* **106** (1991) 63.

48. F. Guidugli, P. Traldi, A. M. Franklin, M. L. Langford, J. Murell, J. F. J. Todd, *Rapid Commun. Mass Spectrom.* **6** (1992) 229.

49. R. F. Bonner, G. Lawson, J. F. J. Todd, *J. Chem. Soc. Chem. Commun.* (1972) 1179.

50. A. G. Harrison, *Chemical Ionization Mass Spectrometry*, CRC Press, Boca Raton, FL, 1983.

51. M. Lausevic, J. B. Plomley, X. Jiang, R. E. March, C. D. Metcalfe, *Eur. Mass Spectrom.* **1** (1995) 149.

52. M. Lausevic, X. Jiang, C. D. Metcalfe, R. E. March, *Rapid Commun. Mass Spectrom.* **9** (1995) 927.

53. J. J. Thomson, *Rays of Positive Electricity and the Application to Chemical Analysis*, Longmans Green, London, 1913, p. 56.

54. T. L. Kruger, J. F. Litton, R. W. Kondrat, R. G. Cooks, *Anal. Chem.* **48** (1976) 2113.

55. K. Levsen, H. Schwarz, *Angew. Chem. Int. Ed. Engl.* **15** (1976) 509.

56. F. W. McLafferty, F. M. Bockhoff, *Anal. Chem.* **50** (1978) 69.

57. W. F. Haddon, Organic trace analysis using direct probe sample introduction and high resolution mass spectrometry, in M. L. Gross (Ed.), *High Performance Mass Spectrometry*, American Chemical Society, Washington, DC, 1978, pp. 97–119.

58. F. W. McLafferty (Ed.), *Tandem Mass Spectrometry*, Wiley, New York, 1983.

59. J. N. Louris, R. G. Cooks, J. E. P. Syka, P. E. Kelley, G. C. Stafford, Jr., J. F. J. Todd, *Anal. Chem.* **59** (1987) 1677.

60. J. B. Plomley, C. J. Koester, R. E. March, *Org. Mass Spectrom.* **29** (1994) 372.

61. J. B. Plomley, C. J. Koester, R. E. March, GC/MS/MS of tetrachlorodibenzo-*p*-dioxins with a quadrupole ion storage mass spectrometer. Proc. 42nd Ann. ASMS Conf. on Mass Spectrometry and Allied Topics, Chicago, IL, May 29–June 3, 1994, pp. 718–719.

62. J. B. Plomley, R. S. Mercer, R. E. March, Optimal ion trap MS/MS parameters for the analysis of dioxins and furans. Proc. 43rd Ann. ASMS Conf. on Mass Spectrometry and Allied Topics, Atlanta, GA, May 21–26, 1995, p. 230.

63. G. Hamelin, C. Brochu, S. Moore, *Organohalogen Compounds* **23** (1995) 125.

64. J. B. Plomley, R. S. Mercer, R. E. March, *Organohalogen Compounds* **23** (1995) 7.

65. M. Splendore, J. B. Plomley, R. E. March, R. S. Mercer, *Int. J. Mass Spectrom. Ion Processes* **165/166** (1997) 595.

66. R. I. Martinez, R. G. Cooks, MS/MS CAD database: instrument design and operation apposite to CAD dynamics, 35th Annual Conference of the American Society for Mass Spectrometry and Allied Topics, Denver, CO, May 24–29, 1987, pp. 1175–1176.

67. H. M. Rosenstock, M. M. Wallenstein, A. L. Wahlhaftig, H. Eyring, *Proc. Natl. Acad. Sci. U.S.A.* **38** (1952) 667.

68. H. I. Kenttämaa, R. G. Cooks, *Int. J. Mass Spectrom. Ion Processes* **64** (1985) 79.

69. D. A. Catlow, A. Clayton, J. J. Monaghan, J. H. Scrivens, Analytically useful collision regimes in MS-MS. Proc. 35th Ann. ASMS Conf. on Mass Spectrometry and Allied Topics, Denver, CO, May 24–29, 1987, pp. 1036–1037.

70. M. Wang, S. Schachterle, G. Wells, *J. Am. Soc. Mass Spectrom.* **7** (1996) 668.

71. J. B. Plomley, R. E. March, R. S. Mercer, *Anal. Chem.* **68** (1996) 2345.

72. R. E. March, M. Splendore, E. J. Reiner, R. S. Mercer, J. B. Plomley, D. S. Waddell, K. A. MacPherson, *Int. J. Mass Spectrom. Morrison Honour Issue* **194** (2000) 235. Republished as an Erratum in *Int. J. Mass Spectrom.* **197** (2000) 283.

73. E. J. Reiner, D. H. Shellenberg, V. Y. Taguchi, *Environ. Sci. Technol.* **25** (1991) 110.

74. J. B. Plomley, C. J. Koester, M. Lauševic, X. Jiang, R. E. March, F. A. Londry, Analytical protocols for environmental substances using an ion trap. Proc. 43rd Ann. ASMS Conf. on Mass Spectrometry and Allied Topics, Atlanta, GA, May 21–26, 1995, pp. 993–994.

75. R. Zimmermann, U. Boesl, D. Lenoir, A. Kettrup, Th. L. Grebner, H. J. Neusser, *Int. J. Mass Spectrom. Ion Processes* **145** (1995) 97–108.

76. R. E. Pedder, R. A. Yost, Computer simulation of ion trajectories in a quadrupole ion trap mass spectrometer. Proc. 36th Ann. ASMS Conf. on Mass Spectrometry and Allied Topics, San Francisco, CA, June 5–10, 1988, pp. 632–633.

77. E. J. Reiner, D. H. Schellenberg, V. Y. Taguchi, R. S. Mercer, J. A. Townsend, T. S. Thompson, R. E. Clement, *Chemosphere* **20** (1990) 1385.

8

ION TRAP MASS SPECTROMETRY/LIQUID CHROMATOGRAPHY

Quadrupole Ion Trap Mass Spectrometry, Second Edition, By Raymond E. March and John F. J. Todd
Copyright © 2005 John Wiley & Sons, Inc.

8.1. INTRODUCTION

Mass spectrometry refers collectively to a broad range of techniques that fall into the categories of ion formation, measurement of ion mass/charge ratio, and ion detection. Each technique has its own set of characteristics, and it is possible to mix and match to obtain a variety of combinations of ion source/mass analyzer/detector; while some combinations are highly compatible, others are not or were thought not to be compatible at first consideration. An example of this latter category is the combination of electrospray ionization (ESI) and the QIT. One clear problem associated with such a combination was that of admission of ions generated externally into the ion trap to which is applied an RF drive potential of some hundreds of volts [1] such that ions would encounter a strong repulsive potential for almost half of each RF cycle. A deterrent to the investigation of such a combination was the uncertainty that this combination would offer any advantage over ESI combined with a QMF. However, one possible advantage of the ion trap was the high trapping efficiency reported for product ions following collision-induced dissociation and the concomitant enhancement of sensitivity.

The early work of McLuckey and co-workers [2] showed in convincing fashion that a modestly modified commercial ion trap instrument having a nominal mass/charge ratio range of merely 650 Th could be combined with ESI and yield information on biomolecules of molecular weight some 100 times the nominal mass range of the ion trap mass spectrometer. Once the compatibility of ESI with the QIT had been established, advantage could be taken of modern separation sciences, such as high-performance liquid chromatography (HPLC) and capillary electrophoresis (CE), online with mass spectrometry. It is now recognized clearly that the QIT is a remarkably powerful instrument when coupled with ESI [2–6]. The compatibility of ion trap mass spectrometry with matrix-assisted laser desorption ionization (MALDI) [7–9] and with CE [10] has been demonstrated also.

8.2. ELECTROSPRAY IONIZATION

Overlapping the impressive developments in ion trap mass spectrometry that were made in the 1990s was the advent of ESI [11, 12], a major development in the area of ionization. The evolution and characteristics of the various "spray" techniques have been reviewed recently [13–19] and its performance as a powerful analytical tool has been demonstrated [20–23]. Elucidation of the ESI processes by which ions from polar compounds arrive in the gas phase free of the cumbersomeness of solvent molecules has been sought [24]. There is general agreement that the ESI process occurs in three main steps: (i) the production of charged droplets from solution by nebulization under the influence of a potential of some kilovolts, (ii) shrinkage of the charged droplets by solvent evaporation and uneven fission of the shrunken droplets, and (iii) generation of gas-phase ions from small highly-charged droplets. For the formation of gas-phase ions in ESI/MS, two models have been proposed; they are the charge residue model (CRM) [25–27] and the ion evaporation model (IEM) [28, 29]; however, as yet, the complicated and dynamic ESI processes are not completely understood. The award of the Nobel Prize in Chemistry for 2002 to Koichi Tanaka and John Fenn "for their development of soft desorption ionization methods for mass spectrometric analyses of biological macromolecules" (Royal Swedish Academy of Sciences, press release), is testament to the eminence of mass spectrometry as an important analytical tool for the investigation of biological molecules. An interesting personal account of the origin and development of ESI is given by John Fenn in the Foreword to Richard Cole's book [19].

Electrospray ionization provides a means for obtaining gas-phase ions from a wide variety of analyte species such as alkali halides [30], salts of polyatomic acid groups and of multivalent metals [31], flavonoid glycosides [32], synthetic polymers [33], and biopolymers [34]. While there are many analytical applications of importance for species that form singly-charged ions, it is the multiply-charged ions that pose new opportunities and challenges in the derivation of structural information. It should be noted that the description of an ESI ion as multiply charged does not mean that the neutral has suffered the loss of several electrons as, for example, in NO^{2+}; rather, multiple charges are bestowed upon the neutral species, M, of the ESI ion by the addition of multiple protons to form $[M + nH]^{n+}$ or by the loss of multiple protons to form $[M - nH]^{n-}$. Examples of other ionizing cations are Na^+ and K^+.

8.3. COMMERCIAL INSTRUMENT MANUFACTURERS

While many people in many laboratories contribute to the overall advancement of the field of mass spectrometry, it is fitting in this field to acknowledge the enormous contributions made by instrument manufacturers to the advancement of science and to the enhanced availability of high-performance instrumentation, particularly instrumentation for the trapping and identification of gaseous ions. In the preface to the first edition of this work (R. E. March, R. J. Hughes, and J. F. J. Todd *Quadrupole Storage Mass Spectrometry*, Wiley, New York, 1989), we wrote of the excitement arising from the announcement of the FINNIGAN MAT ITD gas chromatographic mass detector and the great improvement in ion trap performance (p. vii–viii):

> Yet the excitement was not only for the rather narrow reason of expediting research in gaseous ion chemistry and physics, important though this may be, but also for the much wider possibility of making mass spectral information readily available at greatly reduced cost. There is now abundant evidence of the application to the health services of mass spectrometric techniques with concomitant high sensitivity and resolution for toxicological studies; studies of metabolism and incipient disease; environmental problems; the quality of food, well water, and materials; forensic sciences; and so forth. Thus the advent of the ion trap detector permits a much greater use of mass spectrometric techniques not only in the technically advanced countries but also in those countries which are technically less advanced. The effects of quadrupole storage mass spectrometry through utilisation of the ion trap detector will be considerable.

Since that time, ion traps, particularly combined with a gas chromatograph, have become ubiquitous, and such instruments are almost as common in universities as infrared absorption spectrometers. The cost of mass spectral information has diminished. We can bear testament to the wide distribution of ion trap instruments on the basis of many letters and e-mail messages from first-time users in many countries; as these men and women set out on their ion-trapping adventures, they sought counsel, guidance, and information by way of published papers. We wished them well. In each of the application areas described above, ion trap instruments have played and continue to play a vital role for the general good of mankind.

The manufacturers have continued to improve their ion-trapping wares and we, the users, acknowledge gratefully their contributions to mass spectrometry.

8.3.1. Commercial Instrument Development

The dynamics of commercial instrument development are driven largely by the ability of manufacturers to improve their instruments, reduce their manufacturing costs, patent protection, the market share of sales of a particular instrument, and the extent of competition. When, for a given type of instrument, patent protection is solid and competition is negligible, the resulting high market share may prove to be a disincentive to improvement of instrument performance. On the other hand, the dynamics differ in the case of, say, two manufacturers each of whom has developed an instrument with similar performance. The hypothetical two instruments, though similar in their basic physics, may differ sufficiently so as to justify patent protection for each manufacturer. In such a hypothetical situation, the user community may anticipate an enhanced pace

References pp. 287–290.

of instrument development as each manufacturer strives to increase market share. A somewhat similar situation has arisen recently in the mass spectrometry field due to the advent of the LIT (Chapter 5). The LIT is compatible with ESI [35], and the early details of the performance of the Thermo Finnigan and SCIEX instruments indicate that the ESI–LIT combination will be competitive with the ESI–QIT combination. Because the manufacturer of one of two commercial LITs (SCIEX) does not manufacture a QIT instrument, an enhanced pace of instrument development is anticipated.

8.3.2. Commercial Instrumentation

A similar situation has existed throughout the past decade for the ESI–QIT combination. There have been two principal manufacturers of instruments that combined ESI with a QIT; Thermo Finnigan introduced the LCQ™ instrument and Bruker/Franzen introduced the Esquire ion trap mass spectrometer. The evolution of each of these instruments has been rapid. The Finnigan LCQ Deca XP MAX mass spectrometer is used for rapid metabolite identification and employs multiple stages of mass selectivity (MS^n) in combination with liquid chromatography (LC). The LCQ Deca XP MAX instrument features a universal Ion Max source that allows simple toolless switching of ionization probes. The new source design along with ion sweep gas provides ruggedness and full-scan sensitivity for analysis of mixtures in complex matrices. The speed of fully automated MS^n capabilities and advanced data-dependent scan functions permit operation of the instrument on the time scale of elution of peaks from a liquid chromatograph. The performance of the LCQ Deca XP MAX instrument has been demonstrated, particularly in the areas of metabolite structural elucidation and peptide sequence determination. Powerful Xcalibur Software facilitates data analysis and simplifies generation of results providing a high-throughput platform that accelerates drug discovery and development.

    Bruker Daltonics has recently introduced a high-capacity ion trap, the Esquire (HCT) mass spectrometer, which claims to open a new bioanalytical dimension of ESI–ion trap performance. Its substantially greater ion storage capacity features ultrahigh-performance scan modes for protein sequencing and metabolite research (see below).

8.4. EARLY EXPLORATION OF ESI COMBINED WITH A QIT

The ESI–QIT combination introduces the complexity of ensembles of multiply-protonated ions formed by ESI and the necessity to extend the mass range of the QIT and to enhance mass resolution. In addition, the wide variety of fragment ions formed by CID of multiply-protonated ions of peptides requires a degree of familiarity with the nomenclature proposed for such ions. The approach taken here is one of consideration of the early work in this field, an introduction to the form of ESI mass spectra, and an examination of the limits of performance of a QIT that had been manufactured as a mass detector for compounds separated by gas chromatography! The limits to performance will indicate the directions for further evolution of the QIT for the investigation of biological molecules.

8.4.1. Instrument Configuration

The initial experiments using ESI combined with an ion trap were performed on a modified version of an ITMS (ion trap mass spectrometer) that had been adapted for ion injection from an atmospheric pressure glow discharge ionization source [36]. In Figure 8.1 is shown a side-view schematic of the instrument that is composed of an ESI source at atmospheric pressure, an intermediate pumped region at ~0.3 Torr, and a housing for the QIT with detector at a pressure of 10^{-5} Torr. A solution of the analyte is pumped from a syringe or a high-performance liquid chromatograph through a length of tubing and through a 120-μm stainless steel needle. The outlet side of the needle is located ~5–10 mm from the inlet aperture (diameter 100 μm) to the intermediate pumped region. A potential of 3–4 kV was applied to the needle. Ions are carried through the intermediate pumped region, through two systems of ion lenses, and into the ion trap by the wind of air entering the inlet aperture.

Helium is admitted to the ion trap so as to maintain a pressure of 1–3 mTorr. Helium is used for the dissipation of ion momentum upon entry to the ion trap, for the collisional focusing of ions to the vicinity of the center of the ion trap, and as a collision gas to increase ion internal or vibrational energy during collision-induced dissociation of selected ion species.

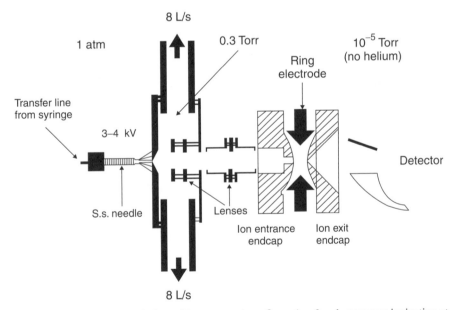

Figure 8.1. Cross-sectional view of instrumental configuration for electrospray ionization at atmospheric pressure and ion injection into QIT. Drawing not to scale. (Reprinted by permission from "Electrical ionization combined with ion trap mass spectrometry," by G. J. Van Berkel, G. L. Glish, S. A. McLuckey, *Analytical Chemistry*, **62** (1990), 1284–1295, Fig. 1. Copyright (1990) American Chemical Society.)

8.4.2. Axial Modulation and Mass Range Extension

Axial modulation (see Chapter 3) is the application of an RF potential, of amplitude $\sim 6\,V_{0-p}$ in dipolar mode, to the end-cap electrodes of the ion trap. The frequency is chosen to be slightly less than half that of the RF drive frequency; that is, for a drive frequency of 1 MHz, axial modulation is carried out at $\sim 490\,kHz$. In this manner, axial ejection occurs at a q_z value slightly less than 0.908 and the mass scanning range is 10–650 Th. When the axial modulation frequency is halved to 244 kHz, the mass range is increased twofold, that is, to 10–1300 Th. Similarly, when the axial modulation frequency is chosen as 70 kHz, the mass range is extended by a factor of 7 to 10–4550 Th.

8.5. ELECTROSPRAY MASS SPECTRUM

In Figure 8.2 is shown an electrospray mass spectrum of horse skeletal muscle myoglobin acquired with a sevenfold extension of the mass/charge ratio range of a standard QIT [37]. For those well versed in infrared spectroscopy of diatomic molecules, the general appearance of the mass spectrum is reminiscent of the R branch of the infrared vibrational absorption spectrum of $H^{35}Cl$ or $H^{37}Cl$. The signal intensities of the major peaks form a Boltzmann-like distribution with the interval between adjacent peaks increasing with increasing mass/charge ratio. The steadily increasing interval between adjacent peaks in the ESI mass spectrum indicates a regular change in one of the ion properties, that is, a regular reduction in the number of ionizing protons

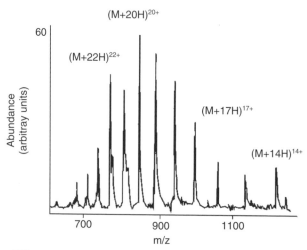

Figure 8.2. An ESI mass spectrum of horse skeletal muscle myoglobin acquired with 7-fold extension of mass/charge ratio range of standard QIT. (Reprinted by permission from "Electrical ionization combined with ion trap mass spectrometry," by G. J. Van Berkel, G. L. Glish, S. A. McLuckey, *Analytical Chemistry*, **62** (1990), 1284–1295, Fig. 13. Copyright (1990) American Chemical Society. The figure has been redrawn for clarit.)

attached to the parent molecule. In Figure 8.2, the base peak is identified as bearing 20 protons, but such is not always the case; one of the puzzles of electrospray mass spectrometry is the variation of the most probable degree of protonation.

8.5.1. Charge State and Molecular Weight

The charge on any ion can be determined from the isotope pattern for any of the 12 major peaks shown in Figure 8.2. Once the charge state and the nature of the ionizing cation are known, the molecular weight can be determined. The isotope spacing, that is, the mass difference arising from the substitution of one ^{13}C atom for a ^{12}C atom in the ion, varies inversely with the number of charges. The mass difference between a ^{13}C atom and a ^{12}C atom is very close to $1u$ and so the isotope spacing for a singly-charged ion is 1 Th (the unit Thomson is defined as 1 Th $= 1u/e_0$, where u is the atomic mass unit and e_0 the elementary charge [38]); the isotope spacing for a positively-charged ion bearing 10 charges is 0.1 Th. Provided the mass spectrometer has adequate mass resolution, the charge state of an ion can be determined directly, but, alas, such is not always the case.

An alternative method is to consider two adjacent peaks, for example, the peak labeled $[M + 20H]^{20+}$ ($\sim m/z$ 851) in Figure 8.2 and the next major peak of lower charge state and higher mass/charge ratio ($\sim m/z$ 894) that we shall identify tentatively as $[M + (19)H]^{(19+)}$. The molecular weight M is unknown; let us assume that the charge state is unknown also. If p is the number of protons in m/z 851 and $p - 1$ is the number of protons in m/z 894, then $[M + pH]^{p+}$, m/z 851, can be represented as

$$\frac{[M(u) + p(u)]}{p(e_0)} = 851\left(\frac{u}{e_0}\right). \tag{8.1}$$

Two simultaneous equations can be obtained, one from each peak:

$$M(u) = \left[p\,(e_0) \times 851\left(\frac{u}{e_0}\right)\right] - p(u) \tag{8.2}$$

$$M(u) = \left[(p-1)\,(e_0) \times 894\left(\frac{u}{e_0}\right)\right] - (p-1)(u). \tag{8.3}$$

Substituting for M, one obtains

$$\left[(p-1)\,(e_0) \times 894\left(\frac{u}{e_0}\right)\right] - p(u) + 1(u) = \left[p(e_0) \times 851\left(\frac{u}{e_0}\right)\right] - p(u) \tag{8.4}$$

which rearranges to

$$894p(u) - 894(u) + 1(u) = 851p(u) \tag{8.5}$$

$$p(e_0) = \frac{894(u) - 1(u)}{894(u/e_0) - 851(u/e_0)} \tag{8.6}$$

from which p is found to be close to $20e_0$. Substitution of $p = 20e_0$ into Eq. (8.2) yields $M = 17,000\,u$. This process can be repeated with other pairs of peaks to obtain an average value for M of $16,960 \pm 30\,u$. Thus a molecular weight of $16,960\,u$, with an experimental uncertainty of $\pm 0.18\%$, has been determined for myoglobin using a modified commercial ion trap mass spectrometer having a nominal mass range of 650 Th.

It should be noted that the mass/charge ratio of a multiply-charged peak maximum, such as in Figure 8.2, reflects neither the lowest isotopomer nor the average of the isotope peaks; rather, it reflects the most abundant isotope. For myoglobin, the mass/charge ratio of the most abundant isotope is 16,949.5 Th while that of the average of the isotope peaks is greater by about 1 Th.

8.5.2. Computer Algorithms

Computer algorithms are available from which a singly-charged representation can be obtained from a summation of the multiply-charged peaks, provided that the ionizing entity is known. Mann et al. [39] published the first algorithm for transforming multiply-charged mass spectra, and Zhou et al. [40] refined the algorithm for the first deconvolution program. Statistical packages such as MAX ENT [41] permit one to obtain high-resolution solutions from data such as are shown in Figure 8.2. New algorithms become available as the field develops [42].

8.5.3. Ion Trap Extended Mass Range Operation

Within the field of application of ESI–QIT, the data shown in Figure 8.2 are almost historic in the sense that much development has occurred in the intervening years. The mass resolution ($m/\Delta m$) reflected in this mass spectrum is between 500 and 1000. The normal scan rate for a QIT is 5555 Th/s. However, when the mass range is extended sevenfold by using axial modulation at a q_z value of $\sim 0.903/7$, that is, at $q_z = 0.129$, the scan rate is also increased sevenfold to 38,885 Th/s. At this scan rate, a mass/charge ratio interval of 1 Th is scanned in 25.7 µs, but because the data system assigns a data point every 28 µs, it is clear that the digitizing rate is limiting the assignment of peak position. Note that the digitizing rate is a function of the software and not a limitation of the performance of the QIT per se. Nevertheless, the time taken to scan the mass range (700 Th) of the mass spectrum in Figure 8.2 is but 18 ms; thus the scan rate used here is compatible with the rate of elution (one isolated compound in 3–8 s) from an LC column. Either the software could be changed, so as to affect an enhanced digitization rate, or the scan rate could be reduced so as to increase the time interval over which a mass/charge ratio interval of 1 Th is scanned. The latter procedure has the additional advantage of affecting an increase in mass resolution [8,43,44] that is an inherent aspect of QIT behavior. Parenthetically, the inadequacy of the relatively low mass resolution of the QIT for the investigation of ESI mass spectra was a spur to further research. The landmark publication of Schwartz et al. [43] concerning high mass resolution in a QIT mass spectrometer opened the door to the investigation of ESI at high mass resolution. Briefly, the mass resolution of a QIT is inversely proportional to the mass-scanning rate; when the mass-scanning rate is reduced, the observed mass resolution is increased, thus permitting a "zoom" operation.

This claim is borne out in the ESI mass spectra of renin substrate shown in Figure 8.3. The principal mass spectrum was acquired at the "normal" scan rate of ~38,885 Th/s and the inset, which shows the dramatic affect of zooming in on the $[M + H]^{4+}$ ion, was acquired with a 200-fold reduction of the scan rate to 194 Th/s. The isotopic distribution of the quadruply-charged ion, that is, the most probable charge state in this mass spectrum, is clear. The separation of the isotopic peaks in the inset is seen to be 0.25 Th, which indicates a 4+ charge state; the mass resolution is 4000. A back-of-an-envelope calculation based only on the triply- and doubly-charged ions at $m/z \approx 596$ and $m/z \approx 878$, respectively, shows that the charge state of $m/z \approx 596$ is, indeed, 3+ and that the molecular weight of renin substrate is approximately 1785.

Reduction in the scan rate is accompanied by an increase in noise in the baseline, as shown in the inset to Figure 8.3. When the bandwidth of the detection system is matched to the scan rate, part of the loss in signal/noise ratio can be retrieved. Reduction of the number of ions in the ion trap enhances mass resolution, and so optimum results are obtained when the ions of interest have been isolated in the QIT and a low scan rate is employed. The inverse relationship between mass scan rate and mass resolution is seen clearly in Figure 8.4. In each of the four parts of Figure 8.4 is shown

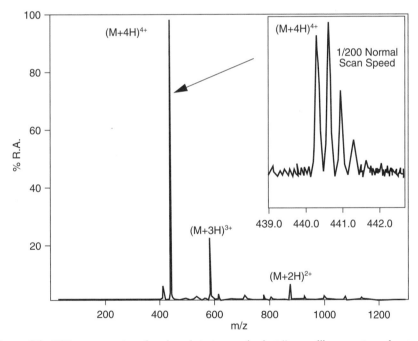

Figure 8.3. ESI mass spectra of renin substrate acquired at "normal" scan rate and narrow scan over 4+ charge state (inset) was acquired with 200-fold reduction of scan rate. (Reprinted from *Practical Aspects of Ion Trap Mass Spectrometry*, Vol. 2, R. E. March, J. F. J. Todd (Eds.), CRC Press: Boca Raton, FL, 1995. Chapter 3, "Electrospray and the ion trap," by S. A. McLuckey, G. J. Van Berkel, G. L. Glish, J. C. Schwartz, Fig. 11. © CRC Press. Reproduced with permission.)

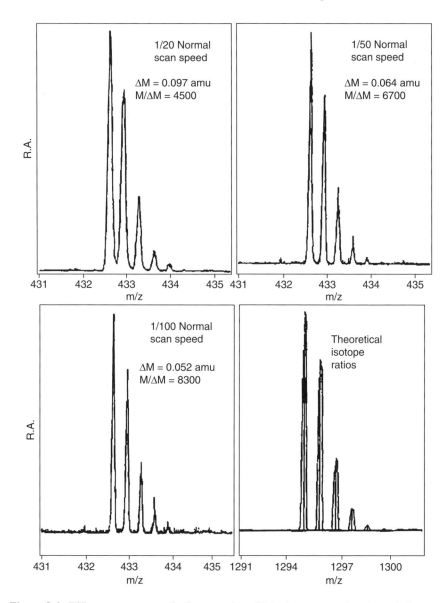

Figure 8.4. ESI mass spectra acquired over region of 3+ charge state of angiotensin I at various scan speeds along with simulated spectrum showing theoretical isotope abundances. (Reprinted from *Practical Aspects of Ion Trap Mass Spectrometry*, Vol. 2, R. E. March, J. F. J. Todd (Eds.), CRC Press: Boca Raton, FL, 1995. Chapter 3, "Electrospray and the ion trap," by S. A. McLuckey, G. J. Van Berkel, G. L. Glish, J. C. Schwartz, Fig. 12. © CRC Press. Reproduced with permission.)

a mass spectrum ~4.5 Th in width of the 3+ charge state of angiotensin I. In the upper left-hand corner, the mass spectrum was obtained at $\frac{1}{20}$ the normal mass scan speed; the full width of a peak at half maximum (FWHM) $\Delta m = 0.097$ Th such that $m/\Delta m = 4500$. When the mass scan rate is reduced to $\frac{1}{50}$ and $\frac{1}{100}$ the normal mass scan speed, it is seen that the FWHM is reduced to 0.064 and 0.052 Th, respectively, and the corresponding mass resolutions are increased to 6700 and 8300, respectively. In the lower right-hand corner of Figure 8.4 are shown the theoretical isotope ratios that are in good agreement with the data obtained at 100 times the normal mass scan speed. McLuckey and co-workers achieved mass resolution in excess of 40,000 for multiply-charged biomolecules at scan rates of the order of a few thomsons per second [2].

8.5.4. Ion/Molecule Reactions

Multiply-charged even-electron polyatomic molecules, particularly biopolymers, constitute a new class of gas-phase ions. Because the peaks in Figure 8.2 are well separated, it is relatively facile to isolate ions of a single charge state, say 20+, and to investigate the charge state–specific ion/molecule reactions of this species with a neutral species added at low pressure to the QIT. Further, because a QIT can, over a limited mass range, confine simultaneously negatively- and positively-charged ions, ion/ion reactions can be investigated. The initial charge state–specific ion/molecule reactions were carried out in a modified version of a commercial ion trap [45] as recently as 1990.

8.5.5. MS^n of Peptides and Proteins

A peptide is a form of biopolymer where two or more amino acids, such as $CH_2(NH_2)COOH$ (glycine) and $CH_3CH(NH_2)COOH$ (alanine), are combined into a linear condensation polymer by elimination of a water molecule from the amino–NH_2 group and the acid–COOH group. Let us consider a hypothetical peptide formed from each of glutamine, isoleucine, threonine, methionine, and serine, as shown in Scheme 8.1. The three-letter code for this peptide is Gln–Ile–Thr–Met–Ser and the

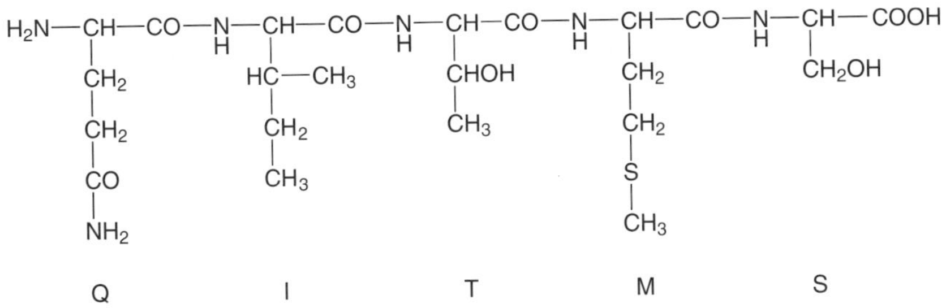

Scheme 8.1. Schematic representation of hypothetical amino acid pentamer, QITMS.

References pp. 287–290.

one-letter code is Q–I–T–M–S. By convention, a peptide is depicted with the amino group to the left and the acid group to the right; the fragmentation nomenclature discussed below is based on this convention. The peptide can be protonated by ESI to form the $[QITMS + H]^+$ ion. It is not anticipated that this peptide will be multiply protonated due to the repulsive force between two protons on this relatively small peptide of MW = 578. Note that the mass/charge ratios of the peaks in Figure 8.2 were in excess of 650 Th; thus the minimum molecular mass necessary to bear two charges must be of the order of 1300 u. In general, the site of protonation is localized on a nitrogen atom in any of the amide groups and, because fragmentation is directed to the site of protonation, virtually any bond in the peptide backbone can be broken, as shown in Scheme 8.2. There are three types of sites where a bond can be broken. The double-headed arrow in Scheme 8.2 at the far left depicts scission of the CHR–CO bond [where R = CH(CH$_3$)CH$_2$CH$_3$]; the upper and lower arrowheads are labeled a_2 and x_3, respectively. Thus, two product ions may be formed as a result of this scission, one with the charge on the N-terminus (a) and the other with the charge on the C-terminus (x). The subscript in a_2 refers to scission occurring in the second amino acid from the N-terminus, and the subscript in x_3 refers to the same scission that occurs after the third amino acid from the C-terminus. The double-headed arrow labeled b_2 and y_3 in Scheme 8.2 depicts scission of the peptide–amide bond (CO–NH) and delineates more clearly that the b_2 ion, having a charge on the N-terminus, is formed by scission of two peptides from the left-hand side of the ion and that the y_3 ion, having a charge on the C-terminus, is formed by scission of three peptides from the right-hand side of the ion. A useful mnemonic is the North Bay Yacht Club, which relates the N-terminus to b ions and the C-terminus to y ions. The double-headed arrow labeled c_2 and z_3 in Scheme 8.2 depicts scission of the amino–alkyl bond (HN–CRH) where R = (CHOH)CH$_3$. Thus three scissions have given rise to six ions; this group of characteristic scissions is repeated along the peptide backbone, as indicated by additional double-headed arrows in Scheme 8.2. This standard

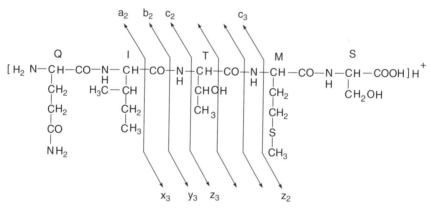

Scheme 8.2. Schematic representation of 6 characteristic scissions, yielding 12 fragment ion species, of protonated QITMS. This standard nomenclature used for the 12 fragment ion species was proposed by Roepstorff and Fohlman [46] and modified by Biemann [47].

nomenclature was proposed by Roepstorff and Fohlman [46] and was modified by Biemann [47].

8.5.6. Positive-Ion MS/MS and MS^n Studies

In one of the earliest tandem mass spectrometric applications of the QIT to electrospray mass spectra, quadruply-protonated mellitin $[M + 4H]^{4+}$, where M is mellitin, the base peak in the electrospray mass spectrum, was examined [37]. The resulting product ion mass spectrum, obtained with an ITMS, is shown in Figure 8.5. The parent ion, $[M + 4H]^{4+}$, is shown at m/z 712.6; hence the molecular weight of mellitin is 2846.4, and a single charge is associated with, on average, $711.6\,u$. The mass/charge range of the instrument was extended by a factor of 6 by using axial modulation. The mass spectrum shown in Figure 8.5 was acquired as the average of 10 scans during a period of 12 s. The amount of material sprayed during this period was $\sim4\,pmol$ of analyte. Thus, even at this early stage, the QIT demonstrated great sensitivity.

The parent ion of m/z 712.6 was outside the normal mass range of the ion trap (650 Th) so that the normal parent ion isolation procedure could not be carried out. However, by using the maximum values of the voltages available, it was possible to isolate the range of m/z 650–m/z 740. Thus the daughter ion assignments in the range of m/z 650–m/z 740 in Figure 8.5 are less reliable than those that fall outside of this region. The major ions in the mass range m/z 500–m/z 1100 are of the Y type, and

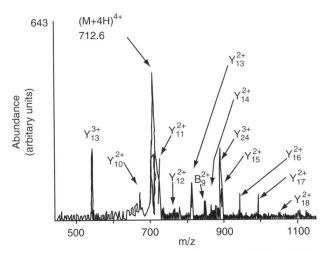

Figure 8.5. Product ion mass spectrum of $[M + 4H]^{4+}$ from mellitin obtained using electrospray ionization combined with a QIT. During data acquisition period, $\sim4\,pmol$ of analyte flowed from capillary needle. (Reprinted by permission from "Electrical ionization combined with ion trap mass spectrometry," by G. J. Van Berkel, G. L. Glish, S. A. McLuckey, *Analytical Chemistry*, **62** (1990), 1284–1295, Fig. 14. Copyright (1990) American Chemical Society.)

most of the Y ions are doubly charged. An ion signal is attributed to each Y ion from Y_{10}^{2+} to Y_{19}^{2+}. Ion signals are attributed to singly-charged B ions from B_9^+ to B_{12}^+. Two triply-charged Y ions are observed clearly and correspond to Y_{13}^{3+} and Y_{24}^{3+}.

Barinaga et al. [48], using a TSQ mass spectrometer, reported the product ion mass spectrum of quadruply-protonated mellitin $[M + 4H]^{4+}$. The qualitative similarity between the mass spectrum of Figure 8.5 and that of Barinaga et al. is remarkable. All of the fragment ions observed in Figure 8.5 were observed with the TSQ mass spectrometer and their relative ion signal intensities were similar. Furthermore, in neither the QIT nor the TSQ mass spectrometer were complementary ions observed in comparable abundance; that is, conservation of charge was not observed. The close similarity in the mass spectra from the two instruments is surprising when one considers that there were significant differences in the respective target species (in that helium was used in the ion trap and nitrogen was used in the TSQ mass spectrometer), center-of-mass collision energies, number of collisions for parent ions, and time scale. A discussion is given by Barinaga et al. [48] of the structural information that can be gleaned from the mass spectrum of Figure 8.5.

8.6. RECENT APPLICATIONS OF ESI COMBINED WITH A QIT

Quadrupole ion traps have become the workhorse for LC/tandem mass spectrometry for a variety of applications, mainly because of their fast-switching capability between MS and MS/MS and their interesting price/performance ratio. Further increases in sensitivity have been limited by the filling capacity of the ion trap and its duty cycle. Bruker Daltonik has introduced a high-ion-capacity (or high-charge) ion trap, HCT, which offers an improved combination of mass scanning rate and mass resolution.

8.6.1. Major Nonlinear Resonances for Hexapole and Octopole

To appreciate the level of sophistication in the HCT, it is necessary to review the action of hexapole and octopole resonances (see Chapter 3) on ion trajectories in the QIT [49]. Low-order multipole fields, that is, hexapole ($n = 3$) and octopole ($n = 4$) give rise to nonlinear resonances that can influence ion motion. It is assumed that ion trap operation is confined to the first stability region, given by

$$0 \leq \beta_r, \beta_z \leq 1. \tag{8.7}$$

The nonlinear resonances are expressed as

$$n_r \beta_r + n_z \beta_z = 2v \quad \text{with} \quad 0 \leq \beta_r, \beta_z \leq 1 \tag{8.8}$$

where n_r, n_z, and v are integer values and n_r is even for all types of multipoles, n_z is even for superposition of even multipoles only, and n_z is any integer for odd multipoles. Only the resonances with $v = 1$ take up energy. When the sum $(n_r + n_z)$ is

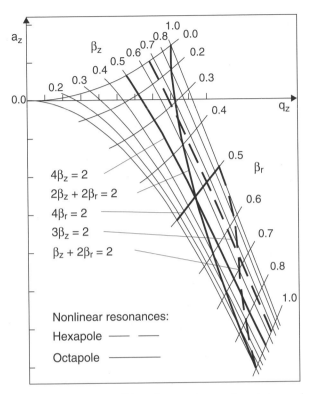

Figure 8.6. Nonlinear resonance conditions for superimposed weak hexapole and octopole fields form curved lines in the a_z, q_z plane. Only lines inside first stability region are shown; some lines (not all) continue into the surrounding instability region. (Reprinted from *Practical Aspects of Ion Trap Mass Spectrometry*, Vol. 1, R. E. March, J. F. J. Todd (Eds.), CRC Press: Boca Raton, FL, 1995. Chapter 3, "Nonlinear ion traps," by J. Franzen, R.-H. Gabling, M. Schubert, Y. Wang, Fig. 11. © CRC Press. Reproduced with permission.)

limited to values not greater than the multipole order n (which is reasonable but not strictly correct), the following resonances can be found for the first stability region:

$$\text{Hexapole:} \qquad 3\beta_z = 2 \qquad 2\beta_r + \beta_z = 2 \qquad (8.9)$$

$$\text{Octopole:} \qquad 4\beta_r = 2 \qquad 4\beta_z = 2 \qquad 2\beta_r + 2\beta_z = 2. \qquad (8.10)$$

The equations for the two hexapole and the three octopole resonance conditions are presented in a somewhat unusual form so as to emphasize the symmetry within the equations. For a rotationally symmetric ion trap structure, Eqs. (8.9) and (8.10) form the full set of hexapole and octopole resonance conditions. These nonlinear resonances are presented in Figure 8.6 for an ion trap with rotationally symmetric superpositions of hexapole and octopole fields. There are only two hexapole resonances by which energy may be taken up, while for the octopole there are three. One should note that the hexapole $3\beta_z = 2$ and the octopole $2\beta_r + 2\beta_z = 2$ intersect just below the q_z axis.

8.6.2. Ejection by a Dipole Field

The motion of ions confined in a QIT is characterized by series of frequencies of which the major frequencies are ω_r and ω_z. Figure 8.7 shows the resonance of an ion with a dipolar AC field, the frequency of which matches the secular frequency ω_z of the ion under study, applied across the end-cap electrodes of an ion trap. The increase in axial amplitude of an ion's trajectory produced by an ion trap resonance can be represented, to a good approximation, by the equation

$$\frac{dz}{dt} = C_{n-1} z^{n-1} \tag{8.11}$$

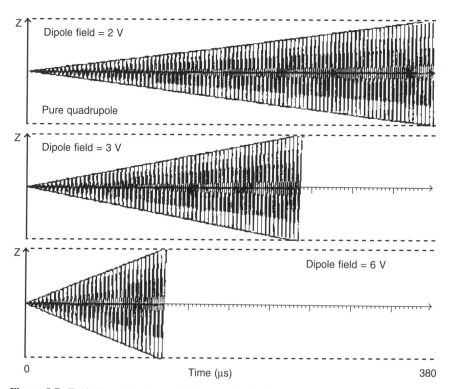

Figure 8.7. Excitation of ion's oscillations by dipolar fields of three different field strengths results in linearly increasing amplitudes. Constant C_0 (see text) for slope of enveloping curve is identical in all three tracks. (Reprinted from *Practical Aspects of Ion Trap Mass Spectrometry*, Vol. 1, R. E. March, J. F. J. Todd (Eds.), CRC Press: Boca Raton, FL, 1995. Chapter 3, "Nonlinear ion traps," by J. Franzen, R.-H. Gabling, M. Schubert, Y. Wang, Fig. 28. © CRC Press. Reproduced with permission.)

n being the order of the multipole. Equation (8.11) holds true for the dipole ($n = 1$), for the quadrupole ($n = 2$), and, approximately, for all odd higher multipoles ($n = 3, 5, 7, \ldots$) such as hexapole, decapole, and so on. The higher even multipoles ($n = 4, 6, \ldots$) such as octopole, duodecapole, and so on, strongly quench their own resonances because of the marked shift of the frequency with increasing amplitude causing the resonant excitation to abate.

The simulation showed in Figure 8.7 for excitation of an ion's oscillations by dipolar fields of three different field strengths reveals the linear increase of the amplitude, in accordance with Eq. (8.11). The constant C_0 is proportional to the AC voltage, V_{AC}:

$$\frac{dz}{dt} = C_0 \qquad C_0 = C_0' V_{AC}. \tag{8.12}$$

The integration of Eq. (8.12) results in a linear increase. When the simulation is repeated for many ions, it is found that equal numbers of ions reach the upper and lower end-cap electrodes; thus, only half of the ions ejected are detected normally.

8.6.3. Ejection by a Hexapole Field

In Figure 8.8 is shown the axial amplitude increase of oscillating ions matching the nonlinear hexapole resonance at $\beta_z = \frac{2}{3}$ for different starting amplitudes of oscillation. The envelope curves are obtained differentially according to the equation

$$\frac{dz}{dt} = C_2 z^2 \tag{8.13}$$

with the same value of the constant C_2 for all three curves. Upon integration, the envelope function has the form of a hyperbolic function

$$z(t) = \frac{C_2'}{z - C_2''} \tag{8.14}$$

with a mathematical pole at $z = C_2''$.

For ions near the center, $z = 0$, this function is weaker than the exponential function; however, for ions outside the center, the increase in amplitude is stronger than with the exponential function because of the mathematical pole of this function.

8.6.4. Octopole Field

It has been recognized for some time that the octopole quenches its own resonance [50]. For positive octopoles, ions are focused to the center of the ion trap by forces that increase more strongly than linearly, leading to a higher oscillation frequency. Thus, with increasing oscillation amplitude, the oscillation frequency is increased so that the ion is no longer in resonance with the applied frequency. Because the ion is no longer in resonance, its oscillation amplitude decreases whereupon it comes, once more, into resonance; this type of behavior is described as beating. The result of an

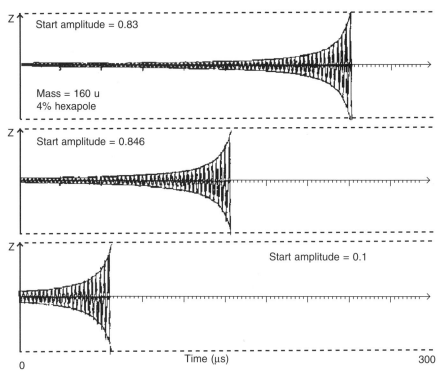

Figure 8.8. Hexapole resonances $\beta_z = \frac{2}{3}$ for three different starting amplitudes. Envelope curve is drawn by Eq. (8.13) with same constant C_2 for all three curves and starting amplitudes as parameters. (Reprinted from *Practical Aspects of Ion Trap Mass Spectrometry*, Vol. 1, R. E. March, J. F. J. Todd (Eds.), CRC Press: Boca Raton, FL, 1995. Chapter 3, "Nonlinear ion traps," by J. Franzen, R.-H. Gabling, M. Schubert, Y. Wang, Fig. 33. © CRC Press. Reproduced with permission.)

octopole superposition on its own z-type resonance at $\beta_z = \frac{1}{2}$ is shown in Figure 8.9. In the upper part of Figure 8.9 is seen the linear increase of the amplitude by resonant excitation in the pure quadrupole field. The conditions are such that the applied dipole frequency is one-quarter of the RF drive frequency, identical to the octopole resonance condition $\beta_z = \frac{1}{2}$ in Figure 8.6. In the lower part of Figure 8.9, the same experiment is carried out in an ion trap with 2% octopole superimposed. The ion does not reach either of the end-cap electrodes. Thus, the acquisition of energy is arrested by the phase shift resulting from the nonlinearity. The octopole superposition arrests its own z-type nonlinear resonance even when it is supported by a matching dipole resonance.

8.6.5. Modified Hyperbolic Angle Ion Traps

An interesting development of the ion trap has resulted in a different design for the ion trap. The modified hyperbolic angle ion trap was developed by Bruker-Franzen

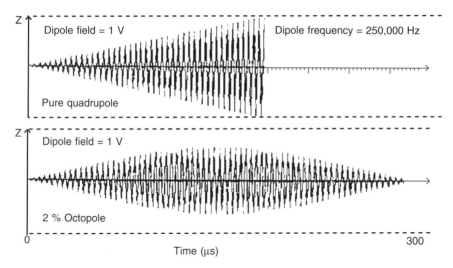

Figure 8.9. Ion resonance due to dipolar excitation in pure quadrupole field (upper) and in field with 2% octopole component (lower). Octopole arrests its own z-type nonlinear resonance. Linear amplitude growth of ion's trajectory in pure quadrupole field (upper) by matching dipolar excitation at quarter of RF drive frequency is stopped for case of octopole superposition, in spite of the fact that excitation frequency used is octopole z-resonance frequency. Resulting beat has a symmetric structure. (Reprinted from *Practical Aspects of Ion Trap Mass Spectrometry*, Vol. 1, R. E. March, J. F. J. Todd (Eds.), CRC Press: Boca Raton, FL, 1995. Chapter 3, "Nonlinear ion traps," by J. Franzen, R.-H. Gabling, M. Schubert, Y. Wang, Fig. 36. © CRC Press. Reproduced with permission.)

Analytik GmbH, under contract, as a chemical–biological mass spectrometer [49, 51]. The hyperbolic angle is defined as the angle α of the asymptote to the radial plane. As discussed in Chapter 2, the ring electrode and the end-cap electrodes must share common asymptotes in order to establish a quadrupolar field. Consequently, the slopes of the asymptotes to the ring electrode, m, and the slopes of the asymptotes to the end-cap electrodes, m', must be equal, as in Eq. (2.59):

$$m = m' = \pm\frac{1}{\sqrt{2}} = \tan\alpha.$$

It is possible to modify the hyperbolic angle such that the new angle has a slope of

$$\tan\alpha = \pm\frac{1}{\sqrt{1.9}}. \tag{8.15}$$

This type of modification of the hyperbolic angle leads to the superposition of different types of higher even multipoles, particularly the octopole. The slit between the electrodes, however, is narrower in this design. The ion storage behavior is much better than that of the stretched ion trap design.

The surfaces of the modified hyperbolic ion trap are hyperboloids of revolution. The equations of the surfaces are

$$r = \pm\sqrt{\theta z^2 + r_0^2} \tag{8.16}$$

for the ring electrode and

$$z = \pm\sqrt{\frac{r^2}{\theta} + z_0^2} \tag{8.17}$$

for the end-cap electrodes. A favorable value for θ is 1.9 such that the corresponding angle $\alpha = 35.96°$, compared to $35.264°$ when $\theta = 2$. The weights of the principal multipoles in the modified hyperbolic angle ion trap with $\theta = 1.9$ are 4-pole (quadrupole), 97.7557%; 8-pole (octopole), 1.4083%; and 12-pole (dodecapole), 0.1593% [49]. Compared to the Paul ion trap truncated at, say, $2r_0$, the octopole component is much larger and has opposite sign. The sign of the octopole is the same as that for the quadrupole field; hence, the octopole is added.

8.6.6. Combined Hexapole and Octopole Fields

In Figure 8.10 is shown ion ejection with combined superimposed hexapole (4%) and octopole (2%) fields. With a dipolar field of 0.6 V, ion ejection is not observed within the duration of the simulation. With a dipolar field in the range 0.7–4 V, unidirectional ion ejection is observed in each case such that all of the ions ejected thus should be detected externally. Unidirectional ion ejection is a consequence of the presence of the superimposed hexapole and octopole fields. The frequency of the dipole was selected to match the z-type nonlinear resonance $\beta_z = \frac{2}{3}$ of the hexapole field superposition. Ion ejection is extremely sharp, indicating that a very good mass resolution may be possible. This double-resonance phenomenon (dipole excitation and nonlinear hexapole resonance, both at precisely the same frequency) appears to be extremely concise and effective. The secular frequency of the ion matches both the dipole resonance with its linear amplitude amplification and the hexapole nonlinear resonance with its hyperbolic growth of amplitude. The influence of the combined hexapole and octopole field makes the beat asymmetric, causing unidirectional ion ejection, which is not the case for pure hexapole or pure octopole superpositions. It should be noted here that the scan speed in the above simulations is extremely fast. The scan speed is chosen such that $1u$ is scanned in only 10 oscillations of the secular motion; based on an RF drive frequency of 1 MHz, the scan speed is $1u/30\,\mu s$, or 33,333 Th/s.

8.7. THE HCT ION TRAP

The HCT ion trap is termed a nonlinear ion trap because an optimized combination of hexapole and octopole nonlinear resonance effects is employed for the ejection of ions [52]. The hexapole resonance is engaged by ion ejection at a hexapole resonance in the stability diagram while the octopole resonance is a consequence of the construction of the HCT ion trap with modification of the hyperbolic angle.

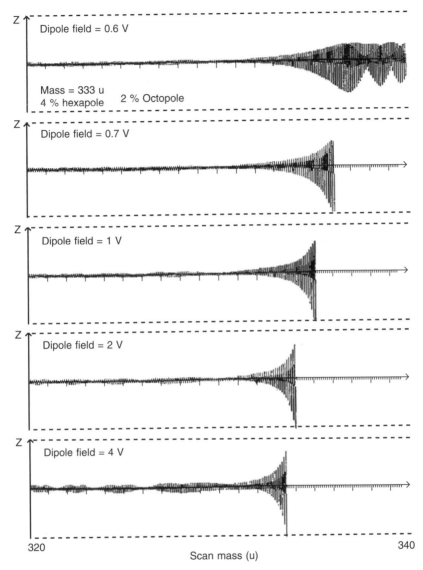

Figure 8.10. Ion ejection with combined hexapole (4%) and octopole (2%) fields. Ion ejection is unidirectional for all dipolar excitation voltages greater than 0.6 V, and the necessary voltages are unusually low. Dipolar excitation with 2 and 4 V results in sharp ejections, causing only very low beating oscillations before point of ejection is reached. (Reprinted from *Practical Aspects of Ion Trap Mass Spectrometry*, Vol. 1, R. E. March, J. F. J. Todd (Eds.), CRC Press: Boca Raton, FL, 1995. Chapter 3, "Nonlinear ion traps," by J. Franzen, R.-H. Gabling, M. Schubert, Y. Wang, Fig. 52. © CRC Press. Reproduced with permission.)

8.7.1. Ion Ejection

In the usual type of QIT, axial modulation is employed at a frequency slightly less than half the RF drive frequency so as to eject ions at a q_z value of ~0.904, that is, just inside the $\beta_z = 1$ boundary of the stability diagram shown in Figure 8.6. Axial modulation is employed with the HCT at a q_z value of ~0.78 where the hexapole resonance at $\beta_z = \frac{2}{3}$ crosses the q_z axis.

A mass scan of the HCT ion trap is affected with pre-excitation of the ion cloud in the z direction by a dipolar frequency of 333 kHz (equal to one-third of the RF drive frequency of 1 MHz) in order to resonate the ions out of the center before they encounter the nonlinear hexapole resonance. From Figure 8.8 it is seen that ion ejection is more rapid when the starting positions of the ions are dispersed away from the center of the ion trap. The cloud of ions of a given mass/charge ratio encountering the nonlinear resonance increases its secular oscillation amplitude in a hyperbolic manner and, after a few subsequent oscillatory swings, leaves the ion trap through small holes in one of the end-cap electrodes. The ions exiting thus form short pulses of ~100 ns in duration that are about 3 μs apart, corresponding to the secular oscillation frequency of the ion cloud. Ion ejection is unidirectional, as shown in Figure 8.10.

8.7.2. Ion Trap Capacity

The ion storage capacity of the HCT has been compared with that of the Bruker Esquire 3000 Plus ion trap instrument in the following manner. The indicated mass/charge ratio for m/z 609 and m/z 610 from reserpine was determined with each instrument as a function of ion signal intensity or, more precisely, ICC count. In Figure 8.11 are shown the determinations of the mass/charge ratio of each ion species as a function of ICC. As the ion charge count increases, the effects of space charge appear, leading to a shift of mass/charge ratio. When a mass shift of ~0.25 u is deemed acceptable (presumably allowance can be made for such a mass shift once the ICC value is known), the ICC limit of the Bruker Esquire 3000 Plus ion trap instrument is about 6×10^4 ions, as shown by the vertical line marked Esquire3000plus. However, the ICC limit of the Bruker HCT, also indicated with a vertical line, is about 6×10^5 ions; thus the HCT has a charge capacity about 10–20 times that of the Esquire 3000 Plus ion trap instrument. The upright triangles represent data points for ions recognized as having a mass/charge ratio of m/z 610. At very high ICC values, the software has difficulties in recognizing the mass/charge ratio of mass-shifted m/z 609 and assigns the mass/charge ratio as m/z 610.

8.7.3. Mass Resolution at High Scan Speeds

The mass scanning rate of the HCT in UltraScan mode is 26,000 Th/s such that unit mass is scanned every 38.4 μs. When allowance is made for ion ejection at $\beta_z = \frac{2}{3}$ rather than close to $\beta_z = 1$, this mass scanning rate is 3.12 times that of the early Finnigan ITD instrument of 5555 Th/s. Despite the increased mass scanning rate of the HCT, a mass resolution, as defined according to the FWHM, of <0.5 Th has been achieved. When the mass scanning rate is reduced to 8100 Th/s, the FWHM is

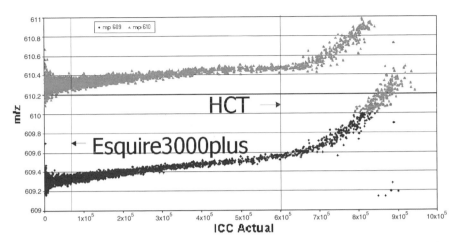

Figure 8.11. Display of ion signal intensity expressed as observed ion charge current (ICC Actual) as function of observed mass/charge ratio for two ion species, m/z 609 and m/z 610 from reserpine, ejected from HCT. In a series of experiments, HCT was filled incrementally with two ion species and mass/charge ratios were observed upon ion ejection. Normal ion capacity of Esquire 3000plus ion trap is indicated at $\sim 6 \times 10^4$ ions. As ion density within HCT is increased, effects of space charge perturbation become apparent in that indicated mass/charge ratio moves to higher values. For m/z 609, once ICC reaches 8×10^5 ions, ion assignment changes from m/z 609 to m/z 610. Effect of space charge on ions of m/z 610 is less marked than with m/z 609 because m/z 609 ions are ejected before ejection of m/z 610. Working capacity for HCT is indicated as 8×10^5 ions, thus ion capacity of HCT, accepting mass shift of ~ 0.2 Th, is some 10–20 times that of Esquire 3000plus instrument. (C. Baessmann, A. Brekenfeld, G. Zurek, U. Schweiger-Hufnagel, M. Lubeck, T. Ledertheil, R. Hartmer, M. Schubert, A nonlinear ion trap mass spectrometer with high ion storage capacity. A workshop presentation at the 51st ASMS Conf. on Mass Spectrometry and Allied Topics, Montreal, PQ, June 8–12, 2003. Reprinted by permission of Bruker Daltonik GmbH.)

reduced further to <0.3 Th. The major implications of a high mass scanning rate combined with high mass resolution are twofold. First, when a compound elutes from a liquid chromatograph in, say, 3 s, several full scans can be made because the full mass range of the HCT (6000 Th) can be scanned in 0.23 s. Under these conditions, there is sufficient time for a full mass scan followed by several selected tandem mass spectrometric (MS/MS) scans. Second, the high mass resolution is sufficiently precise to permit identification of a given peptide ion with a high degree of probability. The combined performances of high mass scanning rate and high mass resolution permit the rapid identification of peptides (see below).

Let us examine the mass resolution near the limit of the mass range. In Figure 8.12 (upper) is shown the isotopic cluster of protonated insulin where the most probable peak is detected at 5738.392 Th. The protonated insulin molecules $[M + H]^+$ were obtained from an atmospheric pressure MALDI source and the mass spectrum

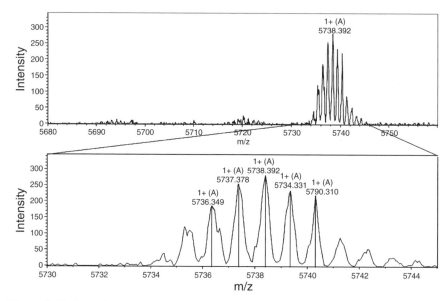

Figure 8.12. Isotopic cluster of protonated insulin, $[M + H]^+$, obtained from atmospheric pressure MALDI source. Upper part of figure shows mass spectrum of isotopic cluster of protonated insulin within mass range of 80 Th near upper limit of mass scanning range of 6000 Th. Mass spectrum was obtained at mass scanning rate of 26,000 Th/s. Lower part of figure shows zoom, obtained at mass scanning rate of 8100 Th/s, of isotopic cluster within mass range of 14 Th. (C. Baessmann, A. Brekenfeld, G. Zurek, U. Schweiger-Hufnagel, M. Lubeck, T. Ledertheil, R. Hartmer, M. Schubert, A nonlinear ion trap mass spectrometer with high ion storage capacity. A workshop presentation at the 51st ASMS Conf. on Mass Spectrometry and Allied Topics, Montreal, PQ, June 8–12, 2003. Reprinted by permission of Bruker Daltonik GmbH.)

was obtained at a mass scanning rate of 26,000 Th/s. The lower part of Figure 8.12 shows a zoom of the isotopic cluster alone obtained at a mass scanning rate of 8100 Th/s. The singly-charged peaks are well separated and the valley between peaks is <10% of peak height.

8.7.4. Sensitivity for a Protein Digest

Peptide ions, derived from a digest of a protein with trypsin, can be separated by LC and directed into an ESI source prior to injection into a QIT such as the LCQ or HCT. Fortuitously, ESI tends to produce a preponderance of doubly-charged tryptic peptides and, by and large, doubly-charged peptide ions tend to fragment more evenly across a given sequence than do singly-charged ion species [53]. Thus a large proportion of de novo sequencing of peptides has been performed on doubly-charged ions.

An example of the sensitivity of the HCT for doubly-charged peptide ions is given in Figure 8.13. Fifty attomoles of Enolase was digested and the peptide ion fragments were separated by LC. The four mass spectra shown in Figure 8.13 were obtained in the UltraScan mode and the principal doubly-charged ion is identified in each mass spectrum. The signal/noise ratio (S/N) is given for each of the doubly-charged ions

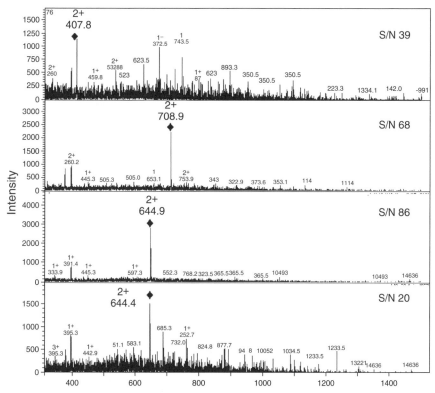

Figure 8.13. Identification of four peptides from a digest of 50 amol of Enolase; the peptide ion fragments were separated by LC. Four mass spectra shown were obtained in the UltraScan mode. Principal doubly-charged ion is identified in each mass spectrum and signal/noise ratio (S/N) is given for each doubly-charged ion identified. (C. Baessmann, A. Brekenfeld, G. Zurek, U. Schweiger-Hufnagel, M. Lubeck, T. Ledertheil, R. Hartmer, M. Schubert, A nonlinear ion trap mass spectrometer with high ion storage capacity. A workshop presentation at the 51st ASMS Conf. on Mass Spectrometry and Allied Topics, Montreal, PQ, June 8–12, 2003. Reprinted by permission of Bruker Daltonik GmbH.)

identified. An average S/N of about 50 for each of the major doubly-charged ions resulting from 5×10^{-17} mol of Enolase augurs well for the application of the HCT for de novo peptide sequencing.

8.7.5. De Novo Peptide Sequencing

During the elution of each of the four peptides, the major doubly-charged ion species are isolated and subjected to CID. From each of the resulting product ion mass spectra, the sequence of constituent amino acids can be determined. The doubly-charged ion of m/z 708.9 in Figure 8.13 has the sequence GNPTVVELTTEK (amino acids 15–27), as shown in bold text in Table 8.1. Similarly, m/z 644.4^{2+} has the sequence

TABLE 8.1. Sequence of 436 Amino Acids of Protein Enolase

10	20	30	40	50	60	70
AVSKVYARSV	YDSRGNPTVE	**VELTTEKGVF**	RSIVPSGAST	GVHEALEMRD	GDKSKWMGKG	**VLHAVKNVND**
80	90	100	110	120	130	140
VIAPAFVKAN	IDVKDQKAVD	DFLISLDGTA	NKSKLGANAI	LGVSLAASRA	AAAEKNVPLY	*KHLADLSKSK*
150	160	170	180	190	200	210
TSPYVLPVPF	LNVLNGGSHA	GGALALQEFM	IAPTGAKTFA	EALRIGSEVY	HNLKSLTKKR	YGASAGNVGD
220	230	240	250	260	270	280
EGGVAPNIQT	AEEALDLIVD	AIKAAGHDGK	VKIGLDCASS	EFFKDGKYDL	DFKNPNSDKS	KWLTGPQLAD
290	300	310	320	330	340	350
LYHSLMKRYP	IVSIEDPFAE	DDWEAWSHFF	KTAGIQIVAD	DLTVTNPKRI	ATAIEKKAAD	ALLLKVNQIG
360	370	380	390	400	410	420
TLSESIKAAQ	DSFAAGWGVM	VSHRSGETED	TFIADLVVGL	RTGQIKTGAP	ARSERLAKLN	*QLLRIEEELG*
430	440					
DNAVFAGENF	HHGDKL					

Source: C. Baessmann, A. Brekenfeld, G. Zurek, U. Schweiger-Hufnagel, M. Lubeck, T. Ledertheil, R. Hartmer, M. Schubert, A nonlinear ion trap mass spectrometer with high ion storage capacity. A workshop presentation at the 51st ASMS Conf. on Mass Spectrometry and Allied Topics, Montreal, PQ, June 8–12, 2003. Reprinted by permission of Bruker Daltonik GmbH.

Note: Sequences identified from 50 amol of sample are shown in boldface.

NVNDVIAPAFVK (amino acids 67–78), m/z 407.8^{2+} has the sequence AADALL-LKVNQ (amino acids 338–348), and m/z 644.9^{2+} has the sequence IGTLSESIK (amino acids 349–357). Using 50 amol of Enolase, the percentage sequence coverage by MS/MS is 7.3% (see below). Associated with these results is a MOWSE score of 150 (see below).

When 1 fmol of Enolase (MW 46,642 Da) was used, 14 peptides were identified such that the sequence coverage was 21.6%. The Δm value for each of the 14 peptides was $\leq$90 mDa corresponding to a root-mean-square (rms) error of 48 ppm. The total Mowse score was 566.

8.7.5.1. *Protein Identification*

A decade ago, when the mass accuracy of reflectron time-of-flight instruments was about $\pm$0.5 Th and that of QIT instruments was somewhat greater than $\pm$0.5 Th, the identity of a protein could be only *suggested* on the basis of mass spectrometric data. Karl Clauser and Peter Baker (http://prospector.ucsf.edu.html) devised a program MS-Fit that used molecular weight data to suggest a protein identity. To establish protein identity, a knowledge of the sequences within the peptides was required; such information can be obtained now using MS/MS [54]. Identification of the sequence GNPTVVELT-TEK (amino acids 15–27) for the ion of m/z 708.8^{2+}, as described above, is associated with a MOWSE score based on a scoring algorithm described by Pappin et al. [55].

8.7.5.2. *MOWSE Score*

The first stage of a MOWSE search is to compare the calculated peptide masses for each entry in a nonredundant protein sequence database (NCBI) with the set of experimental peptide masses (http://www.matrix-science.com.html). Matrix Science offers the Mascot program for peptide sequencing and an example of the application of this program is found in Ref. 56. Each calculated value that falls within a given mass tolerance of an experimental value counts as a match. The matches obtained either from peptide masses or from product ions as a result of MS/MS are handled on a probabilistic basis. The total score is the absolute probability that the observed match is a random event. Because of the wide range of probability magnitudes and the ambiguity of a "high" score being associated with a "low" probability, scores are reported in much the same way as are hydrogen ion concentrations, that is, as a pH value. Probability scores are reported as

$$\text{MOWSE score} = -10 \log_{10}(P) \qquad (8.18)$$

where P is the absolute probability. Thus, a probability of 10^{-20} becomes a MOWSE score of 200. A commonly-accepted threshold is that an event is significant if it would be expected to occur at random with a frequency of less than 5%. One may then find on the master results page for a typical peptide mass fingerprint search the statement "Scores greater than 38 are significant ($p < 0.05$)."

8.7.5.3. *Expectation Value*

In a peptide mass fingerprint, the score for each protein and the score for each product ion in an MS/MS search are accompanied by an expectation value. For a score at the significance threshold, $p = 0.05$, the expectation value is 0.05; when the score is increased 10-fold, the expectation value decreases to 0.005. Thus, the lower the expectation value, the more significant is the score.

8.7.6. De Novo Peptide Sequencing of Thirteen Proteins

In the uppermost part of Figure 8.14 is shown the TIC obtained with an HCT from a mixture of 13 digested proteins eluting from a liquid chromatograph. Note that the peak width is of the order of 3 s. Below the TIC in Figure 8.14 is given mass spectrum 846 obtained at 518.1 s of the LC separation; four doubly charged ions are identified in this mass spectrum. Each of the four doubly charged ions is identified by the software and each is subjected to autoMS/MS in Ultrascan mode in sequence; the corresponding product ion mass spectra, 847–850, respectively, are shown in Figure 8.14. The data are treated by the MS-Fit program to obtain the amino acid sequence in each peptide and to correlate the peptide data to each of the 13 proteins. In this example, 184 peptides were identified from a one-dimensional LC separation over a period of 15 min.

8.7.7. Multidimensional Liquid Chromatography

In the preceding paragraph reference was made to one-dimensional LC separation; it must be emphasised that the QIT is compatible [57] with multidimensional LC [58–60]. Multidimensional LC is used for the separation and identification of a large number of proteins from a protein mixture. Usually the separation is accomplished with a salt step gradient; however, due to the discontinuity in eluting solvent increase between successive salt steps, peptides may be distributed in more than one fraction with a concomitant degradation of the detection limit.

8.8. DIGITAL ION TRAP

In the conventional operation of a QIT instrument, the ion trap is driven by an RF sinusoidal wave voltage applied to the ring electrode whereupon ions may be confined within the electrode assembly. To achieve axial ejection of the trapped ions,

---

Figure 8.14. Identification of 13 proteins. Uppermost part shows TIC obtained with HCT from mixture of 13 digested proteins separated by LC. Below the TIC is given mass spectrum #846 obtained at 518.1 s of LC separation; four doubly-charged ions are identified in this mass spectrum. Below mass spectrum #846 are presented corresponding product ion mass spectra, #847–850, respectively. In this example, 184 peptides were identified from one-dimensional LC separation over period of 15 min. (C. Baessmann, A. Brekenfeld, G. Zurek, U. Schweiger-Hufnagel, M. Lubeck, T. Ledertheil, R. Hartmer, M. Schubert, A nonlinear ion trap mass spectrometer with high ion storage capacity. A workshop presentation at the 51st ASMS Conf. on Mass Spectrometry and Allied Topics, Montreal, PQ, June 8–12, 2003. Reprinted by permission of Bruker Daltonik GmbH.)

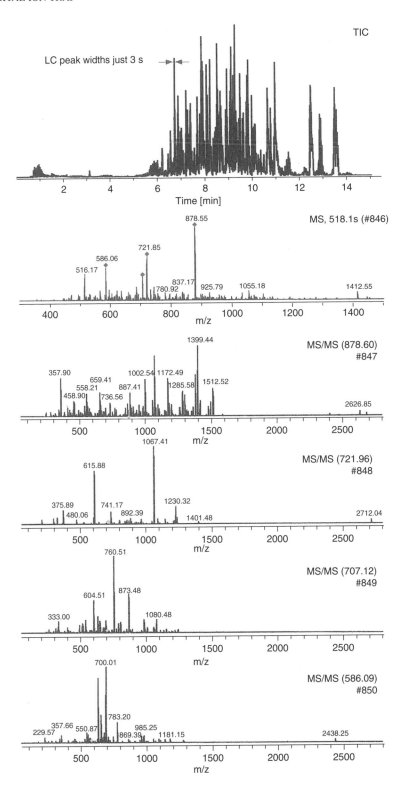

the amplitude of the RF drive potential is ramped in concert with axial modulation (see Chapter 3). In principle, the conditions required for ion trapping can be achieved with any periodic waveform [61].

8.8.1. Introduction

In a patent filed in 1971 Hiller [62], proposed that the potential applied to the electrodes of a QMF "may be formed by a repetitive sequence of segments each one of which is composed of either one or a number of linearly-varying functions of time and/or one or a number of exponentially-varying functions of time where the exponents of each portion are either real functions of time or complex functions of time but not purely imaginary functions of time." This patent covered rectangular and trapezoidal waveforms in addition to other functions. The stability conditions for a pulsed ion-trapping waveform applied to a QIT have been derived by Sheretov and Terent'ev [63] without resorting to the Mathieu equation (see Chapter 2). Contemporaneously, Richards et al. derived stability criteria for a square waveform of fixed frequency and demonstrated a mass scan of a QMF with such a square waveform [64]. In addition, Richards et al. [64] explored the application of rectangular and trapezoidal waveforms to the QIT and described stability diagrams for an ion trap driven by such waveforms [65]. More recently, Sheretov et al. [66], using a switching circuit to generate a frequency scan of a pulsed waveform, operated a QIT in mass-selective instability mode (see Chapter 3). In a digital quadrupole ion trap (DIT), the trapping quadrupole field and the excitation field are generated by a switching circuit and are controlled with an advanced digital algorithm [67].

8.8.2. Concept of the DIT

The concept of the DIT was first introduced through a theoretical study of the secular frequency of ions trapped by a digital waveform [68]. In the DIT, the trapping waveform applied to the ring electrode of a nonstretched QIT (see Chapter 3 for a discussion of the stretched ion trap) is generated by the rapid switching between discrete DC high-voltage levels where the timing of the switch can be controlled precisely by specifically-designed digital circuitry. The switches for producing the high-voltage digital waveform were constructed with power metal–oxide–semiconductor field-effect transistors (MOSFETs). The AC excitation voltages applied across the end-cap electrodes are generated and controlled by the same digital circuitry. The waveform parameters such as period, duty cycle, and relative phase are under digital control. The waveforms for the rectangular drive and the dipole excitation are shown in Figure 8.15. The period of the asymmetric rectangular waveform is T, where Td ($d < 1$) is the duration of the positive voltage V_1 and $T(1 - d)$ is the duration of the negative voltage V_2; the delay of the midpoint of the DC pulse with respect to the rising edge of the rectangular waveform is given by t_d, and the width of the DC pulse of amplitude V_d is w_d. The frequency of the dipolar excitation in Figure 8.15 is one-fifth that of the drive frequency such that excitation is imposed at $\beta_z = 0.4$.

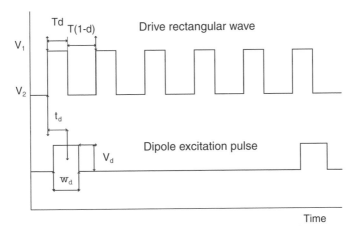

Figure 8.15. Schematic representation of rectangular waveform (upper) of RF drive potential digitized at frequency of 1 MHz and dipolar excitation waveform (lower) of frequency 200 kHz. (Reprinted from the *Journal of Mass Spectrometry*, Vol. 39, L. Ding, M. Sudakov, F. L. Brancia, R. Giles, S. Kumashiro, "A digital ion trap mass spectrometer coupled with atmospheric pressure ion sources," Fig. 2, pp. 471–484 (2004). © John Wiley & Sons Limited. Reproduced with permission.)

8.8.3. Stability Parameters

Ion motion under the influence of a digital waveform can be expressed in terms of the conventional a_z, q_z trapping parameters (see Chapter 2), and thus it is possible to relate the DIT readily with the QIT. A theoretical analysis of the ion-trapping action of the DIT has been presented together with the results of simulations [68–71], and the first demonstration of confinement of ions generated internally has been reported [72]. Stability conditions of ion motion in a pure quadrupole field with a periodic square wave [where $Td = T(1 - d)$] applied have been derived analytically [61, 63]; the trapping parameters are given as a_0, q_0. These trapping parameters can be compared with those that pertain to a sinusoidal waveform, Eqs. (2.74) and (2.75). Initial DIT experiments have been carried out on the q_0 axis ($U = 0$; therefore $a_0 = 0$) in an ion trap in which $r_0^2 = 2z_0^2$, such that q_z for a comparable QIT driven by a sinusoidal voltage V oscillating at a frequency $\Omega = f/2\pi$ is given by

$$q_z = \frac{4eV}{mr_0^2\Omega^2}. \tag{8.19}$$

For a DIT to which a square-wave potential is applied, the trapping parameter q_0 has values that extend, when $a_0 = 0$, from 0 to 0.7125, compared to the corresponding q_z values that extend from 0 to 0.908. The period of the square-wave potential T_{SWF} is given by

$$T_{SWF} = td + t(1 - d) = \frac{1}{f} = \frac{2\pi}{\Omega} \tag{8.20}$$

such that the trapping parameter q_0 for the DIT is expressed as

$$q_0 = \frac{eV}{mr_0^2\pi^2} T_{SWF}^2. \tag{8.21}$$

The LMCO (see Chapter 2) of the DIT is given by

$$\left(\frac{m}{e}\right)_{LMCO} = \frac{V}{q_0 r_0^2 \pi^2} T_{SWF}^2. \tag{8.22}$$

For a fixed trapping voltage V, a linear mass scale is dependent upon the square of the period of the square waveform.

A digitally-controlled square waveform permits the implementation of a frequency scan [73] by change of the period of the square waveform rather than by change of the waveform amplitude. To eject ions at a fixed value of q_0, the frequency of the dipolar excitation voltage must be scanned in concert with the trapping frequency.

8.8.4. Field Adjustment

The electrodes of the DIT are hyperboloidal in form, in accordance with Eqs. (2.57) and (2.58) for the ring and end-cap electrodes, respectively, the radius of the ring electrode, $r_0 = 10$ mm, and the relationship between ring radius and end-cap electrode separation is given above. On the basis of past experience, one would expect to observe with the DIT effects of fields of order higher than quadrupole (see Chapter 3) due to electrode truncation and perforations in the electrodes; such effects are manifested as mass shifts. A conical field-adjusting electrode has been placed adjacent to the entrance end-cap electrode, as shown in Figure 8.16, in order to compensate for higher-order fields. A potential of the order of 1 kV applied to the field-adjusting electrode, during the negative phase of the rectangular waveform, opposes penetration through the aperture of the entrance electrode of the equipotential lines due to the drive potential. Such penetration, if not compensated for, would influence ion secular frequency and lead to mass shifts. A fine mesh covers the aperture of the exit end-cap electrode to reduce penetration of the negative extraction potential during a mass scan.

8.8.5. Trapping Ability

A significant advantage of the DIT is the capability for high mass range at a low trapping voltage and fixed q_0. In Figure 8.17 is shown the MALDI mass spectrum obtained from 5 pmol of horse heart myoglobin at a trapping voltage of ± 1 kV; the singly-protonated molecule $[M + H]^+$ at m/z 16,980.8 and the doubly-protonated molecule $[M + 2H]^{2+}$ at m/z 8502.7 are observed. At a mass scan rate of 40,000 Th/s, the FWHM of the $[M + H]^+$ species is 7.8 Th. Resonant excitation was carried out at $\beta_z = 0.5$ such that the mass range of 3600–18,7000 Th was scanned in 388 ms.

8.8.6. Mass Resolution

High mass resolution can be obtained with the DIT when an appropriate voltage is applied to the field-adjusting electrode. For example, when an optimized V_{FAE} of

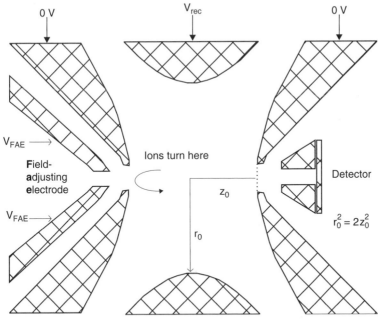

Figure 8.16. Ion trap geometry showing conical field-adjusting electrode to which potential V_{FAE} is applied during negative phase of rectangular waveform, a fine mesh across aperture of exit end-cap electrode, and the extraction electrode. During negative phase of rectangular waveform, ions approach apertures and are turned back. (Reprinted from the *Journal of Mass Spectrometry*, Vol. 39, L. Ding, M. Sudakov, F. L. Brancia, R. Giles, S. Kumashiro, "A digital ion trap mass spectrometer coupled with atmospheric pressure ion sources," Fig. 5a, pp. 471–484 (2004). © John Wiley & Sons Limited. Reproduced with permission.)

~1130 V was applied to the field-adjusting electrode, the MALDI mass spectrum shown in Figure 8.18 of $[M + H]^+$ ions of bovine insulin was obtained at a mass scan rate of 325 Th/s. The mass resolution in Figure 8.18 is at least 17,000.

8.8.7. Pseudopotential Well Depth

It is essential for the tandem mass spectrometric operation of the DIT that the pseudopotential well depth (see Chapter 3) be characterized. For the DIT, the pseudopotential well depth $\bar{D}_{DIT}$ is given by [68]

$$\bar{D}_{DIT} = 0.206q_0V. \tag{8.23}$$

The expression for $\bar{D}_{DIT}$, Eq. (8.23), is of a form similar to that for $\bar{D}_z$, given in Eq. (3.18). For a QIT, a mass scan is carried out by ramping the amplitude of the RF drive potential V with ion ejection at constant q_z. Under these conditions, the magnitude of

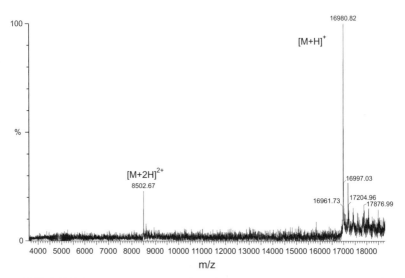

Figure 8.17. MALDI mass spectrum showing singly-protonated molecule $[M + H]^+$ at m/z 16,980.8 and doubly-protonated molecule $[M + 2H]^{2+}$ at m/z 8502.7. Mass spectrum was obtained with 5 pmol of horse heart myoglobin loaded on target. Resonant excitation was carried out $\beta_z = 0.5$ for mass range of 3600–18,7000 Th scanned in 388 ms. Trapping voltage of ± 1 kV was applied. (Reprinted from the *Journal of Mass Spectrometry*, Vol. 39, L. Ding, M. Sudakov, F. L. Brancia, R. Giles, S. Kumashiro, "A digital ion trap mass spectrometer coupled with atmospheric pressure ion sources," Fig. 3, pp. 471–484 (2004). © John Wiley & Sons Limited. Reproduced with permission.)

the pseudopotential well depth $\bar{D}_z$ varies linearly with the mass of the ion, m [Eq. (3.15)], as shown in Figure 8.19. However, a mass scan of the DIT is accomplished by changing the period of the square waveform at constant voltage, V_{rec}, and at constant q_0; thus the magnitude of the pseudopotential well depth $\bar{D}_{DIT}$ for the DIT remains constant during a mass scan, as shown also in Figure 8.19.

The variation in pseudopotential well depth as a function of mass/charge ratio for each DIT and a conventional ion trap is shown in Figure 8.19. During a mass scan where q_z increases, the pseudopotential well depth of a conventional QIT increases also and becomes deeper than that for a DIT. Under the conditions applied for Figure 8.19, the pseudopotential well depth for m/z 612 is the same in each instrument. For $m/z > 612$, the RF ion trap provides a deeper pseudopotential well depth that will permit more energetic CID than does the DIT; for $m/z < 612$ the converse situation holds.

8.8.8. Forward and Reverse Mass Scans

In a commercial ion trap with a stretched geometry, as a result of modifying the relative positions of the end-cap electrodes, the mass resolution of a reverse scan, wherein ions are ejected resonantly in the reverse direction from high mass to low mass, is not comparable to that observed with a forward scan [74, 75]. Brancia et al.

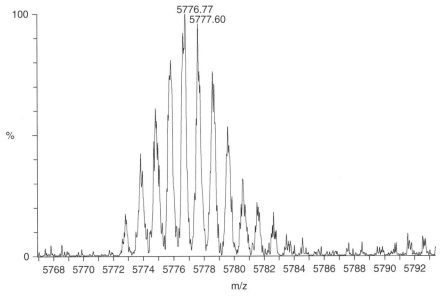

Figure 8.18. High mass resolution scan of $[M + H]^+$ of bovine insulin. Scan speed was 325 Th/s. (Reprinted from the *Journal of Mass Spectrometry*, Vol. 39, L. Ding, M. Sudakov, F. L. Brancia, R. Giles, S. Kumashiro, "A digital ion trap mass spectrometer coupled with atmospheric pressure ion sources," Fig. 3, pp. 471–484 (2004). © John Wiley & Sons Limited. Reproduced with permission.)

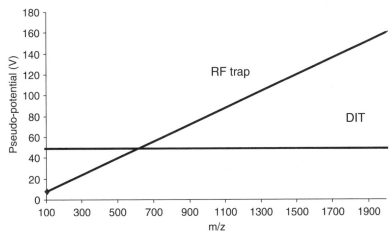

Figure 8.19. Variation in pseudopotential well depth as a function of mass/charge ratio for the DIT and a conventional RF ion trap. Pseudopotential well depth for DIT at $q_0 = q_{0,\,max}/3$ and $V_{rec} = \pm 1\,kV$ is constant; that for RF ion trap at $q_z = q_{z,\,max}/3$ increases with mass/charge ratio. (Reprinted from the *Journal of Mass Spectrometry*, Vol. 39, L. Ding, M. Sudakov, F. L. Brancia, R. Giles, S. Kumashiro, "A digital ion trap mass spectrometer coupled with atmospheric pressure ion sources," Fig. 8, pp. 471–484 (2004). © John Wiley & Sons Limited. Reproduced with permission.)

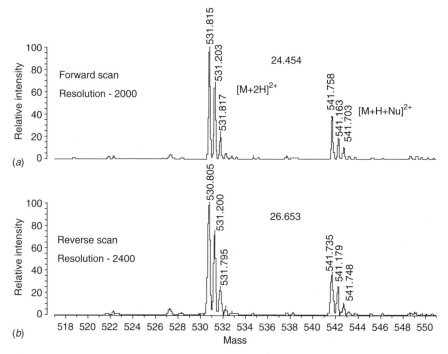

Figure 8.20. Electrospray mass spectra of 5 pmol μL^{-1} of bradykinin acquired with (*a*) forward and (*b*) reverse scans in absence of space charge using ion accumulation period of 10 µs. During resonant ejection at $\beta_z = 0.5$, scan rate of 782 Ths^{-1} was used. Each mass spectrum shown here is sum of 30 scans. Ion abundance, expressed in volts and shown on mass spectra, represents peak height of doubly-protonated bradykinin [M + 2H]$^{2+}$. (Reprinted from the *Journal of Mass Spectrometry*, Vol. 39, F. L. Brancia, R. Giles, L. Ding, "Effect of reverse scan on mass measurement accuracy in an ion trap mass spectrometer," Fig. 2, pp. 702–704 (2004). © John Wiley & Sons Limited. Reproduced with permission.)

[76] have reported reverse and forward mass scans of an electrospray mass spectrum of bradykinin that they obtained using a DIT as described above. In the DIT it is the frequency, rather than the amplitude, that is varied during a mass scan; during the frequency scans employed here, the voltage on the ring electrode was held constant at an amplitude of ±500 V. Ions ejected resonantly from the DIT reach the detector at different times, giving rise to a spectrum in which ion abundance versus ejection time is displayed. Because the frequency scan rate of the DIT is designed to provide a linear relationship between scan time and mass/charge ratio, the time on the abscissa represents directly the mass/charge ratio. The scale is calibrated for both scan directions.

The electrospray mass spectra obtained are shown in Figure 8.20. Data files derived from the mass spectra acquired with a reverse scan have been converted so that all ions are displayed in the same order as those acquired with a forward scan. The voltage applied to the field-adjusting electrode was adjusted to correct for field faults, as described above, then, along with all other parameters except for the direction of scan, was held constant during acquisition of the mass spectra. The mass spectra are remarkably similar with respect to appearance, mass resolution (2800 for forward scan and 2400 for reverse scan), and ion abundance of $[M + 2H]^{2+}$ (24.45 V for forward scan and 26.68 V for reverse scan).

8.8.9. Summation of Digital Ion Trap Development

The development of the DIT constitutes a significant advance in the field of QIT mass spectrometry. With a field-adjusting electrode incorporated into the electrode structure of a QIT having $r_0^2 = 2z_0^2$, compensation of fields of higher order can be achieved such that both forward and reverse mass scans can be performed with satisfactory mass resolution. The versatility of the DIT with MALDI and with ESI has been demonstrated.

8.9. CONCLUSIONS

In this chapter, we have attempted to demonstrate two important aspects concerning the application of QIT mass spectrometry to de novo peptide sequencing: first, the compatibility of the QIT with ion sources wherein the ions of interest are generated externally to the ion trap; second, the not inconsiderable improvements in mass scanning rate, mass resolution, mass accuracy, dynamic range due to enhanced ion capacity, and MS/MS sensitivity that have been made recently.

Although the combination of ESI with the three-dimensional QIT has been emphasised in this chapter, it must be appreciated that the HCT, LCQ, and DIT instruments are compatible with MALDI sources [7–9, 77–79] and with capillary electrophoresis [10] as, indeed, is the two-dimensional LIT instrument that was discussed in Chapter 5.

REFERENCES

1. P. Kofel, Injection of mass-selected ions into the radiofrequency ion trap, in R. E. March and J. F. J. Todd (Eds.), *Practical Aspects of Ion Trap Mass Spectrometry*, Vol. II, CRC Press, Boca Raton, FL, 1995, Chapter 2.

2. S. A. McLuckey, G. J. Van Berkel, G. L. Glish, J. C. Schwartz, Electrospray and the quadrupole ion trap, in R. E. March and J. F. J. Todd (Eds.), *Practical Aspects of Ion Trap Mass Spectrometry*, Vol. II, CRC Press, Boca Raton, FL, 1995, Chapter 3.

3. J. C. Schwartz, I. Jardine, Quadrupole ion trap mass spectrometry, in B. Karger and W. Hancock (Eds.), *High Resolution Separation and Analysis of Biological Macromolecules, Part A: Fundamentals*, Vol. 270, Spectrum Pub. Services, York, PA, 1996. (Chapter 23, pp. 552–586.)

4. B. L. Kleintop, D. M. Eades, R. A. Yost, Liquid chromatography/mass spectrometry, in R. E. March and J. F. J. Todd (Eds.), *Practical Aspects of Ion Trap Mass Spectrometry*, Vol. III, CRC Press, Boca Raton, FL, 1995, Chapter 5.

5. A. Mordehai, H. K. Lim, J. D. Henion, Ion spray liquid chromatography/mass spectrometry and capillary electrophoresis/mass spectrometry on a modified benchtop ion trap mass spectrometer, in R. E. March and J. F. J. Todd (Eds.), *Practical Aspects of Ion Trap Mass Spectrometry*, Vol. III, CRC Press, Boca Raton, FL, 1995, Chapter 6.

6. H.-Y. Lin, R. D. Voyksner, Electrospray/ion trap mass spectrometry: Applications to trace analysis, in R. E. March and J. F. J. Todd (Eds.), *Practical Aspects of Ion Trap Mass Spectrometry*, Vol. III, CRC Press, Boca Raton, FL, 1995, Chapter 14.

7. K. Jonscher, G. Currie, A. L. McCormack, J. R. Yates III, *Rapid Commun. Mass Spectrom.* **7** (1993) 20.

8. J. C. Schwartz, M. E. Bier, *Rapid Commun. Mass Spectrom.* **7** (1993) 27.

9. W. Zhang, A. N. Krutchinsky, B. T. Chait, *J. Am. Soc. Mass Spectrom.* **14** (2003) 1012.

10. L. A. Gennaro, J. Delaney, P. Vouros, D. J. Harvey, B. Domon, *Rapid Commun. Mass Spectrom.* **16** (2002) 192.

11. M. Yamashita, J. B. Fenn, *J. Phys. Chem.* **88** (1984) 4451.

12. M. Yamashita, J. B. Fenn, *J. Phys. Chem.* **88** (1984) 4671.

13. J. B. Fenn, M. Mann, C. K. Meng, C. M. Whitehouse, *Science* **246** (1990) 64.

14. R. D. Smith, J. A. Loo, C. G. Edmonds, C. J. Barinaga, H. R. Udseth, *Anal. Chem.* **62** (1990) 882.

15. J. B. Fenn, M. Mann, C. K. Meng, S. F. Wong, C. M. Whitehouse, *Mass Spectrom. Rev.* **9** (1990) 37.

16. R. D. Smith, J. A. Loo, R. R. Ogorzalek Loo, M. Busman, H. R. Udseth, *Mass Spectrom. Rev.* **10** (1991) 359.

17. M. Mann, *Org. Mass Spectrom.* **25** (1990) 575.

18. E. C. Huang, T. Wachs, J. J. Conboy, J. D. Henion, *Anal. Chem.* **62** (1990) 713A.

19. R. B. Cole (Ed.), *Electrospray Ionization Mass Spectrometry*, Wiley, New York, 1997.

20. K. Biemann, *Protein Sci.* **4** (1995) 1920.

21. J. F. Banks, Jr., S. Shen, C. M. Whitehouse, J. B. Fenn, *Anal. Chem.* **66** (1994) 406.

22. T. Li, Y. Ohashi, *Carbohydr. Res.* **273** (1995) 27.

23. B. Yeung, P. Vouros, M. L. Siucaldera, G. S. Reddy, *Biochem. Pharmacol.* **49** (1995) 1099.

24. P. Kebarle, L. Tang, *Anal. Chem.* **65** (1993) 972A.

25. M. Dole, L. L. Mack, R. L. Hines, R. C. Mobley, L. D. Ferguson, M. B. Alice, *J. Chem. Phys.* **49** (1968) 2240.

26. F. W. Röllgen, E. Bramer-Wegner, L. Buttering, *J. Phys. Colloq.* **45**(Suppl. 12) (1984) C9.

27. G. Schmelzeisen-Redeker, L. Buttering, F. W. Röllgen, *Int. J. Mass Spectrom. Ion Processes* **90** (1989) 139.

28. B. A. Thomson, J. V. Iribarne, *J. Chem. Phys.* **64** (1976) 2287.

29. J. V. Iribarne, B. A. Thomson, *J. Chem. Phys.* **71** (1979) 4451.

30. C. Hao, R. E. March, T. R. Croley, J. C. Smith, S. P. Rafferty, *J. Mass Spectrom.* **36** (2001) 79.

31. C. Hao, R. E. March, *J. Mass Spectrom.* **36** (2001) 509.

32. R. E. March, X.-S. Miao, C. D. Metcalfe, M. Stobiecki, L. Marczak, *Int. J. Mass Spectrom.* **232** (2004) 171.

33. S. F. Wong, C. K. Meng, J. B. Fenn, *J. Phys. Chem.* **92** (1988) 546.

34. C. K. Meng, M. Mann, J. B. Fenn, *Z. Phys. D* **10** (1988) 361.

35. B. A. Collings, J. M. Campbell, M. Dunmin, D. J. Douglas, *Rapid Commun. Mass Spectrom.* **15** (2001) 1777.

36. S. A. McLuckey, G. L. Glish, K. G. Asano, B. C. Grant, *Anal. Chem.* **60** (1988) 2220.

37. G. J. Van Berkel, G. L. Glish, S. A. McLuckey, *Anal. Chem.* **62** (1990) 1284.

38. R. G. Cooks, A. L. Rockwood, *Rapid Commun. Mass Spectrom.* **5** (1991) 93.

39. M. Mann, C. K. Meng, J. B. Fenn, *Anal. Chem.* **61** (1989) 1702.

40. J. X. G. Zhou, I. Jardine, Computer programs for interpretation of mass spectra of multiply charged ions of mixtures. Proc. 38th Ann. ASMS Conf. on Mass Spectrometry and Allied Topics, Tucson, AZ, June 3–8, 1990, pp. 134–135.

41. A. G. Ferrige, M. J. Seddon, B. N. Green, S. A. Jarvis, J. Skilling, *Rapid Commun. Mass Spectrom.* **6** (1992) 707.

42. H. Zheng, P. C. Ojha, S. McClean, N. D. Black, J. G. Hughes, C. Shaw, *Rapid Commun. Mass Spectrom.* **17** (2003) 429.

43. J. C. Schwartz, J. E. P. Syka, I. Jardine, *J. Am. Soc. Mass Spectrom.* **2** (1991) 198.

44. J. D. Williams, K. A. Cox, J. C. Schwartz, High mass, high resolution mass spectrometry, in R. E. March and J. F. J. Todd (Eds.), *Practical Aspects of Ion Trap Mass Spectrometry*, Vol. II, CRC Press, Boca Raton, FL, 1995, Chapter 1.

45. S. A. McLuckey, G. J. Van Berkel, G. L. Glish, *J. Am. Chem. Soc.* **112** (1990) 5668.

46. P. Roepstorff, J. Fohlman, *Biomed. Mass Spectrom.* **11** (1984) 601.

47. K. Biemann, *Biomed. Environ. Mass Spectrom.* **16** (1988) 99.

48. C. J. Barinaga, C. G. Edmonds, H. R. Udseth, R. D. Smith, *Rapid Commun. Mass Spectrom.* **3** (1989) 160.

49. J. Franzen, R.-H. Gabling, M. Schubet, Y. Wang, Nonlinear ion traps, in R. E. March and J. F. J. Todd (Eds.), *Practical Aspects of Ion Trap Mass Spectrometry*, Vol. I, CRC Press, Boca Raton, FL, 1995, Chapter 3.

50. N. R. Whetten, P. H. Dawson, *J. Vac. Sci. Technol.* **6** (1969) 100.

51. Contract DAAA15-87-C0008 awarded to Teledyne CME for the U.S. Army ERDEC (Aberdeen Proving Ground), Chemical- Biological Mass Spectrometer, CBMS.

52. C. Baessmann, A. Brekenfeld, G. Zurek, U. Schweiger-Hufnagel, M. Lubeck, T. Ledertheil, R. Hartmer, M. Schubert, A nonlinear ion trap mass mass spectrometer with high ion storage capacity, A workshop presentation at the 51st Ann. ASMS Conf. on Mass Spectrometry and Allied Topics, Montreal, PQ, June 8–12, 2003.

53. D. L. Tabb, L. L. Smith, L. A. Breci, V. H. Wysocki, D. Lin, J. R. Yates III, *Anal. Chem.* **75** (2003) 1155.

54. J. K. Eng, A. L. McCormack, J. R. Yates III, *J. Am. Soc. Mass Spectrom.* **5** (1994) 976.

55. D. J. C. Pappin, P. Hojrup, A. J. Bleasby, *Curr. Biol.* **3** (1993) 153.

56. S. Laugesen, P. Roepstorff, *J. Am. Soc. Mass Spectrom.* **14** (2003) 992.

57. Y. Wagner, A. Sickmann, H. E. Meyer, G. Daum, *J. Am. Soc. Mass Spectrom.* **14** (2003) 1003.

58. A. J. Link, J. Eng, D. M. Schieltz, E. Carmack, G. J. Mize, D. R. Morris, B. M. Garvik, J. R. Yates III, *Nat. Biotechnol.* **17** (1999) 676.

59. R. Aebersold, D. Figeys, S. Gygi, J. Corthals, P. Haynes, B. Rist, J. Sherman, Y. Zhang, D. Goodlett, *J. Protein Chem.* **17** (1998) 533.

60. S. B. Ficarro, M. L. McCleland, P. T. Stukenberg, D. J. Burke, M. M. Ross, J. Shabanowitz, D. F. Hunt, F. M. White, *Nat. Biotechnol.* **20** (2002) 301.

61. M. G. Floquet, *Ann. Ecole Norm. Sup. Paris* **12** (1883) 47.

62. J. Hiller, UK Patent 1,346,393, filed March 8, 1971.

63. E. P. Sheretov, V. I. Terent'ev, *Sov. Phys.-Tech. Phys.* **17** (1972) 755.

64. J. A. Richards, R. M. Huey, J. Hiller, *Int. J. Mass Spectrom. Ion Processes* **15** (1974) 417.

65. J. A. Richards, Ph.D. Thesis, University of New South Wales, Sydney, 1972.

66. E. P. Sheretov, O. W. Rozhkov, D. W. Kiyushin, A. E. Malutin, *Int. J. Mass Spectrom.* **190/191** (1999) 103.

67. L. Ding, R. Giles, M. Sudakov, F. L. Brancia, S. Kumashiro, Development of a digital ion trap mass spectrometer with an electrospay ion source, 16th International Mass Spectrometry Conference, Edinburgh, UK, August 31–September 5, 2003.

68. L. Ding, S. Kamashiro, *Chinese Vac. Sci. Technol.* **21** (2001) 176.

69. M. Sudakov, E. Nikolaev, *Eur. J. Mass Spectrom.* **8** (2002) 191.

70. N. V. Konenkov, M. Sudakov, D. J. Douglas, *J. Am. Soc. Mass Spectrom.* **13** (2002) 597.

71. L. Ding, M. Sudakov, S. Kumashiro, *Int. J. Mass Spectrom.* **221** (2002) 117.

72. L. Ding, A. Gelsthorpe, J. Nuttal, S. Kumashiro, Rectangular wave quadrupole field and digital QMS technology. Proc. 49th Ann. ASMS Conf. on Mass Spectrometry and Allied Topics, Chicago, IL, May 27–31, 2001, CD.

73. U. P. Schlunegger, M. Stoeckli, R. M. Caprioli, *Rapid Commun. Mass Spectrom.* **13** (1999) 1792.

74. J. D. Williams, K. A. Cox, R. G. Cooks, S. A. McLuckey, K. J. Hart, D. E. Goeringer, *Anal. Chem.* **66** (1994) 725.

75. G. Dobson, J. Murrell, D. Despeyroux, F. Wind, J. C. Tabet, *Rapid Commun. Mass Spectrom.* **17** (2003) 1657.

76. L. F. Brancia, R. Giles, L. Ding, *J. Mass Spectrom.* **39** (2004) 702–704.

77. C. S. Creaser, J. C. Reynolds, D. J. Harvey, *Rapid Commun. Mass Spectrom.* **16** (2002) 176.

78. S. C. Moyer, R. J. Cotter, A. S. Woods, *J. Am. Soc. Mass Spectrom.* **13** (2002) 274.

79. S. C. Moyer, L. A. Marzilli, A. S. Woods, V. V. Laiko, V. M. Doroshenko, R. J. Cotter, *Int. J. Mass Spectrom.* **226** (2003) 133.

9

AN ION TRAP TOO FAR?
THE ROSETTA MISSION TO
CHARACTERIZE A COMET*

9.1. INTRODUCTION

At precisely 07:17 Greenwich mean time (GMT) on Tuesday March 2, 2004, an Ariane-5 rocket carrying the Rosetta "comet chaser" was launched at Kourou in French Guyana: the mission, to characterize the comet Churyumov-Gerasimenko, otherwise known as 67P. Although this event captured the fleeting attention of the media at the time, what was not particularly apparent is that one of the instruments with which this survey will be carried out is a mini–chemical laboratory, MODULUS,

---

*The authors are greatly indebted to Simeon J. Barber, of the Planetary and Space Sciences Research Institute, Open University, Milton Keynes, United Kingdom, for his considerable help in preparing the material for this chapter.

Quadrupole Ion Trap Mass Spectrometry, Second Edition, By Raymond E. March and John F. J. Todd
Copyright © 2005 John Wiley & Sons, Inc.

that includes a GC/MS system incorporating an ion trap mass spectrometer specially designed for isotope ratio measurements.

The purpose of this chapter is to give an account of this highly unusual application of the ion trap and, in particular, to explore some of the technical and design considerations of a system that is fully automated yet is not due to reach its sample until 2014! While reading the description that follows, the reader may care to reflect on the fate of this lonely ion trap during its journey covering hundreds of millions of miles, effectively frozen in time since the earliest stages of the development program began in 1994! Most mass spectroscopists would expect to have utilized three or four new generations of instruments in a 20-year period and would not normally have to wait two decades to (hopefully) see their first analytical mass spectrum! Nor would they expect to have to incorporate an age distribution table into the initial funding application in order to demonstrate that at least some members of the original team, who know how to control the system and to interpret the data signals, will still be in place when the analyses are carried out.

9.2. THE ROSETTA MISSION

The name Rosetta was taken from the Rosetta stone, a slab of volcanic rock, now in the British Museum, that was found in the village of Rashid (Rosetta) in the Nile delta in Egypt by French soldiers in 1799. The stone is covered with carved inscriptions in ancient Greek, together with Egyptian hieroglyphic and Demotic. Only Greek could be translated at the time and, by comparing the three sets of characters, scholars were able to decipher the meaning of the various symbols, thereby unlocking the secrets of 3000 years of ancient history. In the same way, this space mission seeks to take a range of physical and chemical measurements from a comet that is 4600 million years old, even older than the planets in our solar system, in order to try and find clues as to how our Earth was formed. Both the Rosetta project and the Rosetta stone provide the key to unravelling history by comparing the known with the unknown. Hopefully, this twenty-first-century endeavor will offer some of the answers to the question of how life on Earth started.

Rosetta resembles a large aluminum box of dimensions $2.8 \times 2.1 \times 2.0\,m$ and has two 14-m-long solar panels with a total area of $64\,m^2$. It comprises two main components: an Orbiter and a Lander. On arrival at its destination in May 2014, Rosetta will orbit 67P in order to map the surface (which is approximately equal in area to that of London Heathrow Airport!) in order to determine a suitable site for the Lander to target during its descent in November 2014. In addition to containing the main command and communications module, the *Orbiter's* payload includes the following 11 experiments, all controlled by different research groups.

- The *Ultraviolet Imaging Spectrometer (ALICE)* will analyze gases in the coma and tail and measure the comet's production rates of water and carbon monoxide or dioxide. It will provide information on the surface composition of the nucleus.

- *Comet Nucleus Sounding Experiment by Radiowave Transmission (CONSERT)* will probe the comet's interior by studying radio waves that are reflected and scattered by the nucleus.
- The *Cometary Secondary Ion Mass Analyser (COSIMA)* will analyze the characteristics of dust grains emitted by the comet, such as their composition and whether they are organic or inorganic.
- The *Grain Impact Analyser and Dust Accumulator (GIADA)* will measure the number, mass, momentum, and velocity distribution of dust grains coming from the comet nucleus and from other directions (reflected by solar radiation pressure).
- The *Micro-Imaging Dust Analysis System (MIDAS)* will study the dust environment around the comet and provide information on particle population, size, volume, and shape.
- The *Microwave Instrument for the Rosetta Orbiter (MIRO)* will determine the abundances of major gases, the surface outgassing rate, and the nucleus subsurface temperature.
- The *Optical, Spectroscopic and Infrared Remote Imaging System (OSIRIS)* is a wide-angle camera and narrow-angle camera that will obtain high-resolution images of the comet's nucleus.
- The *Rosetta Orbiter Spectrometer for Ion and Neutral Analysis (ROSINA)* will determine the composition of the comet's atmosphere and ionosphere, the velocities of electrified gas particles, and reactions in which they take part.
- The *Rosetta Plasma Consortium (RPC)* will measure the physical properties of the nucleus, examine the structure of the inner coma, monitor cometary activity, and study the comet's interaction with the solar wind.
- *Radio Science Investigation (RSI)*, using shifts in the spacecraft's radio signals, will measure the mass, density, and gravity of the nucleus, define the comet's orbit, and study the inner coma.
- The *Visible and Infrared Mapping Spectrometer (VIRTIS)* will map and study the nature of the solids and the temperature on the surface of the nucleus as well as identify comet gases, characterize the physical conditions of the coma, and help to identify the best landing sites.

The *Lander* structure consists of a baseplate, an instrument platform, and a polygonal sandwich construction, all made of carbon fiber. Some of the instruments and subsystems are beneath a hood that is covered with solar cells. An antenna transmits data from the surface to Earth via the Orbiter. The Lander carries the following further nine experiments, including a drilling system to take samples of subsurface material; the payload of the Lander is about 21 kg.

- The *Alpha Proton X-ray Spectrometer (APXS)*, when lowered to within 4 cm of the "ground," will detect α particles and X-rays that will provide information on the elemental composition of the comet's surface.

References p. 307.

- The *Rosetta Lander Imaging System (ÇIVA/ROLIS)* is a charge-coupled device (CCD) camera that will obtain high-resolution images during descent and stereo panoramic images of areas sampled by other instruments. Six identical microcameras will take panoramic pictures of the surface, and a spectrometer will study the composition, texture, and albedo (reflectivity) of samples collected from the surface.

- *Comet Nucleus Sounding Experiment by Radiowave Transmission (CONSERT)* will probe the internal structure of the nucleus: radio waves from CONSERT will travel through the nucleus and will be returned by a transponder on the Lander.

- *Cometary Sampling and Composition experiment (COSAC)* is one of two evolved gas analyzers. It will detect and identify complex organic molecules from their elemental and molecular composition.

- The *Evolved Gas Analyser (MODULUS Ptolemy)* is the second evolved gas analyzer and will obtain precise measurements of stable isotope ratios of the light elements H, C, N, and O in their various forms within material sampled from the comet subsurface, surface, and near-surface atmosphere.

- The *Multi-Purpose Sensor for Surface and Subsurface Science (MUPUS)* will use sensors on the Lander's anchor, probe, and exterior to measure the density and thermal and mechanical properties of the surface.

- The *Rosetta Lander Magnetometer and Plasma Monitor (ROMAP)* is a magnetometer and plasma monitor that will study the local magnetic field and the comet/solar wind interaction.

- The *Sample Drill and Distribution system (SD2)*, which will drill more than 20 cm into the surface, will collect samples and deliver them to different ovens for evolved gas analysis by COSAC and Ptolemy and microscope inspection by CIVA/ROLIS.

- The *Surface Electrical, Seismic and Acoustic Monitoring Experiments (SESAME)* use three instruments to measure properties of the comet's outer layers: the Cometary Acoustic Sounding Surface Experiment will measure the way sound travels through the surface, the Permittivity Probe will investigate its electrical characteristics, and the Dust Impact Monitor will measure dust falling back to the surface.

Integration of these 20 very different experiments into a single operation is clearly a highly complex matter. Each separate system must be capable of functioning under automated control and of being brought into or out of use according to a strictly predetermined schedule when the cometary encounter commences (see also below). Furthermore, communication with the instrumentation has to be coordinated through a scientific command center and then through mission control using special software and cannot be carried out directly by the individual research groups from their own institutions. In the case of MODULUS Ptolemy (which receives its instructions via command systems on the Lander, which in turn is instructed from the Orbiter), the scientists in their laboratories at the Open University at Milton Keynes in the United

TABLE 9.1. Time Sequence for Rosetta Mission

Date	Event
March 2, 2004	Launch of Rosetta
March 2005	First Earth gravity–assisted fly-by
February 2007	Mars gravity–assisted fly-by
November 2007	Second Earth fly-by
November 2009	Third Earth fly-by
May 2014	Comet Churyumov-Gerasimenko rendezvous maneuver
November 2014	Landing on comet
December 2015	Escorting comet; end of mission

Kingdom are almost five steps removed from direct control of their mass spectrometer and associated equipment. Added to this complexity, the transfer time for signals over the 500 million miles separation between Earth and Rosetta at the time of the encounter will be of the order of 50 min!

Following the launch into a tightly specified flight path, the timeline for the mission is as summarized in Table 9.1. The fly-by stages are a means by which the space module gains speed, rather like a child swinging around a lamp post when running down a street, so that Rosetta will eventually reach the velocity of the comet (up to 135,000 km/h). The landing process itself will present some significant hazards to the mission: because the comet is so small, and hence its gravitational field so weak, there is a danger that the Lander will simply bounce off the surface. The legs of the Lander contain a damping system to absorb most of the kinetic energy on contact, and harpoons will be fired into the surface in order to anchor the system.

Further details of Rosetta's journey may be found at http://www.esa.int/export/esaMI/Rosetta/, and information about the actual launch is available at http://www.arianespace.com/site/news/mission_up_153.html. It should, perhaps, be noted that originally it was intended that Rosetta would target another comet, 46P/Wirtanen, in 2011, with a launch date in January 2003. However, because of the failure of the preceding Ariane-5 flight in December 2002 the operation was delayed and the precise time window required for the trajectory was thereby missed. Consequently, a new target (i.e., 67P) was chosen and the whole project reprogrammed; the additional cost of the delay has been estimated at 70 million Euros!

9.3. THE MODULUS PTOLEMY EXPERIMENT

The name MODULUS stands for Methods Of Determining and Understanding Light elements from Unequivocal Stable isotope compositions. It was concocted by Colin Pillinger and his coinvestigators of the Planetary and Space Sciences Research Institute at the Open University in the United Kingdom in honor of Thomas Young, the English physician turned physicist who was the initial translator of the Rosetta Stone and whose name is best known by the measure of elasticity, Young's modulus. In the original research plan it was intended to develop two versions of the MODULUS instrument, Ptolemy and Berenice: as noted above, the former is part of the package of experiments on the Lander, while the latter was intended to examine (and thus provide a comparison with) the volatile species surrounding the comet as part of the research conducted by the Orbiter. In the event, only Ptolemy ultimately flew, and elements of the proposed Berenice science were incorporated into the Ptolemy instrument. Genealogists may be interested to note that Berenice and Ptolemy were subjects of the inscriptions on the Rosetta stone.

9.3.1. Stable Isotope Ratio Measurements for Light Elements

The underlying aim of the Ptolemy experiment is to determine the degree of isotopic enrichment (or depletion) of D (i.e., ^{2}H), ^{13}C, ^{15}N, ^{17}O and, ^{18}O in cometary samples relative to specified standard reference materials. These measurements will yield data on the respective degrees of isotopic fractionation that have occurred, which should in turn provide information about the temperature regime within which the samples were formed as well as give indications of the sources from which the samples were derived. To determine relative isotopic abundances, samples of solid material taken by the SD2 system, mentioned earlier, from the body of the comet (which has been described as being like a dirty snowball!) will be subjected to stepped pyrolysis in ovens according to preprogrammed protocols so as to generate evolved gases at predetermined temperatures that will then be converted (if necessary) chemically into compounds such as H_2, O_2, CO_2, and N_2. The resulting mixtures will then be separated and isotopically assayed by GC/MS using an ion trap mass spectrometer.

The isotope ratios are measured as the differential values according to the delta notation of Urey [1]:

$$\delta(\text{rare isotope}) = \left[\frac{R_{sample} - R_{standard}}{R_{standard}} \right] \times 1000\%o \qquad (9.1)$$

where R is the ratio of intensities for (D/H), (^{13}C/^{12}C), (^{15}N/^{14}N), (^{17}O/^{16}O), (^{18}O/^{16}O), and so on. As a result, all the determinations on the sample data must be directly compared with contemporary measurements on appropriate standard "onboard" reference materials that have been calibrated to an agreed international standard. In this way, compensation can be made for any systematic fractionation effects in the instrumentation in order to maximize the accuracy and precision of the data.

9.3.2. The Ion Trap Mass Spectrometer as the Instrument of Choice

Normally, to obtain the most accurate and precise isotope ratio measurements on a single gaseous compound, one would choose a magnetic-sector instrument, preferably with a dual inlet system designed for contemporaneous assays of the sample and of the standard reference material. However, since the total payload and physical space available were severely limited, given all the other instrumentation being carried by Rosetta and the associated power requirements, alternative instruments had to be considered. The small size and simplicity of construction of the ion trap, the fact that it functions on the basis of a single parameter control (i.e., the amplitude of the RF drive potential), and its tolerance to moderately high pressures ($\sim 10^{-3}$ mbar) of helium (which is employed as the GC carrier gas as well as for the actuation of the pneumatic valves of the GC injection valves) made this the obvious mass analyzer of choice. Fortuitously, the working pressure of the ion trap is of the same order as the maximum value anticipated in the region of the comet: at initial encounter the ambient pressure is expected to be around 10^{-7} mbar, but this is predicted to rise to $\sim 10^{-3}$ mbar at point of closest solar approach. A further advantage is that the ion accumulation time can be adjusted to allow the buildup of ions from the minor isotopes, thus increasing the precision of their measurement. In the final design, a nonstretched ion trap ($r_0 = 8.0$ mm; $2z_0 = 11.3$ mm) having grounded end-cap electrodes has been employed, operating at a nominal RF drive frequency of 0.6 MHz and amplitude variable between approximately 25 and 300 V_{0-p}; the exact frequency will be determined by a self-tune feature, which selects the most appropriate frequency depending upon the ambient temperature and hence tuning of the RF circuit. These parameters offer a mass/charge ratio range of 12–150 Th, allowing both general sample characterization (e.g., from water to xenon) and isotope ratio measurement. The scan function is under software control and is built up segment by segment, allowing the scientist to tailor scan functions for each of the planned analyses. To reduce complexity and power demands and indeed mass, there is no provision for DC isolation or resonant excitation experiments. Similarly, though the ionization time is preselectable in the range 0.1–5 ms, there is no provision for AGC (see Chapter 3). Although this may appear to suggest a very basic ion trap system, reminiscent of the original Finnigan ITD 700 instrument, compared to current state-of-the-art instruments, it should be remembered that this is a highly specific application upon which considerable research effort has been expended in terms of determining the precise operating conditions, sample amounts, and so on, to achieve the desired level of performance. It should also be noted that the Ptolemy ion trap is operating at rather low mass/charge ratios (isotopic analyses are conducted at $m/z < 50$) compared to laboratory instruments targeted at organic analyses. Perhaps it is therefore not surprising that the optimum design for fulfilling this unique application did not follow conventional wisdom derived from a knowledge of analytical organic mass spectroscopy.

The electrodes are fabricated from aluminum, cut away to reduce the mass, and the overall external dimensions of the analyzer are 60 mm diameter $\times$ 70 mm height (see Figure 9.1). To minimize power consumption and provide some redundancy in this vital area, ionization is effected by a beam of electrons generated from a 3×1 array of microstructures etched from a silicon wafer; each of the microfabricated units comprises an array of 40×40 nanotips [2] and ion detection is accomplished

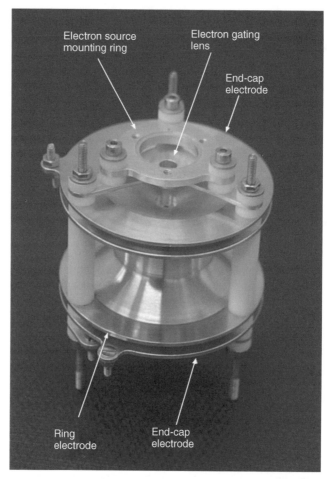

Figure 9.1. Photograph of assembled electrodes of flight model (FM) of Ptolemy ion trap. The electrodes are fabricated from aluminum; the electron source and detector are each mounted on supports, one of which is shown above upper end cap. Total mass of electrode assembly together with ion source and detector is 75 g. (Copyright The Central Laboratory of the Research Councils, reproduced with permission.)

using a novel type of spiral ceramic electron multiplier operating in the pulse-counting mode and developed by the Max-Planck-Institut für Aeronomie (MPAe), Lindau, Germany. The mass of the analyzer assembly (electrodes plus ion source and detector) was 75 g; including electronics and structural items the ion trap weighed less than 500 g. An illustration of the flight version of the ion trap with its associated ionizer, detector, and electronic circuits is shown in Figures 9.2a and b.

The supply of helium, used variously as the carrier gas, actuator for the pneumatic valves, and ion trap buffer gas, is Grade 6 helium (i.e., 99.9999% pure) admixed with argon (Grade 6) to a dilution ratio of 100 ppm (Ar/He). The reason

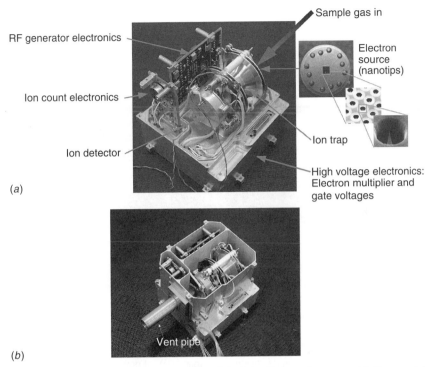

Figure 9.2. (*a*) Isometric photograph of the FM of the Ptolemy ion trap with its associated electronic circuits but with shielding container removed. (Copyright The Central Laboratory of the Research Councils, reproduced with permission.) (*b*) Isometric photograph of the FM of the Ptolemy ion trap with its associated electronic circuits, showing the lower half of shielding container and gas vent pipe. (Copyright The Central Laboratory of the Research Councils, reproduced with permission.)

for the inclusion of argon is twofold: first, it is used to aid mass calibration of the ion trap (providing a well-defined signal at m/z 40) and, second, it is employed in the measurement of D/H ratios (see below). The helium/argon supply is contained within two independent gas tanks of a "sealed-for-life" design fabricated from titanium using an all-welded construction. Each tank has an internal volume of $30 \, cm^3$ and is filled to a pressure of 50 bars, giving a total volume of gas of 30 liters at standard temperature and pressure (STP). Once Ptolemy arrives on site, the gas will be released in the gas management system by puncturing each vessel using a frangible pillar and Shape Memory Alloy (SMA) actuator and in-line particulate filter, all built into the base of each pressure vessel. Although under the conditions of use there will be no need for conventional vacuum pumping (see earlier), chemical "getter" pumps will be employed, for example, calcium oxide to remove carbon dioxide. The entire Ptolemy instrument, comprising the sample collection system,

Helium tanks   Electronics box

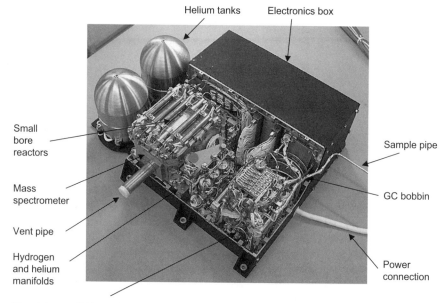

Small
bore
reactors

Mass
spectrometer

Vent pipe

Hydrogen
and helium
manifolds

Sample pipe

GC bobbin

Power
connection

Chemistry manifolds and large bore reactors

Figure 9.3. Isometric photograph of the FM of the complete MODULUS Ptolemy system with
one cover removed to reveal the components of the "mini-laboratory." The complete system
(with cover) has mass of 4.5 kg and consumes less than 10 W of power. (Copyright The Central
Laboratory of the Research Councils, reproduced with permission.)

gas-handling and sample-processing units, gas chromatograph, ion trap mass spectrometer, electronic units, and control/data management system, occupies a volume
of approximately $250 \times 330 \times 110$ mm, weighs 4.5 kg, and consumes less than
10 W of electrical power. Figure 9.3 shows the flight model with the gas tanks fitted
but with the cover removed. As with all the instruments carried on Rosetta, Ptolemy
has been vibration tested and designed to withstand temperature variations of -55
to $+70°C$ during its flight to the comet. All the control routines for the entire operation of the mass spectrometer and associated analytical procedures are preprogrammed into EEPROMS, since once the measurements commence, there will be
no opportunity for interactive real-time interpretation and response to the data being
obtained. However, the EEPROMS can be reprogrammed in flight prior to the
encounter or between experimental periods should this be necessary (see also
below).

For the mission there are essentially three versions of Ptolemy: the actual flight
model (FM) that forms part of the Rosetta package now in space, an identical
qualification model (QM) mounted in a high-vacuum system (10^{-7} mbar) in the laboratory and upon which analytical procedures can be checked and replicated prior
to sending signals to the FM version in Rosetta, and a ground-based electronic simulator reference model with which one can test the transfer of signals prior to their
being sent to ensure that the correct instructions are being transmitted.

9.3.3. Sample Processing and Isotope Ratio Measurements

9.3.3.1. Ion Trap Operation   From the preceding discussion we have seen that early in the planning stage it was determined that the QIT mass spectrometer offered considerable advantages over other alternative analyzers, for example, magnetic-sector, time-of-flight, and quadupole mass filter. However, at this time (1995–1998) there were no literature reports on the use of the ion trap for isotope ratio determinations. An intensive program was undertaken, therefore, to fully characterize and evaluate this novel application [3, 4].

Key figures of merit in relation to isotope ratio determinations are the abundance sensitivity, accuracy, and precision associated with the measurements. The *abundance sensitivity* is a measure of the extent to which the mass spectral peak tail from a lighter major isotope contributes to the height of the adjacent peak arising from the minor isotope, for example, the contribution made by the peak at m/z 44 (^{12}C $^{16}O_2^{+\cdot}$) to the peak at m/z 45 (^{13}C $^{16}O_2^{+\cdot}$) in carbon dioxide. Ideally, for a minor isotope of 1% abundance, the contribution to the measured intensity from the major isotope should be $<0.01\%$. This parameter is clearly determined by the resolution of the mass spectrometer. The *accuracy* is a measure of the ability of the instrument to determine the "true" isotopic ratio of the sample. In the present application, this factor is not regarded as being highly critical since, as noted above, comparisons are being made contemporaneously with calibrated reference samples. Of much greater importance is the *precision*, which is determined by the reproducibility of the measurements. This parameter may be determined by the zero-enrichment (ZE) technique, whereby many repeat measurements are made of the same isotopic ratio and then Eq. (9.1) is applied to each pair of consecutive ratios, in which the nth ratio is taken as being $R_{standard}$ with respect to the $(n+1)$th ratio as R_{sample}. Ideally the value of δ(ZE) should be zero. Thus for, say, 50 repeat measurements of R one would obtain 49 values of δ(ZE), and the precision can then be evaluated from the standard error of the mean of the set of consecutive measurements. While with a terrestrial magnetic-sector isotope ratio mass spectrometer one would hope to achieve precisions corresponding to one standard deviation of much better than $\pm 1‰$, for the Rosetta mission a precision of $\pm 5‰$ was specified as being acceptable. Hence the characterization and optimization of the ion trap mass spectrometer were carried out with $\pm 5‰$ as the target level of performance.

The initial experiments were carried out with a standard Finnigan MAGNUM ion trap combined with a Varian Model 3400 gas chromatograph equipped with a Model 1075 split/splitless injector. "Pure" samples of gases were admitted via a "sniffer" system at a rate of approximately 1 nmol/s, and the peak intensity ratios determined for mass/charge ratio values 29/28 for nitrogen, 33/32 and 34/32 for oxygen, and 45/44 and 46/44 for carbon dioxide. Numerous experiments were carried out to explore the effects of helium buffer gas pressure and space charge on the quality of the data obtained, but essentially the measured isotope ratios showed extremely low accuracy, with some values being in error by more than 100% compared to the

References p. 307.

expected values! The problem appeared to stem from at least three fundamental issues: insufficient resolution of the QIT, insufficient number of digital steps (DAC steps) controlling the RF drive amplitude across the mass spectral peaks, and the occurrence of ion/molecule reactions within the ion trap.

To overcome the first two, instrumental inadequacies, a new instrument control system and associated software were developed, called ACQUIRE (Advanced Control of the Quadrupole ion trap for Isotope Ratio Experiments); this allowed the scan speed to be reduced to 2000 Th/s (i.e., about three times slower than that of the standard trap) and the number of DAC steps to be increased from around 8 to 23 for each "mass unit."

While these modifications gave substantially improved mass spectral resolution and peak shapes, they did not address the remaining serious problem, namely the fact that the ions being studied were effectively "changing mass" through the occurrence of ion/molecule reactions with background gases, especially water, within the ion trap. For example, the $CO_2^{+\cdot}$ ion will react with H_2O according to

$$CO_2^{+\cdot} + H_2O \rightarrow CO_2H^+ + OH^\cdot \tag{9.2}$$

with an ergicity $\Delta_r H = -65\,kJ/mol$ [5]. As a result, depending upon the degree of conversion, the ion peak at m/z 45 will contain unknown proportions of $^{12}CO_2H^+$ and $^{13}CO_2^{+\cdot}$. Inevitably, without being able to resolve the "doublet" at m/z 45, the "measured" proportion of ^{13}C in the sample will be significantly exaggerated. In analogous reactions with H_2O, $CO^{+\cdot}$ forms COH^+ ($\Delta_r H = -137\,kJ/mol$) and $N_2^{+\cdot}$ forms N_2H^+ ($\Delta_r H = -186\,kJ/mol$). On the other hand, the corresponding reaction of $O_2^{+\cdot}$ with H_2O is not thermodynamically favorable ($\Delta_r H = +224\,kJ/mol$).

One possible solution to this problem in the Ptolemy instrument would be to attempt to remove all hydrogen-containing species from the trap. But given that the major constituent of the comet is ice, this was regarded as being impractical. The alternative approach, which was adopted in the flight system, has been to attempt to force complete protonation of each of the isotopic species by adding hydrogen gas to the helium buffer gas stream. Isotope ratios may then be measured on the protonated ions, thereby eliminating the isobaric interferences.

Reactions analogous to those described in reaction (9.2) with H_2O occur with H_2:

$$M^{+\cdot} + H_2 \rightarrow MH^+ + H^\cdot \tag{9.3}$$

for $M^{+\cdot} = CO_2^{+\cdot}$ ($\Delta_r H = -128\,kJ/mol$), $CO^{+\cdot}$ ($\Delta_r H = -200\,kJ/mol$), and $N_2^{+\cdot}$ ($\Delta_r H = -249\,kJ/mol$); with $O_2^{+\cdot}$ the reaction is again thermodynamically unfavorable ($\Delta_r H = +161\,kJ/mol$).

In the initial investigations using the modified MAGNUM ion trap mass spectrometer controlled by the ACQUIRE system [3], pure hydrogen was used in place of helium as the buffer gas at a flow rate of $\sim$1 ml/min. Substantially improved isotope ratio data were obtained: isotope ratios within +18‰ for (m/z 46)/(m/z 45) and −55‰ for (m/z 47)/(m/z 45) of the "theoretical" values for CO_2 were obtained using 47-nmol amounts of sample, with precisions determined by the ZE technique of ±4‰ and ±5‰, respectively.

In the flight model of the system, hydrogen gas from a reservoir cylinder is added to the helium buffer gas stream when nitrogen, carbon monoxide, and carbon dioxide are being assayed. Because of their lack of reactivity toward hydrogen, isotopic measurements using oxygen ions are carried out on the nonhydrogenated $O_2^{+\cdot}$ species. To conserve both power and space, the dimensions and frequency of the RF drive potential were reduced from the standard values employed on the MAGNUM instrument ($r_0 = 10.00\,\text{mm}$, $2\pi\Omega = 1.05\,\text{MHz}$) to those noted above (8.00 mm and 0.6 MHz, respectively).

So far this account has not included reference to the fourth element of interest to the Rosetta mission, namely the determination of isotopic ratios in hydrogen; this is, of course, a measurement that is of special importance in characterizing cometary water. Normally, in a terrestrial laboratory using magnetic-sector instruments for iso-topic work, the ratio of D/H is found by converting the water into hydrogen gas and then measuring the intensities of the m/z 2 and 3, corresponding to $H_2^{+\cdot}$ and $HD^{+\cdot}$, respectively. However, the sensitivity of ion traps falls off at low mass/charge ratios; furthermore, hydrogenation reactions of the kind noted above lead to the facile for-mation of H_3^+ and H_2D^+ at m/z 3 and 4, respectively. Since the helium buffer gas will also yield ions at m/z 4 (i.e., $^4He^{+\cdot}$), there is clearly scope for considerable inaccura-cies in such measurements, and an alternative approach must be adopted when using the ion trap mass spectrometer.

Hence a second novel technique for utilizing ion/molecule reactions to aid iso-tope ratio measurements was proposed. Fortunately, the reactions

$$Ar^{+\cdot} + H_2 \rightarrow ArH^+ + H^{\cdot} \quad (\Delta_r H = -144\,\text{kJ/mol}) \qquad (9.4)$$

and

$$Ar^{+\cdot} + H_2O \rightarrow ArH^+ + OH^{\cdot} \quad (\Delta_r H = -81\,\text{kJ/mol}) \qquad (9.5)$$

are both thermodynamically permitted, so that deuterium and hydrogen can be assayed in terms of the ratio of the intensities of the ArD^+ and ArH^+ ions at m/z 42 and 41, respectively. It should be noted that ions from the minor isotopes of argon, namely $^{36}Ar^{+\cdot}$ and $^{38}Ar^{+\cdot}$, will form adducts with H and D at m/z 37, 38, 39, and 40, but these will not cause isobaric interferences with the measured ions at m/z 41 and 42. As mentioned above, in Ptolemy the source of argon used for these experiments is the high-purity helium/argon buffer gas mixture.

9.3.3.2. Sample Processing and Analysis
The analytical procedures make use of a series of chemical reactors connected by a compact manifold containing miniature solenoid-activated shut-off valves and pressure transducers. The reactors are essen-tially ceramic tubes containing solid-state chemical reagents and have a heating ele-ment coiled round the outside capable of reaching 1000°C using 5 W of power. The

system is highly compact in order to minimize mass and to eliminate dead volumes. A reaction may typically be the oxidation of carbonaceous material to form carbon dioxide using a supply of oxygen generated on-board by heating a mixture of CuO/Cu_2O; essentially this would take the form of a stepped combustion analysis to determine the isotopic compositions of organic materials, such as polymers and macromolecules. Other reactors contain adsorbent materials (see below), drying agents or reagents for generating fluorine (see below), and carbon dioxide. Cylinders containing reference gases are also connected to the manifold. A schematic diagram of the experimental system is shown in Figure 9.4.

A typical analytical procedure might be as follows. A solid sample acquired via the SD2 drilling system is placed in an oven, evacuated and heated (with or without added oxygen) to the first temperature step, and held constant for 5 min. Evolved sample gases are released into the static gas manifold system, excess oxygen removed reactively, and the volatiles exposed optionally to a drying agent to remove water. Further treatments are possible, for example, selective removal of active gases using a "getter," and the remaining gases then admitted directly to the ion trap mass spectrometer or passed into one of three parallel GC column systems for separation prior to isotope ratio analysis. Following this mass analysis step, the sample oven is once again evacuated and the temperature increased to that of the next step and the above procedure repeated.

Two further modes of analysis are possible: near-surface volatile measurement and fluorination to release oxygen from silicate-rich material (possibly the remains of a sample that has been previously pyrolyzed as above). For the measurement of the near-surface volatiles, ambient gases are "trapped" using a carbon molecular sieve, Carbosphere, contained within one of the ovens that is then heated to release the gases for analysis as indicted previously. For the fluorination experiments, a supply of fluorine is generated by means of the inert solid compound Asprey's Salt, $K_2NiF \cdot KF$, which when heated to 250°C yields F_2; this in turn is admitted to the oven containing the solid sample whereupon it reacts to displace [O] from the silicate to produce O_2.

As noted above, there are three GC channels whose operation may be summarized as follows:

Channel A: A Varian Chrompack Ultimetal CP-Sil 8CB column whose function is to provide general analysis of evolved gases.

Channel B: A Varian Chrompack Ultimetal PoraPLOT Q column in line with two reactors containing Rh_2O_3 reactor molecular sieve drying agent. This arrangement separates gases such as CO_2, CO, CH_4, and N_2, while CO and CH_4 are converted to CO_2; a complex set of procedures allows carbon, oxygen, and nitrogen isotope ratios to be determined.

Channel C: A Varian Chrompack Ultimetal Molsieve 5-Å column plus associated reactors which is used to determine the isotopic composition of water. The H_2O is converted to H_2 and CO: The CO gives the $^{16}O/^{17}O/^{18}O$ composition of water and the H_2 provides the D/H ratio, as described above.

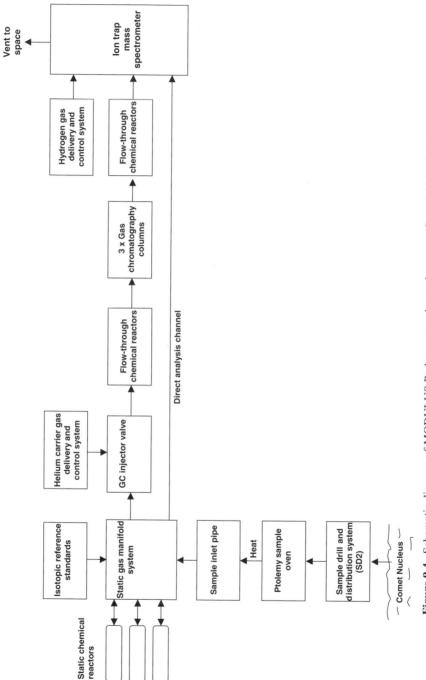

Figure 9.4. Schematic diagram of MODULUS Ptolemy experimental system. (Copyright by the Open University, reproduced with permission.)

9.3.3.3. Operational Sequence at Cometary Encounter   Shortly after landing, the Control and Data Management System (CDMS) on the Lander instructs Ptolemy to undergo a series of operational and survival checks before entering the *Safe Mode*. The time sequence of the encounter is divided into two major mission phases, where the aim is to take those measurements judged to be of the highest levels of scientific priority.

During the *primary mission phase* immediately after landing, which corresponds to the period T (=0) to $T + 65$ h, Ptolemy has been allocated two 6-h operational windows: the first in the period $T + 15$ min to $T + 12$ h, the second in the period from $T + 55$ h to $T + 65$ h. The first time window will be used to perform the highest priority science, namely analysis of a cometary surface sample using a sequence called the *Science* 1 mode (see below). The second time window will be used for the second priority science project, that is, the analysis of the comet's atmosphere (using the *Science* 2 mode). The detailed modes comprise a sequence of instructions in a look-up table stored within the Ptolemy software and are outlined below.

The *secondary mission phase* corresponds to $T + 65$ h to $T + 100$ h, in which a sample will be taken at the greatest possible depth below the surface of the comet and analyzed using the *Science* 1 mode (see below). When all the other instruments on board the Lander have met their objectives, the *Science* 3 mode will be implemented, namely oxygen isotope analysis of silicates using the fluorination procedure mentioned previously. Further experiments may then be carried out according to the nature of the initial results, and it is planned to continue the analysis of the cometary atmosphere at approximately weekly intervals, depending upon the rate of usage of helium from the storage tanks.

Typical operational modes are summarized as follows, although it is possible that the precise details of the experiments will be modified during the flight period as work continues on the QM version of Ptolemy based in the laboratory.

The *Science* 1 sequence:

Load sample into oven.

Heat oven to $-50°C$: Dry sample and analyze CO, CO_2, and N_2.

Heat oven to $+100°C$: Analyze H_2O and dry sample, CO, CO_2, and N_2.

Heat oven to $+400°C$: Dry sample, CO, CO_2, and N_2.

Prepare oxygen, admit to oven, and heat to $+400°C$: Analyze CO_2.

Heat oven to $+800°C$: Dry sample and analyze CO, CO_2, and N_2.

Prepare oxygen, admit to oven, and heat to $+800°C$: Analyze CO_2.

The *Science* 2 sequence:

A sample from the cometary atmosphere is adsorbed on to Carbosphere contained within one of the ovens.

Heat oven to $+200°C$.

Analyze N_2 isotopes.

Analyze water isotopes.

Dry sample.

Analyze reference gas isotopes.

GC analysis.

Analyze isotopes of CO and CO_2.

9.3.4. Summary and Conclusions

The aim of this chapter has been to show how the ion trap mass spectrometer has been adapted to obtain precise stable isotope ratio measurements. Furthermore, the instrument has been combined with a miniaturized and ruggedized automated chemical laboratory capable of working entirely under automated control in the ultimate remote hostile environment. The world of mass spectrometry will watch with bated breath to see whether Ptolemy is successful in performing its mission; in the meantime much of the technological spin-off from this program must surely have an abundance of practical applications in real-world applications, where small-scale portable instruments have a vital role to play?

ACKNOWLEDGMENT

The authors are extremely grateful to C. T. Pillinger and I. P. Wright together with their colleagues S. J. Barber, S. T. Evans, M. Jarvis, M. R. Leese, J. Maynard, G. H. Morgan, A. D. Mores, and S. Sheridan of the Planetary and Space Sciences Research Institute, Open University, Milton Keynes, United Kingdom, and L. P. Baldwin, J. N. Dominey, D. L. Drummond, R. L. Edeson, S. C. Heys, S. E. Huq, B. J. Kent, J. M. King, E. C. Sawyer, R. F. Turner, M. S. Whalley, and N. R. Waltham of the Space Science Department, Rutherford Appleton Laboratory, Didcot, United Kingdom, for access to certain unpublished material upon which this account has been based. One of us (JFJT) wishes to express his sincere appreciation for being invited to join the much smaller team that undertook the initial appraisal of the suitability of the ion trap for this application and explored its potential and limitations as a means of determining isotope ratios.

REFERENCES

1. H. C. Urey, The thermodynamic properties of isotopic substances, *J. Chem. Soc.* (1947) 562–581.

2. B. J. Kent, E. Huq, J. N. Dominey, A. D. Morse, N. Waltham, The use of microfabricated field emitter arrays in a high precision mass spectrometer for the Rosetta missions, paper presented at the Third Round Table on Micro/Nano Technologies for Space, ESTEC, May 15–19, 2000.

3. S. J. Barber, A. D. Morse, I. P. Wright, B. J. Kent, N. R. Waltham, J. F. J. Todd, C. T. Pillinger, Development of a miniature quadrupole ion trap mass spectrometer for the determination of stable isotope ratios, in E. J. Karjalainen, A. E. Hesso, J. E. Jalonen, U. P. Karjalainen (Eds), *Advances in Mass Spectrometry*, Vol. 14, Elsevier Science, Amsterdam, 1998.

4. S. J. Barber, Development of a quadrupole ion trap mass spectrometer for the determination of stable isotope ratios: Application to a space-flight opportunity, Ph.D. Thesis, Open University, Milton Keynes UK, 1998.

5. D. R. Lide (Ed.), *Handbook of Chemistry and Physics*, CRC Press, Boca Raton, FL, 1995.

AUTHOR INDEX

SUBJECT INDEX

CHEMICAL ANALYSIS

A SERIES OF MONOGRAPHS ON ANALYTICAL CHEMISTRY
AND ITS APPLICATIONS

Series Editor
J. D. WINEFORDNER